The Evolution of Lateral Asymmetries, Language, Tool Use, and Intellect

The Evolution of Lateral Asymmetries, Language, Tool Use, and Intellect

John L. Bradshaw

Department of Psychology
Monash University
Clayton, Victoria, Australia

Lesley J. Rogers

Department of Physiology
University of New England
Armidale, New South Wales, Australia

Academic Press, Inc.
Harcourt Brace Jovanovich, Publishers
San Diego   New York   Boston   London   Sydney   Tokyo   Toronto

Cover backdrop: Parietal art from northern Australia (see Figure 8.14).
Front cover inset: Photo by J. H. van Lawick Gooodall. Reprinted by permission of the publishers from The Chimpanzees of Gombe: Patterns of Behavior, by J. Goodall, Cambridge, Mass., The Belknap Press of Harvard University Press,©1986 by the President and Fellows of Harvard College (see Figure 8.2).
Back cover inset: Photo by L. J. Rogers and G. Kaplan (see Figure 5.2).

This book is printed on acid-free paper. ∞

Academic Press, Inc.
1250 Sixth Avenue, San Diego, California 92101-4311

United Kingdom Edition published by
Academic Press Limited
24–28 Oval Road, London NW1 7DX

Library of Congress Cataloging-in-Publication Data

Bradshaw, John L., date.
    The evolution of lateral asymmetries, language, tool use, and
intellect / John Bradshaw and Lesley Rogers.
      p. cm.
    Includes bibliographical references.
    ISBN 0-12-124560-8
    1. Laterality.   2. Comparative neurobiology.   I. Rogers, Lesley.
II. Title.
QP385.5.B695   1992
156'.2334–dc20   92-26556
                                                        CIP

PRINTED IN THE UNITED STATES OF AMERICA
92 93 94 95 96 97   EB   9 8 7 6 5 4 3 2 1

The fragrant Honeysuckle spirals clockwise to the sun
And many other creepers do the same
But some climb anticlockwise; the Bindweed does, for one,
Or Convovulus, to give her proper name.

Rooted on either side a door one of each species grew
And raced towards the window-ledge above;
Each corkscrewed to the lintel in the only way it knew,
Where they stopped, touched tendrils, smiled, and fell in love.

Said the right-handed Honeysuckle
To the left-handed Bindweed:
'Oh, let us get married
If our parents don't mind; we'd
Be loving and inseparable,
Inextricably entwined; we'd
Live happily ever after,'
Said the Honeysuckle to the Bindweed.

To the Honeysuckle's parents it came as a shock.
'The Bindweeds,' they cried, 'are inferior stock,
They're uncultivated, of breeding bereft,
We twine to the right—and they twine to the left!'

From *The Songs of Michael Flanders
and Donald Swann*, with permission.

Contents

3.   Asymmetries in Rats and Mice

4.   Asymmetries in Nonprimate Mammals

8.   Culture, Tool Use, and Art

9.   Brain, Language, Intellect, and the Evolution of the Self

Preface

More years ago than either of us cares to remember, we were colleagues, in different departments at Monash University, Melbourne. LJR worked in the Physiology Department on lateralization of structure and function in the avian brain and the lateralization of memory. JLB studied the lateralization of perceptual, language, and cognitive functions in normal and brain-lesioned humans. Since then, not only have we gone our separate ways—or at least LJR has moved to the University of New England—but our research interests have also expanded and moved on. LJR now spends much of her research time studying lateralization of behavior in marsupials and primates and investigating factors that influence lateralized development of neural pathways in birds. JLB's areas of expertise have similarly widened to encompass attention and movement in normal and dysfunctional brains, including stroke patients and individuals suffering from unilateral neglect and from movement disorders such as Parkinson's and Huntington's disease.

The genesis of this book probably occurred during a workshop on the evolution of language—its origins and development—in March 1989 at the University of New England, where we renewed old acquaintance. JLB had long had an interest in archaeology, having participated in "digs" in the United Kingdom and New Zealand, and even having once been a member of the British Institute of Archaeology at Ankara, Turkey. He had also published on human evolution. This book provided us with the opportunity to combine our complementary (human and nonhuman) research interests on an issue, the evolution of lateral asymmetries, which underlies one of the three "great mysteries"—human uniqueness—the other two, of course, being the origins of matter and life.

Human uniqueness (consciousness, praxis, tool use, language, and art) is very recent, at most a matter of one or two million years, and maybe less than 100,000 years, depending on how we define or link these attributes; we can contrast that period with the duration of complex multicelled life (over half a billion years, with fossil evidence for behavioral asymmetries right at the beginning), of life itself (several billion years), and of the galaxy (ten or twenty billion years, depending on the true value of Hubble's Constant). Human uniqueness, moreover, is highly controversial, and the attributes listed above all seem somehow to involve aspects of cerebral asymmetry. It can be addressed from a variety of viewpoints, including molecular biology, comparative zoology, ethology, sociobiology, paleontology and archaeology, anthropology, experimental and comparative psychology, physiology, and neurology. We do not, of course, claim professional expertise in more than a fraction of these areas and apologize to anyone whose work we have omitted or misunderstood, but between us we believe we have sufficient background, either from first hand research experience or long-standing interest, to offer some sort of synthesis. At least three of these areas are currently very active—neurology, molecular biology, and human prehistory. Readers will certainly dispute many of our claims and interpretations, and many will be aware of radically new discoveries made since we wrote these words, which will still further change how we think about our evolutionary past, unique present, and doubtful future. If we can hazard a guess, these discoveries will continue to narrow any remaining Rubicons between ourselves and other species in such hotly contested issues as the uniqueness of language, tool use, and consciousness. As we find out more about our remote past and evolutionary heritage, whatever remains to be considered uniquely human seems to be ever more recent in its origins.

This book provides a detailed overview of the origins and evolution of lateral asymmetries in the animal kingdom, how they have had impact on a wide range of behaviors, the constraints under which they manifest, and their impact on our own cognitive, linguistic, tool using, and aesthetic capacities. We hope it proves useful for senior undergraduates (it is abundantly illustrated and contains useful reading lists), graduate students, instructors, researchers entering the field and requiring a "state of the art" overview, established workers who wish to check up on specific areas, and specialists, all in a variety of fields. These fields include biologists, physiologists, behavioral scientists of a variety of persuasions (psychologists and neurologists, primatologists, anthropologists, students of human prehistory), and the general reader. We hope that at least for the next few years this text will offer some kind of overview. The provision of a Summary and Conclusions section at the close of each chapter is aimed to provide both a general survey and a balanced judgment of any controversial aspects previously discussed in detail. Each chapter commences by introducing the topics to be covered. We have drawn together and critically reviewed many diverse areas of information, but, if at times we appear uncritical, it is because so much of the information is so new and controversial that it cannot yet be properly assessed.

We commence with an introductory chapter on asymmetries in the natural and artificial human worlds, in physics, chemistry and the cell, and in evolution up to and including the reptiles. There follows a detailed chapter on behavioral and anatomical asymmetries in birds and the role of environmental and hormonal influences. This theme is extended in the third chapter, which deals with analogous if not homologous asymmetries in rats and mice, and the organizing effects of the cerebral commissures. Following a chapter on asymmetries in other mammals (excluding primates), primates are dealt with in two further chapters, one addressing manual asymmetries and the other, cognitive asymmetries and asymmetries in brain morphology. We next devote a chapter to human evolution; a further chapter to culture, tool use, and art in human prehistory; and close with a final chapter which addresses the nature, evolution, and lateralized mediation of speech, language, praxis, and art. This final chapter explores how these essentially human attributes relate to the appearance of intelligence, intellect, and self-awareness. To date, such a defined, systematic, and up-to-date synthesis has not been attempted and is, we believe, long overdue.

We are extremely grateful to the many individuals from a wide range of disciplines who so readily made available illustrative material for reproduction and hope that we have been able to acknowledge everyone adequately. Our grateful thanks also to Judy Bradshaw, Sundran Rajendra, and Gisela Kaplan for assistance, to our drafting artists Rosemary Williams and Ivan Thornton, to Helene Dawson and Di Davies, to photographers Vlad Kohout and David Elkin, and above all to our partners, friends, and students who for so long endured our preoccupation with the book and assisted so fundamentally in its generation.

Finally, we hope our readers do not find the book as difficult to read as we found it to write!

John L. Bradshaw
Lesley J. Rogers

Asymmetries to the Level of the Lower Vertebrates

Introduction

We regard the concept of three dimensions of terrestrial space almost as cliché. They are realized in the x, y, and z axes of mathematics, geometry, and computer graphics. As we shall see, however, all three are more or less arbitrary or relative with respect to the Earth, an observer, or a directional object. We have also long been fascinated by our own orientation and handedness, as witnessed by language, myth, and our interest in systematic asymmetries in particle physics, crystals, organic molecules, biology, and the brain.

In this chapter we shall ask the following: What is special about the left-right dimension, why are most organisms *superficially* symmetrical, why is there an underlying asymmetry that seems nevertheless to prevade all living things and how is it realized, at a macro level, in vertebrates from the dawn of organized life, through chordates and all the vertebrate radiations.

Dimensions in the Physical World

Ignoring time and the considerations of theoretical physics, our world may be thought of as three-dimensional. The vertical dimension is easily defined by reference to gravity, as long as we are not enclosed in a free-falling capsule, or in a space ship, or floating freely in the dark, submerged in a body of water. While supported by substrate (e.g., on the ground, or at the bottom or the surface of the sea), we can easily recognize up and down. We also apply the corresponding terms "top" and "bottom," by extension, to those surfaces of an object which normally or by convention occupy the upper or lower extremities of the vertical dimension; thus a container, even when on its side, has a top and a bottom. In this context, these terms are being used with reference

1

to an object's preferred orientation, rather than directly with respect to gravity.

A second axis, orthogonal to the first, vertical axis, can be easily specified in terms of near and far, or front and back. The dimension of depth, horizontally conceptualized, is necessarily relative to an observer, or to an observed object, and lacks the apparent objectivity of gravitational up and down. If the observer, possessing a leading edge probably with receptors, is moving in a certain direction, "far" is definable as the other end of the continuum which commences proximally with the observer. Similarly, "front" relates either to the leading edge of the observer or, alternatively, to the surface of an observable object that is facing toward that observer. We can, of course, also characterize an object in terms of its front or back surfaces without direct reference to an observer, just as a box on its side still has a top and bottom. Thus "front" can refer to the major surface of that object, for example, a book's front (versus back) cover, and the front (back) of a television set, house, or car.

As Corballis and Beale (1971, 1983) observe, when we view an object beside us in a mirror, its vertical dimension (up–down) undergoes no significant alteration; top and bottom stay in the same place. While we can still meaningfully talk about the object's front and back, or the dimension near and far, the object's orientation relative to the observer has reversed. Indeed, when we view *ourselves* in the mirror, we see our fronts, not our backs; our image *faces* us, rather than away from us looking in the same direction as we are looking. We can now perhaps understand why the third dimension, left–right, is apparently reversed in a mirror image (see Figure 1.1). Thus when we raise our right hand, the image in the mirror raises its *left* hand. The third dimension, left–right, exists with reference to a real or potential observer who possesses a notional "left" and "right" side, and can only be specified in the context of the other two dimensions, that is, after they have been established. Should you walk up to a flat mirror, your top and that of your image will fully coincide at the top edge of the mirror; however, your front surface and that of your image will confront each other, rather than face toward the same point in real space. If you raise your right hand, your image will raise the hand that is on the same side of the mirror, which will appear as a movement of its left hand. If your image did not face *toward* you, but in the *same direction* as you, you would be happy to say that you both raised the same hand. The crucial point is that mirrors seem to confuse only left and right, and not the other two dimensions, because left–right can only be specified after the other dimensions are established. It is made entirely with reference to the observer's own laterality or orientation. This dimension lacks the appar-

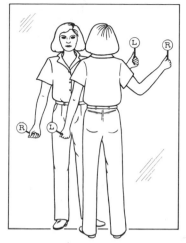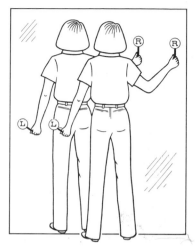

Figure 1.1 Why does a mirror seem to reverse left–right, but not up–down? It would not seem to do so if both my image and I faced in the same direction, rather than facing each other. "Left" and "right" are labels we *apply* with reference to *ourselves*.

ently independent objectivity of gravitationally defined up and down; there is no "true right" like there is "true north" or "up."

Why do humans, and many species of animals, possess left–right asymmetry, which *permits* us to nominate one side as "right" or "left"? Symmetry (or asymmetry), however, need not be exclusively *lateral*; spherical and radial symmetry frequently occur (see Figure 1.2), and lateral symmetry may be secondarily distorted, as with flatfish and the coiling of a snail. The segmental organization of annelids, arthropods,

Figure 1.2 A sphere is the best packaging format for a small free-floating organism that is not exposed to unidirectional forces. A sessile creature may adopt a flattened, pie-shaped, radial symmetry, whereas an animal that moves along a substrate can only evince lateral symmetry.

and even some vertebrates represents *translational* symmetry (Chapple 1977a); each segment is primitively more or less a replica of the one preceding it, with progressive ontogenetic differentiation along the longitudinal axis. Thus the appendages of decapod crustacea perform a progressive variety of functions, from feeding to grasping, to walking and swimming; all are serial homologues of each other. However, with respect to left–right asymmetry, as Corballis and Beale (1971, 1983) again observe, in the natural world few events occur consistently more often on one side than the other, and in such an unbiased world it might be disadvantageous to be laterally asymmetrical. Thus it would probably be more useful to be able to generalize events occurring on either side, than to have to distinguish between them. One example of a possible exception is a migrating bird traveling along a meridian of longitude between pole and equator. Under these circumstances the morning sun will consistently strike it on one side, and the evening sun on the other. Such knowledge, implicit or explicit, together with a circadian "clock," would permit accurate navigation in any direction, not just north and south, although for climatic reasons such migration might be expected generally to occur between high and low latitudes.

Animal navigation probably involves considerably more cues and mechanisms than just the position of the sun (Baker 1989; Able and Bingman 1987; Gallesten 1989; Walcott 1989). Navigational problems aside, species that must navigate might in fact benefit from lateral asymmetries, and might therefore exhibit them in exaggerated form, if only at the level of neural information processing. *Symmetry*, or lack of a left–right bias, in a largely symmetrical world might be generally more advantageous for freely moving animals; in this way they would avoid developing turning biases and would respond the same way to appetitive or adversive objects irrespective of the side on which these are encountered. Indeed, there may be only five circumstances where we might expect an organism to display lateral biases:

1. Where, at a macro level, the external world is in fact laterally biased and the organism is responsive to it.

2. Where the organism possesses subtle lateral asymmetries at a micro level, for example, in its amino acids or protein structure.

3. Where an animal navigates using a directional constant such as the sun. Even so, such navigation, or use of an internal compass like crystals of magnetite which in bees or pigeons may sense the Earth's magnetic field (Walker and Bitterman 1989), only requires that the creature remembers to turn toward, or away from, an indicated heading; it does not need to distinguish and remember left and right per se.

4. Where an animal travels around a home base. If it "remembers" always to turn in the same direction it will remain within a smaller radius than when it randomly alternates left–right turns. Conversely, if dispersal is important, any turns should regularly alternate. Again, however, it does not need to be able to distinguish between left and right.

5. Where for reasons of economy of space or function, initially duplicated structures on the two sides, or a single midline structure, become laterally specialized. Thus in decapod crustacea we shall see that this often happens with the two claws (chelae); one claw becomes specialized for cutting and the other for crushing. Similarly in our own species, shortage of cerebral processing space may explain why language and certain other functions have tended to reside in the left hemisphere, and spatial and emotional behaviors in the right (Bradshaw 1989).

As Corballis and Beale (1971, 1983) observe, a perfectly symmetrical organism could not tell left from right (see Figure 1.3); it would not be able to determine whether its view of the world was normal or left–right mirror reversed (e.g., as viewed indirectly via a mirror or as a transparency projected the wrong way round). As they point out, it is logically the same whether the view of the world is reversed either by mirror or slide, or whether the viewing organism is instead somehow laterally transposed; however the latter might in fact be achieved, a totally symmetrical organism would not under such circumstances be changed in any way. They note that we can in fact determine whether a creature can tell left from right in two ways. It can be required to move a limb on one side, or to turn in one direction in response to commands which do not

Figure 1.3 Mirror reversal of a completely symmetrical organism does not affect it in any way. Therefore such an organism would never know if it had entered a mirror-reversed "Alice" world, since this is formally equivalent to the world staying unreversed and the viewer instead undergoing left–right reversal.

directly indicate direction. Thus a green light, or the word "left!" (or "ugh!") could require one response, and a red light, or the word "right!" (or "ooh!") require another. Alternatively, to left–right mirror image stimuli (e.g., visually presented > or <, or as a touch on one or the other side), the subject is forced to make non-mirror-image responses; for example, it must advance or stay put, emit a high or low note, say "left" or "right", "ugh" or "ooh," and so on. It cannot simply turn toward (or away from) the stimulated side, or the direction indicated, as this direct stimulus-response mapping would pose no problem for symmetrical creatures. Thus the subject must *remember* how to make nonsymmetrical responses to symmetrical stimuli, or symmetrical responses to intrinsically asymmetrical stimuli. Such a situation, as we saw, is unlikely to occur in nature, and may in fact be restricted to the peculiarly artificial human world of unidirectional or asymmetrical writing, screws and bolts, water faucets and roads (see Figure 1.4). Even the migrating bird, flying perhaps with the sun on its right in the morning, and on its left in the evening, has only to remember to fly slightly toward (or away from) the sun at certain times of day.

Left and Right in Culture, Myth, and the Body

While an asymmetrical world may be a uniquely human artifact, we humans are not unique within the animal world in terms of our lateral asymmetries. Nor are we likely to be completely alone in experiencing at least some degree of conscious awareness (Griffin 1981, 1984) in making and using tools (Beck 1980), and in learning to communicate via a system which shares at least some of the characteristics of human language. Indeed, it is probably no coincidence that handedness, conscious awareness, tool use, and language have coevolved to significant levels in our species. Nevertheless, whereas laterality (and self-awareness, tool behavior, and communication) may indeed be more prominent in humans, the comparative frequency of sinistrality (left-handedness) in our midst cautions us against making too strong an evolutionary argument on the adaptive significance of dextrality.

The fact that left-handers have everywhere always been very much in the minority (around 10% of the population depending partly on criteria used, see, e.g., Bradshaw 1989, for review) may account for the denigration of sinistrality, for example, in the Bible. Hardyck and Petrinovich (1977) note that the Bible says the right hand of the Lord "is full of righteousness" (Ps. 48:10) and "shall take hold of me" (Ps. 139:10); it "shall spread out the heavens" (Isa. 48.13), while the Son of Man shall

Figure 1.4 Natural scenes look much the same whether viewed as normal photographs, or left–right reversed. The same is certainly not true of human creations which include, for example, writing and road traffic.

"set the sheep on his right hand, but the goats on the left. Then shall the king say unto them on his right hand, Come, ye blessed of my Father, inherit the kingdom prepared for you . . . Then shall he say also unto them on the left hand, Depart from me ye cursed, into everlasting fire" (Matt. 25:33–34, 41). Indeed, pejorative associations with sinistrality are almost universal (the term itself, associated with "sinister," not excepted). Thus it is no accident that "right" also means "correct" in many languages (e.g., German *recht,* in French *droit*), while the sinister left is found in the words gauche, gawk (lout, from English dialect for left-handed), cack-handed (faeces, the Indo-European root is **kak*), *lyft* (weak, Celtic), *laevus* and *scaevus* (ill-omened, Latin from Greek), and *link-isch* (clumsy or wrong, German). In many cultures, the left is the "unhygienic" hand. The political left and right date from the prerevolutionary French assembly where aristocrats sat on the right and the peoples' leaders sat on the left—the left and right wings. During church celebrations of Christian weddings, the bridegroom's family is usually seated on the right, and that of the bride on the left; there is a pervasive tradition linking male with right and good, and female with left and bad, a tradition which Corballis (1980, 1982) claims even extends to the Maories of New Zealand and the Gogo of Tanzania.

The link between handedness and the sexes is codified in the Pythagorean Table of Opposites, a curious mixture of number magic and mythology, which is cited by Aristotle (*Metaphysics*, A5, 985 b 23). Again, left is linked to female, cold, dark, bad, and crooked, whereas right is associated with all of the opposites. The sixth century B.C. Greek philosopher Parmenides thought that the sex of the embryo was determined by the side of the womb in which it lay, males to the right, females to the left. His contemporary, Anaxagoras, intuitively and, as we now know, correctly believed that the father instead was responsible for an offspring's sex. However, he believed that semen from the right testis generates sons, and that from the left, daughters (see Lloyd 1966). For centuries afterwards, potential fathers even tied off one testicle so as to "determine" their offspring's sex (nowadays centrifuging the semen can partially sort by weight the differentially heavy XY and XX sperm, which respectively generate sons and daughters), and pregnant mothers with an enlarged left breast were believed to be carrying a daughter (Mittwoch 1977, 1985). It is interesting to note that infants conceived at different times in the changing temperature gradient which accompanies ovulation differ slightly in male/female ratio, though the mediating factor is likely to be differences in acidity in the cervix acting on XX and XY sperm (Harlap 1979). In the rare eventuality of hermaphroditism in which there is also differentiation of testis and ovary, the single testis is usually found on the right and the single ovary on the left (Mittwoch 1977). In *normal* humans the right gonad is generally heavier, especially in females. In fact, well over 2000 years ago Greek sculptors were well aware of testicular asymmetry (see Figure 1.5), where the left hangs lower than the right (McManus, 1976), although they wrongly believed that the left is heavier. Our right arm (and that of monkeys, too) is also more massive than our left. The umbilical cord is also asymmetrical (Neville 1976). In a cleft palate, which like hermaphroditism exhibits bilateral asymmetry, the cleft occurs more often on the left side than on the right side (Mittwoch 1990). Hoogmartens (1986) reports a tendency for preferring to chew on the right side of the mouth, especially in females, who generally tend to exhibit stronger motor asymmetries than males. It is tempting to re-

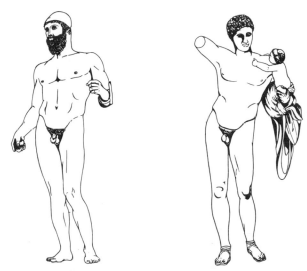

Figure 1.5 As shown by these sketches of classical Greek sculpture, it has long been known that the male's left testicle typically hangs lower than the right. From J. L. Bradshaw, Hemispheric Specialization and Psychological Function. Chichester: Wiley, 1989, with permission.

late such an effect (which first should be confirmed and extended) to the stronger asymmetries on the right side of the mouth that appear during speech movements, and which again are more pronounced in females.

If we look into a double reversing mirror, that is, via two mirrors placed at an angle to each other so that we can see ourselves just as others see us, not left-right reversed as a single normal mirror, we may be struck by the asymmetries in our own face, asymmetries with which we have become so familiar through years of seeing our own mirror image that we are no longer sensitive to them; they are now revealed by the double reversal of our image, which initially opposes our habituating mechanisms. We can circumvent the need for double mirrors and highly familiar faces such as our own if we resort to another trick, the photographic construction of left–left (L⅃) and right–right (ЯR) composite faces (Kolb, Milner, and Taylor 1983). Thus we can take a full-face protrait and prepare its mirror image by printing through a reversed negative (see Figure 1.6). If we cut the original and its mirror image vertically down their exact midlines and juxtapose the two left halves (L⅃), that is, creating a complete composite face made entirely from the original left side, and similarly create a right–right (ЯR) composite, the two composites are immediately seen as somehow dissimilar. Moreover the L⅃

Figure 1.6 Our faces are asymmetrical. Originals appear at the top, and the other two rows are composites composed from photographically abutting two left or two right halves. When one is asked which composite seems more similar to the original, the one corresponding to the left half (as viewed) of the original is generally chosen. Here, this is the middle item for the first two columns, and the bottom item for the last two.

composite (using the term L and R here to refer to what is perceived, from the observer's rather than the owner's viewpoint) always seems to resemble its original (LR) more closely than the Я R composite. The fact that this occurs both with normal and mirror-reversed originals indicates that the effect is perceptual; we take more notice of the half-face seen on the left.

We can simplify the task even further. Instead of performing the above three comparisons, that is, L Ⅎ versus original, Я R versus original, and the results of both comparisons versus each other, to determine which of the two composites more closely resembles an original, we can create two composite happy–sad faces, one the mirror image of the other (see Figure 1.7), and ask which of the two is happier (or sadder). Again, people typically choose the face that has the left side (as seen) of the mouth showing the target emotion, and only a single comparison is needed. This situation arises because the left side of a display, whether or not we steadily fixate on its center (in this case, the nose) is preferentially

Figure 1.7 The bottom face is identical to the top face, merely left–right reversed. If asked which is the happier, people tend to choose the face whose left half has the happy expression, the bottom photograph in this case. The two faces on the left were constructed from the owner's original left and right sides, and the two on the right from the owner's original right and left sides, as we can see by comparing them horizontally.

processed by the contralateral right side of the brain, the right hemisphere (RH), which is generally superior at judgments involving patterns, faces, and emotion (Borod 1989). Indeed, these asymmetries tend to be stronger without fixation, and they may occur with inverted faces (a procedure that tends to destroy the images' face-like qualities), and perhaps even with composites of laterally symmetrical nonfacial stimuli, for example, with images of trees.

Not only is the perceived left side of a face more salient because of its direct access to the viewer's right cerebral hemisphere, but the *owner's* left side (perceived by the viewer on the right side) tends to produce stronger emotional responses, being more directly under control of emotional centers in the RH. Because we face each other when observing one another, we see the "weaker" right side of our opponent's face to our left, primarily via our "stronger" RH. Conversely, mouth asymmetries during *speech* are stronger on the owner's *right* side, in terms of both measured amplitude and velocity of movement. Thus the right side of the mouth opens faster and wider than the left, as it is connected (again contralaterally) to the left hemisphere (LH) which controls speech in almost everyone (see Figure 1.8). Interestingly, such speech asymmetries tend to be greater for more complex movements, and for females, as so often is the case with motor asymmetries, whereas males tend to be more laterally asymmetrical ("lateralized") with cognitive tasks.

Asymmetries in β Decay, the Carbon Molecule, Amino Acids, and Proteins

In 1848, Louis Pasteur microscopically examined crystals of sodium-ammonium tartrate (Hegstrom and Kondepudi 1990; Mason 1984); one sample had an organic origin. He noted that there were two types of crystals which were mirror images of each other in appearance. He manually separated the two forms (now known as enantiomers, which match each other as do the left and right gloves of a pair), and dissolved them in water to form two solutions. He found that a beam of polarized light when shone separately through each solution was rotated, in one case to the left, and in the other case to the right. The sample with the organic origin was dextrorotatory. Later, in 1857, Pasteur discovered that microorganisms could change a previously optically inactive solution to one that rotated polarized light. We now know that many organic molecules exist in such left- and right-handed (chiral) forms as a consequence of the arrangement of the four bonds of a saturated carbon atom (see, e.g., Mason 1984). The bonds point toward the corners of a tetrahedron, a pyramid with equilateral faces (see Figure 1.9). This

Figure 1.8 When we speak, the right side of the mouth (i.e., the left *observed* in the photograph), is typically more mobile than the left. From M. E. Wolf and M. A. Goodale, Oral asymmetries during verbal and nonverbal movements of the mouth. *Neuropsychologia,* 1987, 25:375-396, with permission.

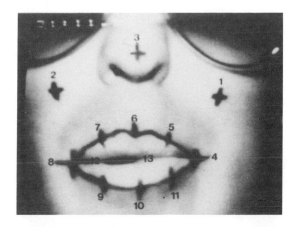

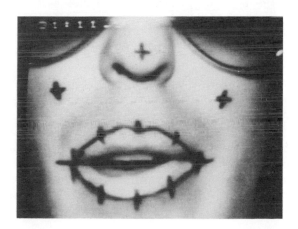

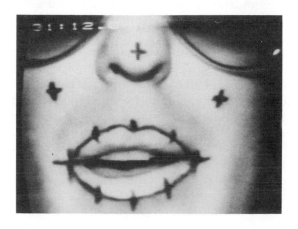

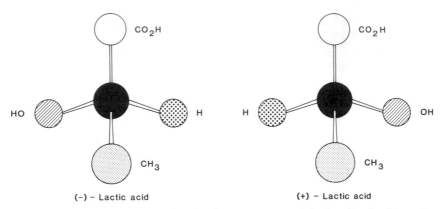

Figure 1.9 The three-dimensional molecular arrangement of the four bonds of the carbon atom can lead to two possible configurations (enantiomers), one the mirror image of the other, as shown here in the two possible forms of lactic acid. From J. L. Bradshaw, Hemispheric Specialization and Psychological Function. Chichester: Wiley, 1989, with permission.

tetrahedron can assume two possible configurations, one the mirror image of the other. However, the absolute positions of atoms and groups of atoms around a chiral carbon atom are not related in a simple way to the direction in which the enantiomer rotates the plane of polarized light, which may itself vary with the wavelength of the light used to measure it, and with the solvent, concentration, temperature, and pH when in solution (Kean, Lock, and Howard-Lock 1991). There are therefore several systems of classification and numerous symbols to assign, depending on whether reference is made to laboratory measurement of the rotation of polarized light, or to the spatial configuration of a chiral molecule.

An even more fundamental asymmetry exists in the natural world in the world of particle physics, which may interact with the chirality of organic carbon compounds. In the 1950s, contrary to all previous expectations, it was discovered that during radioactive β decay, an unstable nucleus ejects either an electron and an antineutrino or a positron and a neutrino. Previously there had been no reason to believe in the possibility of a systematic bias in subatomic chirality, but it was found that β particles (electrons or positrons) generally spin in one direction with respect to their direction of movement, thus violating the principle of parity invariance. Their chirality may in fact differentially affect organic molecules, which in living substances are virtually exclusively composed from amino acids of one and the same chirality. Thus in an important sense, all atoms are chiral, and a chiral molecule must exist in a higher or lower energy state than its enantiomer; the one with the lower energy state will tend to be more stable in the presence of radiation and will survive longer than its counterpart.

Kondepudi, Kaufman, and Singh (1990) report an unexpected observation with the familiar substance sodium chlorate. Its individual molecules are symmetrical, but when crystalizing they align themselves into either a right- or a left-handed structure. An undisturbed solution slowly develops an equal number of crystals of either chirality. If, however, it is stirred, the rate of crystal formation greatly increases, and as soon as one crystal forms spontaneously it apparently catalyzes the formation of many secondary crystals of *the same chirality*, suppressing the formation of other primary crystals. Consequently, all crystals in the solution are descended from the one that forms first. As Kondepudi et al. note, these phenomena of autocatalysis and competition may help explain how the molecules of life originally settled upon a single chirality.

While we can artifically synthesize an equal mixture of right-hand (D-, dextro) and left-hand (L-, levo) chiral molecules, living organisms are virtually exclusively synthesized from 20 L-amino acids (only glycine is nonchiral), and in turn from D-sugars. Mirror image forms are found only in the cell walls of some primitive bacteria, while some fungi can produce D-amino acids, which act as antibiotics by interfering with the construction of bacterial cell walls. It is these L-amino acids which Garay (1968) found were destroyed slightly more slowly during exposure to radioactive β particles as compared with their D-isomers, a phenomenon which perhaps led to the evolutionary selection of L-amino acids at the dawn of life. In this context, Engel, Macko, and Silfer (1990) have found that in addition to the nonprotein amino acids that are uncommon in living systems, several amino acids common in living systems also occurred abundantly in the Murchison meteorite. This meteorite landed in 1969. The amino acids were found to be rich in ^{13}C, which indicates an extraterrestrial origin. In particular, alanine was found to occur with an excess of the L-enantiomer. This further supports the hypothesis that β decay may have resulted in the preferential destruction of D-amino acids on Earth (and elsewhere) before the origin of life, with a consequent enrichment of L-amino acids. Indeed, according to Maurus (1991), reviewing recent thinking on this matter by Bonner and Rubenstein at Stanford University, intense ultraviolet light emanating from electrons, accelerated by the powerful magnetic field of a rapidly rotating neutron star, would have strong circular polarization. If any organic compounds strayed into range, the circularly polarized light would selectively destroy molecules of one handedness while sparing the other.

Proteins, living substances, are long polymeric chains, therefore, of L-amino acids. Enzyme proteins catalyze biomolecular reactions, including the synthesis of other proteins; the catalytic ability of enzymes depends on their three-dimensional structure, which in turn depends on their L-amino acid sequence. Synthetic chains of amino acids made from both L-and D-(dextro having the opposite chirality) enantiomers do not twist in a fashion compatible with efficient catalytic activity. Parenthetically, had the drug thalidomide been constructed with entirely left-handed molecules, there would have been no ill effects; only the right-handed form is damaging. There are even suggestions (see Lesser and Bown 1991) that antiasthma drugs like Ventolin, the efficacy of which depends on the action of left-handed molecules, may show certain long-term adverse effects from the simultaneous presence of right-handed molecules.

The nucleic acid macromolecules DNA and RNA are themselves chiral polymers, composed of four bases, each incorporating a D-enantiomer sugar group; hence nucleic acids only form right-handed helices. L-amino acids form D-sugars because such a configuration requires far less rearrangement than does an L-configuration (Neville 1976). Thus when chains of L-amino acids form an α-helix, the most stable configuration is right-handed, like a corkscrew, as the hydrogen bonds formed between the different turns of the helix are stronger than a left-hand configuration. In life, as Neville observes, all α-helices are right-handed; as in a hierarchical structure like a rope, the "sense" of a superhelix is the opposite of that of constituent, component subhelices. Biological structures are also hierarchical, like a thick rope, at many levels—molecules, macromolecules, microfibrils, fibrils, fibers, cells, organs, and organisms. Lower level asymmetries may partly determine the (often complementary) asymmetries at *higher* levels. (See, e.g., Holliday, 1990, who suggests how maternal DNA may encode secondary and tertiary RNA structures in either left- or right-handed forms, with consequent impact on protein synthesis and cytoplasmic structure. Likewise, Corballis and Morgan (1978) and Morgan and Corballis (1978) note that if a molecule possesses handedness, it may undergo differential directional rates of diffusion in the formation of embryonic cytoplasm, thereby leading to left–right maturational gradients.) Thus, while a biological compound or structure may appear quite symmetrical at a macroscopic level, like a basket made from plaited string or rope, at a molecular (or fiber) level, both structures will be asymmetrical. However, as Geschwind and Galaburda (1987) note, compounds on opposite sides of a biological structure will twist in the same direction, not as mirror images; otherwise, for example, if the taste receptors on opposite sides of the tongue were mutually mirror-aged in construction, then steric compounds like glucose would bind only to one side and would therefore only taste sweet unilaterally.

Cytoplasmic Asymmetries

As Geschwind and Galaburda (1985) observe, cell cytoplasm contains many internal structures and organelles; it is not just an amorphous gel or fluid, but consists of a cytoskeletal network of protein structures which define the cell's form. The same characteristic shape of a cell is preserved in its descendants, including any configurational asymmetry. The cytoskeletal architecture is determined by the arrangement of fibrils formed by structural proteins such as actin, the synthesis of which may depend on cytoplasmic genes (though their *expression* may depend on nuclear genes). Cytoplasmic genes, although associated with such organelles as mitochondria, are not just involved, like the mitochondria, with energy metabolism; they also help control the structural architecture of the cells (Davidson, Hough-Evans, and Britten 1982). Moreover, these genes are derived from the mother via the ovum (Birky 1983), and may in principle be traced back through many maternal generations for assessing the dates of the earliest common ancestors of different present-day populations, subspecies, species, or even genera. Certain structural fibrous compounds grow in the form of a coil; as it grows forward, its tip will rotate clockwise or counterclockwise. Likewise, the flagella of sperm exhibit a definite directionality, and bronchial cilia all beat in the same direction to perform their scavenging efficiently. Occasionally, abnormalities in sperm flagella and bronchial cilia occur simultaneously with gross anomalies of internal organ lateralization (e.g., Kartagener's syndrome with the phenomenon of *situs inversus viscerum*, Afzelius and Eliasson, 1979). Unicellular ciliates when swimming rotate around the long axis of the cell, each species in its own direction, although usually to the right (Schaeffer 1928). Likewise, the helical flagella of bacteria, containing many turns of left-handed spirals, avoid entanglement by rotating clockwise as viewed from the body; any points of entanglement travel outwards towards the tree ends, rather than back up to the body and thereby jamming (Neville 1976).

Bacillus subtilis is a common rod-shaped gram-positive bacterium. Certain mutants grow not in terms of increased numbers of individual bacteria, but in long thread-like clones forming a double helix which fold repeatedly to form helical "macrobes" which twist either left or right, depending on the temperature (Galloway 1990a,b). The twist direction reflects structure-determining architectural features of the individual cell walls. At 40°C there is no twisting, whereas below that temperature, right-handed clones appear, and above that temperature left-hand ones appear (see Figure 1.10). Similarly, the foraminiferans *Globigerina truncatulinoides* and *G. pachyderma* tend to show left-hand coiling of the shell with low water temperature and right-hand coiling with higher tempera-

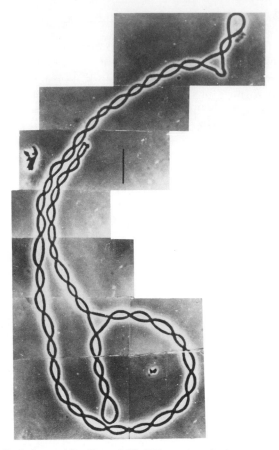

Figure 1.10   Helical clone of *Bacillus subtilis* B1S produced after spore germination and outgrowth in fluid culture at 22°C. Size bar, 10 m. Courtesy of N. H. Mendelson.

tures (Galloway, 1987). Plants, too, may exhibit asymmetries. In legumes of the genus *Medicago*, which includes alfalfa, different alleles, one from each parent at the same locus, determine the direction of pod coiling, the right coiling allele being dominant. It is also a matter of common observation that creepers twine in a preferred direction.

Asymmetries Early in Evolution

The earliest known examples of behavior asymmetries were reported by Babcock and Robison in 1989. In the Cambrian period, half a billion years ago, trilobites, an extinct arthropod with a superficial resemblance to a woodlouse, first appeared; they disappeared at the end of the Paleozoic era. Many fossil specimens bear healed scars which, from their form and appearance, could often be safely attributed to sublethal predation; they appear as broken areas or embayments of the exoskeleton that have become callused. These scars, in contrast to those which were apparently due to "indefinable causation", were found to occur significantly more frequently on the right side. Out of 158 trilobite specimens with injuries that were probably due to "sublethal predation," 69% had injuries only on the right side, 27% only on the left, and 4% had damage to both sides (see Figure 1.11). The authors concluded that predators preferred to attack from a specific direction, although in subsequent correspondence suggestions were made that either the prey might have tended to face their attackers asymmetrically or to have asymmetrically coiled them-

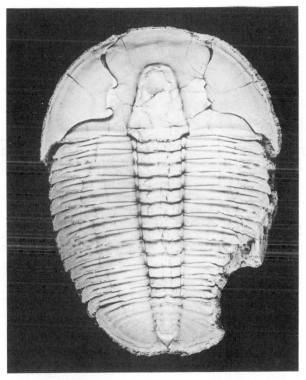

Figure 1.11   The earliest known behavioral asymmetry from around half a billion years ago. A middle Cambrian fossil trilobite *Elrathia kingii* showing a healed arcuate scar which resulted from injury during predation. Such injuries seem to have occurred predominantly on the right side. From L. E. Babcock and R. A. Robison, Preferences of Palaeozoic predators. *Nature*, 1989, 337:695-696, with permission.

selves for protection, or that vital organs were themselves asymmetrically located, leading to differential survival rates. However, Babcock (in personal communication, 1991) dismisses the last possibility. Interestingly, superficial scars on whitefish attributed to the modern lamprey are more common on the left side of the fish (Reist et al. 1987).

The Cambrian period, from which came the earliest trilobites examined by Babcock and Robison, is in fact the oldest system of rocks containing fossils of reasonably complex life forms that can be used for dating and correlation. This and later systems also contain fossils of a controversial group, the carpoids, which Jefferies (1986; Jefferies and Lewis 1978) refers to as the calcichordates, though others (e.g., Ubaghs 1975) regard it as a long extinct group of echinoderms. According to Jefferies, calcichordates (i.e., chordates with an endoskeleton of calcite similar to the echinoderms) are ancestral to the vertebrates. The more conventional view is that vertebrates evolved from "tadpole" larvae of tunicates by paedomorphosis (and see Gee 1989; Thomson 1987), whereas Jefferies claims that adult ancestral chordates were free living, much like the tunicate tadpole larvae, but with a calcite endoskeleton. He redefines a group of primitive carpoids (cornutes) as the stem group of chordates, and defines a relatively derived group of carpoids (mitrates) as the crown group of chordates retaining calcite endoskeletons. He proposes three stem groups within the mitrates, leading, respectively, to acraniates (e.g., the amphioxus *Branchiostoma*), the tunicates, and the vertebrates. Despite problems (e.g., no modern chordates possess a calcite endoskeleton, and the only living animals with such a calcichordate-like skeleton are echinoderms), the calcichordate theory is still held by many, including its critics, as the best and the most self-consistent way of explaining chordate and vertebrate origins (Gee 1989). Whether or not the cornute calcichordates lie on the direct chordate (and ultimately vertebrate) line of evolution, they nonetheless seem (according to the chordate interpretation) to have possessed gill slits only on the left, like the larval amphioxus; conversely, mitrates had gill slits symmetrically located on either side, like the adult amphioxus, though they retained considerable internal asymmetry in the region of the head. According to the echinoderm interpretation, the skeleton (theca) instead would have been asymmetrical.

Neuroanatomists have long pondered the evolutionary significance of the decussation evident in the sensory and motor functions in vertebrates. There is no obvious functional advantage in the left side of the brain monitoring sensorimotor events in the contralateral right space, or for the right side of the brain being largely responsible for what happens on the opposite side of the body. In a scenario different to that of Jefferies,

Kinsbourne (1978) notes that invertebrates naturally lack a rigid notochord or vertebral column running along the dorsal longitudinal axis, even though they may have an anterior specialization of the nervous system analogous to a vertebrate or chordate brain. As in vertebrates, this brain-like ganglion is dorsally located in relation to the more ventral oropharynx, but in invertebrates, the posterior extension of this structure sweeps sharply ventrally, then runs longitudinally along the ventral surface in a posterior direction. Thus, according to Kinsbourne, a change from ventral to dorsal positioning of the main length of the neuraxis may have laid the foundation for, and perhaps necessitated the evolution of, the notochord. The ventral to dorsal displacement of the neuraxis is, he notes, a fundamental innovation of the invertebrate-vertebrate transition, as is a change in the direction of blood flow within the main longitudinal vessels. In invertebrates, blood flowing posteriorly does so ventrally, and returns anteriorly through dorsally located channels. The exact opposite occurs in vertebrates. Kinsbourne speculates on the possibility that early vertebrates underwent a 180-degree rotation of the body beginning behind the anterior head, reversing the position of the neuraxis and blood flow, resulting in decussation. However, as Kolb and Whishaw (1990, p. 29) note, a problem with this theory is that it predicts that the sensory and motor pathways of the vertebrate brainstem should be reversed in a dorso–ventral direction with respect to the spinal cord. However, this is not the case, because the dorsal portion of the midbrain is sensory, as is the dorsal portion of the spinal cord. Moreover, the theory fails to explain why a 180-degree rotation would prove adaptive.

On the other hand, torsional rotation is observed in the abdomens of male flies (Bickel 1990). In most flies belonging to the suborders Nematocera (e.g., crane flies, midges, blackflies, and mosquitoes) and Orthorrhapha (e.g., horseflies, beeflies, and robber flies), the male genitalia at the end of the abdomen undergo torsional rotation about the axis either permanently after emerging from the pupa or temporarily during mating. This causes the internal organs to loop around the gut, dragging with them the reproductive tract, trachea, and plates of the exoskeleton. In the robber fly family, males belonging to some genera twist through 180 degrees when necessary, others are permanently twisted through 180 degrees, and some twist only 90 degrees while coupling. This suggests that inversion may have arisen independently several times, probably to accomodate postural changes during end-to-end coupling. In higher flies (the suborder Cyclorrhapha, e.g., houseflies, fruitflies, and blowflies), the ancestral inversion is supplemented by an additional 180-degree rotation, producing a full 360-degree twist. It is hard to see how such a full 360-degree rotation differs from the unrotated state, unless by doing so constrictions and zones of weakness are produced in the exoskeleton, making the rear part of the male's abdomen more flexible.

Asymmetries in Extant Invertebrates

Neville (1976) documented many of the currently known asymmetries in his short, appropriately titled monograph, *Animal Asymmetries*. He mentions how coelenterates like the Portugese Man-o'-War (*Physalia physalis*) and Jack Sail-by-the-Wind (*Velella velella*) exist in two enantiomorphic forms, their gross structures resembling each other as mirror images, like a pair of gloves. (Conversely, according to Larson, 1991, the jellyfish *Linuche unguiculata* swims clockwise while moving upward through sinking water and simultaneously rotating counterclockwise.) In the case of insects, some species overlap their wings predominantly left over right, and other species vice versa. Male bedbugs usually possess an enlarged genital clasper on the left. In the praying mantis and cockroach the oothecal glands which secrete the egg case exhibit a chemical asymmetry; the large gland on the left produces structural proteins, the glucoside of a tanning agent precursor, and phenoloxidase, whereas the smaller gland on the right produces a glucosidase. On mixing, oxidative cross-linking occurs to harden the egg case, just like the two commercial resin components which similarly harden on mixing.

Turning to molluscs, the bivalve scallop, pecten, swims horizontally, the right side up, and the balancing organ (statocyst) on the right is vestigial. The right eye of the cephalopod *Calliteuthis reversa* is smaller and alone possesses a surrounding ring of photophores (Neville 1976). The molluscan two-lobed liver is usually better developed on the left (Vermeij 1978). In the gastropod class of the mollusca, larvae are generally symmetrical; later, at metamorphosis while undergoing torsion, the posteriorly located viscera, mantle cavity, and shell swing counterclockwise, in most species, 180 degrees with respect to the head and foot to face anteriorly, with loss of gills, auricle, and nephridia occuring on the right side. Most commonly found snail shells do in fact curl to the right, like *Lymnaea peregra*, although there are occasional sinistral individuals in predominantly dextral species. However, *Laciniaria biplicata* usually curves to the left, whereas *Partula suturalis* and *Liguus poeyanus* have a roughly 50/50 dextral–sinistral distribution (Neville 1976; Vermeij 1978). In *Lymnaea*, the direction of torsion is ultimately determined by asymmetries in the egg cytoplasm prior to fertilization and by the pattern of spiral cleavage of the fertilized egg during development, that is, by the plane of orientation of the mitotic spindles when the fertilized egg is undergoing cleavage from the two to the four-cell stage. Inheritance is therefore maternal (Johnson 1982), and dextrality appears to be dominant. Freeman and Lundelius (1982) showed that dextrality, but not sinistrality, requires the presence of a certain protein. Thus cytoplasm from dextral eggs injected into uncleaved sinistral eggs causes them to cleave dextrally, while cytoplasm from sinistral eggs has no effect upon the cleavage pattern of dextral eggs. Consequently, the maternal dextral gene specifies a cytoplasmic product that influences the pattern of cleavage; sinistral eggs cleave in the opposite manner because of the absence of this product.

In the case of mammalian *situs inversus viscerum*, individuals may be born with their normal visceral asymmetry reversed, together with (as we just saw) abnormalities in structure and function of sperm flagella and bronchial cilia. The condition affects less than 1 in 10,000 people. In mice, it is the result of an autosomal recessive gene. In homozygotes, the ratio of normal to reversed chirality is 1:1, suggesting that the normal allele specifies normal chirality, its absence leading to the random allocation of normal and reversed chiralities (see Galloway 1990a,b). An essentially similar model has been proposed by Annett (1985) for the genetics of left-handedness. She argues that a right-determining allele produces dextrality, and a *neutral* (not a nonexistent *left*-determining allele), if present from both parents, gives offspring a 50/50 chance of dextrality-sinistrality. Thus both *situs inversus* and left-handedness may reflect loss of genetic control resulting in a random allocation of chirality. Morgan and Corballis (1978) claim that such a model in fact operates for most biological asymmetries; one allele apparently codes for directionality, and the other remains neutral, with even the coding alleles perhaps not intrinsically coding for directionality, but instead working on a preexisting gradient of asymmetry at the level of the cytoplasm in the fertilized egg. The coding alleles in this case are "left–right agnosic." This situation may in fact explain the slightly higher maternal (presumably extrachromosomal) rather than paternal inheritance of human handedness (for review, see Bradshaw 1989, Ch. 8). Indeed, neural tube defects (e.g., spina bifada) in humans are often transmitted via the mother (Nance 1969), and parents of such offspring are often sinistral (Fraser 1983).

As discussed at the beginning of this chapter, left- and right-handedness is ultimately definable only with respect to the other two dimensions, up–down and front–back; that is, the anteroposterior and the dorsoventral axes. If the determinants of handedness reside in something performed by some factor such as a protein, then that factor must somehow be oriented to or maintain a gradient to these two axes (see Galloway 1990a,b). In conjoined human monozygotic twins (Siamese twins), the viscera are typically normal in one twin, but may be reversed in the other (*situs inversus*). There seem to be no reports of reversal occurring in both; furthermore, one could possibly explain *situs inversus* in a single offspring as representing the solitary survivor of a twin pair, the "normal" member having been lost in a very early embryo stage. Perhaps cleavage very early in the single-cell stage of the embryo and occurring in a particular plane, ultimately results in two clusters of daughter cells. One cluster would be derived from the half cell in which the aforementioned chirality factor was originally confined along a concentration gradient (such that that cluster goes on to develop normal chirality), with the other cluster derived from the half in which the factor was absent. This cluster will then go on to develop normal or abnormal chirality (including *situs inversus*) on a random 50/50 basis (see Galloway 1990a,b).

To return to invertebrates, the decapod crustacea are fundamentally symmetrical and invariably so in their larval stages, although later abdominal asymmetries (torsion and reduction in the number of abdominal appendages—pleopods—on one side) occur in hermit crabs. These creatures inhabit coiled gastropod shells which (as previously discussed) typically coil dextrally; however, asymmetries develop *before* they find a gastropod shell. When the fully grown (up to 5 kg) *Birgus latro* (the coconut robber crab) leaves its borrowed home, presumably in the absence of any other shell big enough or because it can now cope with most eventualities, it does not regain symmetry (Davis 1978). Heterochely, differences in the size and shape of the first pair of walking legs, is also a common development early in adult life, including in *Birgus latro*. Females develop three extra pairs of swimmerets on the left margin of the abdomen. The claw is also larger on the *left;* however, if its growth is suppressed, the right claw does not grow to take over dominance (compare with what happens in certain other species discussed next). The stronger left claw is used for dragging heavy items, breaking open hard coconut shells, combatting adversaries, or steadying an already split coconut so that the more adept right claw can scrape out the meat (compare with skilled dextral behavior in humans).

In fiddler crabs (*Uca* species) one claw (the smaller) may be used for grooming, grasping, cutting, or crushing, and the other larger one for courtship and agonistic displays (Govind 1989). Only in *Uca vocans* is there a population bias, where all the males are *right*-claw dominant (Davis 1978). Claws may differentiate as cutters or crushers, as in the lobster *Homarus americanus* (Govind 1989), which splits 50–50 into cutter on the left and crusher on the right, or vice versa. Here, different kinds of muscle fibers are found in the two claws: massive, long, and slow acting or slender, short, and fast acting (Chapple 1977a,b; Govind 1989).

Initially there is no differentiation; then, slow-acting fibers largely transform to fast-acting fibers in the cutter, and fast-acting fibers entirely transform to slow ones in the crusher (see Figure 1.12). This occurs during a critical period that is preprogrammed, during which differential activity of the paired claws triggers an asymmetry first in the central nervous system and then in the periphery. Thus an interaction between the claws and the environment influences the lateralization of the developing lobster. Removal of one claw late in the larval sequence leads to the intact claw becoming the crusher, while the regenerating claw becomes the cutter. Removal of a claw later, when asymmetry has become naturally established, does not alter it—intact and regenerate claws retain their original configurations. If the creature is reared without a substrate that can be grasped and manipulated, two symmetrical cutters develop. If, while a juvenile, one side is exercised, it becomes the crusher. Govind (1989) invokes a seesaw model; when activation remains below

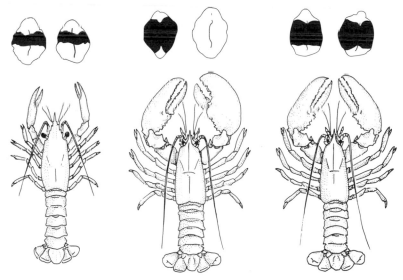

Figure 1.12   The juvenile lobster in its fourth stage (left) has a pair of symmetrical claws in which the compositions of the closer muscles are also symmetrical. A center band of fast fibers is sandwiched by slow fibers (lighter stain). The paired claws and muscles later develop into a cutter claw, with predominantly fast fibers and a ventral band of slow fibers, and a crusher claw with exclusively slow fibers (middle figure). This is the normal condition for adult lobsters reared in the wild, or in the laboratory with a graspable substrate. When lobsters are reared in the laboratory without a graspable substrate, paired, symmetrical cutter claws result (right figure). From C. K. Govind, Asymmetry in lobster claws. *American Scientist*, 1989, 77:468-474, with permission.

a threshold for forming a crusher, the seesaw stays horizontal and two cutters form (see Figure 1.13). With asymmetrical activation, the seesaw tilts, with a crusher developing on the side most activated. The opposite side is then inhibited from becoming a crusher. Paired crushers cannot appear without unbalancing the seesaw. In snapping shrimps removal of the large snapper claw induces the smaller pincer to transform into a snapper, while a new (small) pincer regenerates at the site of the old (large) snapper. The same occurs with the Florida stone crab, *Mennipe mercinaria*, according to Davis (1978); whereas in the case of the Spanish fiddler crab *Uca tangeri*, amputation does not stimulate enlargement of the smaller claw, but instead the amputated claw regenerates to its normal size.

Roundworms and Flatworms

Caenorhabditis elegans is a nematode that exhibits superficial bilateral symmetry, but is in fact asymmetrical at all developmental stages (Wood 1991). Many of its contralaterally analogous cells arise from different lineages on the two sides of the embryo. The embryo is markedly asym-

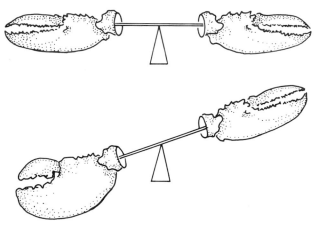

Figure 1.13 A seesaw illustrates how asymmetry is determined in juvenile lobsters. When reflex activity from the two claws is equal to or below a threshold level for forming a crusher claw, the system remains horizontal and both claws develop as cutters. With the occurrence of differences in reflex activity, the system tilts and a crusher develops on the side with higher activity. Simultaneously, the other claw is prevented from becoming a crusher. From C. K. Govind, Asymmetry in lobster claws. *American Scientist*, 1989, 77:468-474, with permission.

metrical, becoming less so during development. Bilateral asymmetry appears between the four- and six-cell stage, though left–right polarity apparently cannot be fixed until after dorsal and ventral polarity is established between the two- and three-cell stages. The four-cell embryo is planar and apparently bilaterally symmetrical. In the next round of cell division there is a skewing of cleavages in an counterclockwise direction, so that eventually the two sides of the embryo differ substantially, not only in the positions of linear homologues, but also in the cell lineages which produce progeny cells of similar fates on either side of the creature. Wood could accomplish a reversal of chirality by micromanipulation at the six-cell stage, resulting in a mirror image, but otherwise normal, healthy, and fertile animal.

According to Eakin and Brandenberger (1981), larvae of certain marine flatworms have two symmetrically situated eyes (ocelli). In *Pseudoceros canadensis*, the two eyes are unlike each other (see Figure 1.14). The right eye is composed of one cup-shaped pigmented cell (eyecup) with a laterally directed cavity and three sensory cells, each extending an array (rhabdomere) of straight, cylindrical, tightly-packed microvilli into the eyecup. One rhabdomere projects dorsally, one ventrally, and the central one projects towards the base of the pigmented cup. The left eye is similarly oriented, and constructed of one pigmented and four sensory cells, three of which bear microvilli like those of the right eye. However the central sensory cell sends unusually large arching cilia into the eyecup among the three rhabdomeres. Eakin and Brandenberger speculate that there may be two evolutionary lines of photoreceptors, ciliary and rhabdomeric; the former is found among phyla which include chordates and vertebrates, whereas the latter occurs among arthropods, annelids, molluscs, and flatworms. Flatworms may, in fact, lie near the divergence between those invertebrate phyla in the rhabdomere line, and those in which ciliary receptors predominate. The larvae of *P. canadensis* maybe retain a ciliary photoreceptor, inherited from more primitive ancestors, together with an innovation, microvilli inherited from ancestral flatworms.

Protochordates

Larvae of the amphioxus (*Branchiostoma lanceolatum*) are markedly asymmetrical; gill slits appear first on the left, having migrated from the right where they were formed (compare with the cornute calcichordates just reviewed). The mouth begins on the left but migrates to the center. Young larvae, swimming by epidermal flagella, rotate clockwise as seen

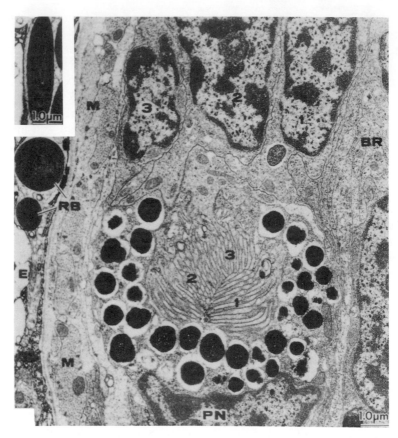

Figure 1.14   Longitudinal sections of asymmetrical ocelli of marine turbellarian (flatworm) *Pseudoceros canadensis*. The section on the left is of a right ocellus showing (towards top) nuclei (1-3) and (beneath) corresponding rhabdomeres (1-3) of the *three* sensory cells. The section on the right is of a left ocellus composed (in the lower half) of the *four* sensory cells: 1-3 bearing rhabdomeres (1-3) of microvilli and one cell bearing cilia (4). M, muscle cells; BR, brain; RB, rhabdites (in cross section); E, epidermis; PN, nucleus of pigmented cells; CS, circumciliary space; BB, basal body. From R. M. Eakin and J. L. Brandenberger, Fine structure of the eyes of *Pseudoceros canadensis* (Turbellaria, Polycladida). *Zoomorphology*, 1981, 98:1–16, with permission.

from the front, so that the mouth on the left is always in front (Neville 1976).

Flatfish

Flatfish are the most conspicuously asymmetrical members of the class Osteichthyes. However, males of *Anableps anableps*, the viviparous tooth

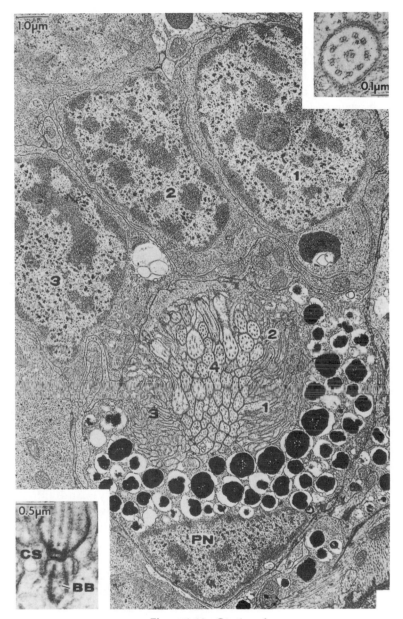

Figure 1.14 Continued

carp of South America, have an anal fin modified as an intromittent organ (gonopodium), which can only be moved to one side (Neville 1976). Females have a genital opening also only on one side, so that males with right-moving gonopodia can only mate with females possessing a genital opening on the left, and vice versa. Apparently there are roughly equal numbers of the two types. In laterally asymmetrical flatfish, of course, it is also advantageous for mating mechanics if most members of a species share the same laterality.

Skates and rays preserve left–right lateral symmetries, undergoing flattening along a vertical dimension. On the other hand, many species of bottom-dwelling flatfish exhibit a multitude of departures from the normal vertebrate body plan of bilateral symmetry, the most conspicuous being the presence of both eyes on the same side of the head. According to Policansky (1982a), the flatfish are a group of over 500 species, using the current taxonomic order. They hatch as externally symmetrical cod-like larvae, which undergo a metamorphosis involving migration of the eyes across the head at ages that range from a few weeks to many months, and at sizes ranging from a few millimeters to more than 12 centimeters. Size, however, in *Platichthys stellatus* may be the important determinant (Policansky 1982b). There is a rotation of the neocranium before ossification is complete, which involves the brain and eye orbits alone, with some resorbtion of the soft bone which would otherwise impede eye travel. The dorsal fin remains behind the head while the eye migrates, and then the fin grows forward almost to the snout tip. (In some species it does so before the eye has completed its journey, and the eye must then move through the fin.) In some species, almost all adult fish lie on the left side (e.g., Soleidae) with both eyes on the right side of the head. In other left-eyed species (e.g., Scophthalmidae and Cynoglossidae) they lie on the right, despite the fact that all probably evolved from the same Percoid ancestor. In still other species (e.g., Psettodidae), about equal numbers of left- and right-eyed individuals are found; nor, apparently, is there any relationship between the side they choose to lie on first and the side on which the eyes end up. In the case of the starry flounder, *Platichthys stellatus,* its nearest relatives are mostly right-eyed, whereas *P. stellatus* in Japanese waters are nearly all left-eyed, with the ratio progressively dropping as a geographical cline to a 50–50 distribution off the west coast of the United States. Whereas in symmetrical fish the left optic nerve crosses the right dorsally or ventrally to it with equal frequency, in flatfish the optic nerve of the *normally* migrating eye almost always crosses dorsally to the other. Thus an individual with an atypical eyedness would suffer possible disadvantage from the need for a double crossing of the optic nerve.

As Policansky (1982a,b) observes, because of the distribution of dextral and sinistral fish, the starry flounder are the ideal flatfish to use for crosses in genetic experiments investigating the inheritance of lateral asymmetries, and for testing the theory of Morgan and Corballis (1978). They too propose that genes do not code for the *direction* of asymmetries, for example, one for left and one for right, but instead work for or against preexisting cytoplasmic asymmetries in the oocyte (and compare with Annett 1985). In particular, they propose that whereas one gene will determine a particular phenotype, its alleles will instead simply interfere with the normal mechanism, perhaps resulting in a chance left–right distribution. (They also propose, on other grounds, that left-biased asymmetries will be in the majority.) Policansky notes that in terms of the Morgan and Corballis model, *P. stellatus* flounders are homozygous for left-eyed alleles in the Japanese population (they receive an allele for left-eyedness from both parents), whereas Californian individuals should be homozygous for neutral alleles. However, the results of currently incomplete breeding experiments were contrary to the prediction of the Morgan and Corballis model that the offspring of all Californian mating patterns would conform to a 50–50 distribution. In fact, of right-eyed Californian males and females, 18% of the offspring were left-eyed; of left-eyed Californians, 83% of the offspring were left-eyed; and the sex of the parents seemed to influence the proportion of left- and right-eyed offspring. In addition, an environmental factor seemed to operate. Cytoplasmic or maternal inheritance seemed to be ruled out by the fact that fathers as well as mothers influenced the proportion of dextral and sinistral offspring in the crosses. Finally, there was no support, looking at the laterality of the various families of flatfish, for the idea of a majority of left-biased asymmetries. Policansky calculates that there are two predominantly dextral families, three predominantly sinistral ones, one family contains a 50–50 split of individuals, and one family consists of two dextral and three sinistral genera.

Habenular Asymmetries and the Parietal Eye of Lower Vertebrates

Braitenberg and Kemali (1970) reported on marked asymmetries between the left and right habenular nuclei which may play a role in reproduction (see Kemali, Guglielmotti, and Fiorino 1990). These epithalamic structures lie in the anterior dorsal diencephalon beneath the epiphysis and to either side of the third ventricle. In frog, newt and eel brains the left side is consistently the more lobate. Indeed according to Morgan, O'Donnell, and Oliver (1973), the left nucleus in tadpoles and young frogs is partially divided by a vertical septum, lacks cells adjoining the third ventricle along part of its length, and contains fewer free-cell bodies in its lumen than the right nucleus. The fully grown adult frog has two distinct habenular nuclei on the left side and only one on the right; the second nucleus on the left is in a position lateral to the original nucleus (see Figure 1.15). Braitenberg and Kemali (1970) speculate that early in evolution two originally paired and bilaterally symmetrical organs, such as two parietal eyes, each connected to the corresponding half of the epithalamus, may have shifted their relative positions. They might have rotated around each other to achieve positions on the medial plane, ending up either one rostral and one caudal, or one dorsal and one ventral. If in this process the connections with the epithalamus were preserved, the fibers would exhibit bilateral asymmetry. One of the two parietal organs may then have lost its function, or assumed a radically different function from its opposite.

The connectivities of the parietal eye of the lizard *Uta stansburiana* were described by Engbretson, Reiner, and Brecha (1981). Again, greater development was found in the medial habenular nucleus on the left, which was subdivided into two parts: *pars dorsolateralis* and *pars ventrome-dialis*. The structure on the right was not so divided, and was found to resemble the *pars ventromedialis* of the left medial habenular nucleus (see Figure 1.16). It seems that centripetal fibers of ganglion cells in the parietal eye project only to the *pars dorsolateralis* of the medial habenular nucleus on the left. The authors speculate that this habenular asymmetry in lizards may be a specialization associated with the unilateral projection from the parietal eye. Reptiles lacking a parietal eye, or a parietal eye and pineal gland (a photosensory complex of the epithalamus present in some anamniotes and reptiles which way be important in reproductive behavior, endocrine function, and thermoregulation) may, however, have symmetrical habenulae. Engbretson et al. conclude that the parietal eye in *Uta stansburiana* may develop ontogenetically from an invagination of the left side of the roof of the diencephalon, whereas the pineal gland may stem from a similar invagination on the right side.

Summary and Conclusions

Here on Earth we can easily define the vertical axis in terms of gravity. Front and back, and left and right are determined at least partly in relation to up–down, partly in relation to an observed object which at least potentially has a preferred orientation or direction of locomotion, and partly in relation to an observer who also is asymmetrical. It is a different matter of course in the void of outer space.

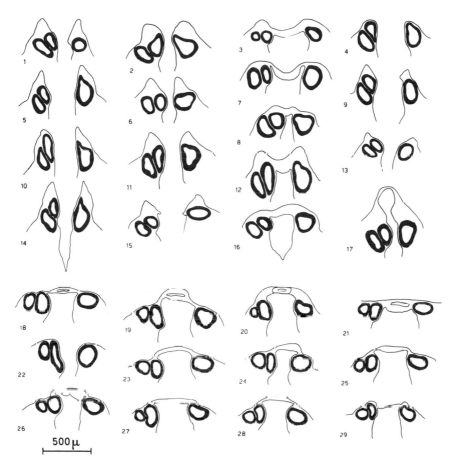

Figure 1.15 Outlines of the nuclear shells of the left and right habenular nuclei in 29 frogs. One to 17: from horizontal sections. Eighteen to 29: from frontal sections. Enlarged 40:1. Note that there are two nuclei on the left and one on the right. From V. Braitenberg and M. Kemali, Exceptions to bilateral symmetry in the epithalamus of lower vertebrates. *Journal of Comparative Neurology*, 1970, 138:137–146, with permission.

While at a superficial level the natural world does not possess any obvious or systematic asymmetries, and most organisms are, superficially at least, bilaterally symmetrical and are thus largely free from consistent turning biases, at a micro level all manner of subtle asymmetries may be present. Indeed, a perfectly symmetrical organism would be unable to tell left from right or to make a consistently biased response.

Human beings have traditionally held in suspicion the sinistral minority, as evidenced by language and myth. Curiously, handedness has long been associated with sex, not only in mythology and prejudice, but

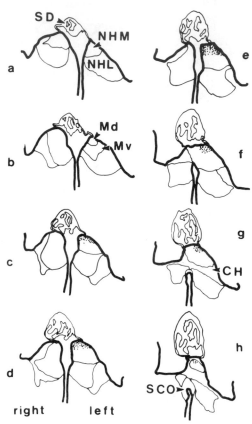

Figure 1.16 Outlines of transverse sections from rostral (a) to caudal (h) of the habenula of the lizard, *Uta stansburiana*. In (a), the left and right medial habenular nuclei (NHM) and left and right lateral habenular nuclei (NHL) are similar, but in (b) the left NHM has become subdivided into pars dorsolateralis (Md) and pars ventromedialis (Mv). The right NHM is not subdivided. By (e), the NHLs are no longer apparent. In (G) the Md appears dorsal to the habenular commissure (CH), while Mv and the entire right NHM remain ventral to CH. SCO, subcommissural organ; SD, dorsal sac. From G. A. Engbretson, A. Reiner, and N. Brecha, Habenular asymmetry and the cerebral connections of the parietal eye of the lizard. *Journal of Comparative Neurology*, 1981, 198:155-165, with permission.

even in the biology of reproduction. However, lateral asymmetries are also found in particle physics, the three-dimensional structure of the carbon molecule, organic molecules, ribonucleic acid, amino acids, and cell cytoplasm. In living creatures we encounter the phenomenon in coiling asymmetry in flagella and cilia, in bacteria (where the direction may be heat dependent), Foramenifera, and of course in creeping plants.

The first behavioral asymmetry was evident half a billion years ago in the wounds inflicted on trilobites, and similar marks of asymmetrical predation are today found on the fish prey of lampreys. Indeed we can trace the evolution of asymmetrical structure and function through all the living phyla. Enantiomorphic or chiral forms are found in coelenterates, molluscs (the gastropod coil has been subject to particular study), arthropods, and flatfish. In amphibians and reptiles we encounter asymmetries in the central nervous system.

A recurrent finding is inheritance via maternal cytoplasm and inheritance of either a positive bias in a single direction or, instead, in the absence of any such inherited factor, chance determination of chirality. Use and environmental factors may modify the expression of "handedness," whereas in certain species of roundworm, asymmetry may appear between the four- and six-cell embryonic stage. In subsequent chapters we shall describe how lateral asymmetries manifest in birds, rodents, and primates, and their relationship to human tool use and language.

Further Reading

Freeman, G., and Lundelius, J. J. The developmental genetics of dextrality and sinistrality in the gastropod *Lymnaea peregra. Wilhelm Roux's Archives of Developmental Biology,* 1982, 191:69 83.

Govind, C. K. Asymmetry in lobster claws. *American Scientist,* 1989, 77:468-474.

Hegstrom, R. A., and Kondepudi, D. K. The handedness of the universe. *Scientific American,* 1990, January, 98 105.

Kemali , M.; Guglielmotti, V.; and Fiorino, L. The asymmetry of the habenular nuclei of female and male frogs in spring and winter. *Brain Research,* 1990, 517:251-255.

Lloyd, G. E. R. Polarity and Analogy: Two Types of Argument in Early Greek Thought. Cambridge: Cambridge University Press, 1966.

Morgan, M. J., and Corballis, M. C. On the biological basis of human laterality: II. The mechanisms of inheritance. *Behavioral and Brain Sciences,* 1978, 2:270-277.

Neville, A. C. Animal Asymmetry. London: Edward Arnold Ltd., 1976.

Policansky, D. The asymmetry of flounders. *Scientific American,* 1982, 246 (5), 96-102. (b)

Asymmetries in Birds

Introduction

The avian brain has clear asymmetry or lateralization of function at several levels of neural organization, ranging from lateralized perceptual input through processing of higher-order information to lateralized control of motor output. Lateralized use of the feet by parrots in manipulating objects has been a subject of interest and study for over a century (Harris 1989). Even though it was originally considered in the same context as handedness in humans, this possible association was later forgotten. Renewed interest in laterality in birds began with the discovery of lateralization of the motor control for singing in songbirds (Nottebohm 1971). Lateralization has now been shown for a range of visual functions, beginning with the report by Rogers and Anson (1979) that in the domestic chicken there is lateralized control by the forebrain hemispheres of a number of visually guided behaviors.

It is not our aim to trace the historical path of research on laterality in birds, but rather to consider laterality in its likely evolutionary context. The requirements of visual perception may have been a major impetus to the evolution of lateralization of brain function. This came about as a consequence of laterally placed eyes, and clearly predates the evolution of birds. In birds, however, with their highly developed visual abilities, visual processing and lateralization may have reached an elaborate form, achieved by mechanisms somewhat different from those used by mammals.

Organization of the Avian Brain and Testing for Lateralization of Visual Behavior

Birds are excellent subjects to test for lateralization of visual function because the optic nerves of the bird cross over almost completely in the optic chiasma (Cowan, Adamson, and Powell 1961; Weidner et al. 1985), and therefore by restricting visual input to one eye or the other, it is possible to relay visual information directly to the contralateral side of the brain only. Thus, lateralization of many visual functions has been revealed simply by testing birds monocularly. In medially eyed humans and primates, by contrast, the left visual field of *each* eye (not the left eye per se) projects to the contralateral right hemisphere, and vice versa. Therefore, it is only by tachistoscopic presentation of a stimulus in the extreme peripheral left or right visual fields that one can achieve in humans the same lateralization of visual input as is possible in birds by simply testing birds monocularly. Some laterally eyed mammals, including albino strains of rats, do have, however, almost complete crossing over of the optic nerve fibers at the optic chiasma and so can be tested monocularly like the chick (see Chapter 3).

In addition, the avian brain lacks the large corpus callosum of the mammalian brain (see Figure 2.1), which interconnects the left and right hemispheres. Both birds and mammals evolved from a common reptilian ancestor, but each diverged along a different path of evolution and only in mammals did large interhemispheric tracts evolve, the corpus

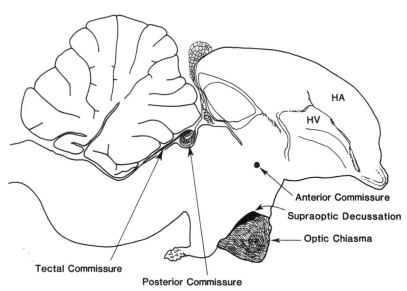

Figure 2.1 A midsagittal section through the chicken brain showing left side–right side connections. Note the absence of the corpus callosum of mammals. There are four commissures, the anterior, posterior, tectal, and pallial (not marked), and one decussation, the supraoptic, which connects nonhomologous regions in the left and right sides. HA, hyperstriatum accessorium; HV, hyperstriatum ventrale. Adapted from Kuenzel and Masson (1988).

callosum in placental mammals and the anterior commissure in marsupials (see Chapters 3 and 4). A number of smaller commissures are present in the avian brain (the pallial, anterior, posterior, and tectal commissures; Cuénod 1974) and left–right connection of nonhomologous brain regions occurs via the supraoptic decussation, but, broadly speaking, the avian brain has fewer pathways that connect the left and right sides, particularly at higher levels of processing in the forebrain. Information can, however, be transferred from one side of the avian brain to the other as interocular transfer for some tasks demonstrates (e.g., visual discrimination learning, Goodale and Graves 1982; Gaston 1984; Watanabe 1986; Zappia and Rogers 1987: imprinting, Horn 1985; but also see, Benowitz 1974) and, similar to the corpus callosum in mammals, those commissures present in the avian brain do appear to have a role in lateralization. Nevertheless, even though interhemispheric and intertectal interactions occur in the avian brain (Watanabe, Hodos, and Bessette 1984), relative to the mammalian brain, birds appear to have less left–right information transfer particularly at higher-order levels of processing.

Thus, not all, but a large amount of information may be processed unihemispherically (i.e., in the hemisphere contralateral to the open eye in monocularly tested birds). When we also take into account that most species of birds have laterally placed eyes (as is also the most common arrangement among vertebrates as a whole; Martin, 1986), we may hypothesize that there were strong evolutionary pressures for birds to either develop, or maintain and elaborate upon, lateralization of brain function. Birds with laterally placed eyes have only a small degree of binocular vision in the frontal field (in pigeons and chickens it is around 16° on either side of the beak; Jahnke 1984; and see Figure 2.2). The majority of their entire visual field is monocular. Hence, most stimuli are perceived by only one eye, at least initially, and the visual input is processed to a large extent by the side of the brain contralateral to that eye. Additionally, with the exception of owls, those avian species which have been studied have been found to frequently use independent movements of the two eyes in order to scan the environment (Walls 1942; Wallman and Pettigrew 1985). Thus each eye often examines different visual environments and stimuli even in the "binocular" field. Furthermore, as Schaeffel, Howland, and Farkas (1986) found in the chicken, each of the eyes can focus completely independently in the lateral field of vision and in part also in the frontal field. Each eye, therefore, acts as an independent unit, and so too must the higher processing centers which receive input from the eyes. Consequently, the presence of hemispheric specialization in birds may be most important to prevent conflicting responses elicited by stimuli perceived by the left and right eyes. By

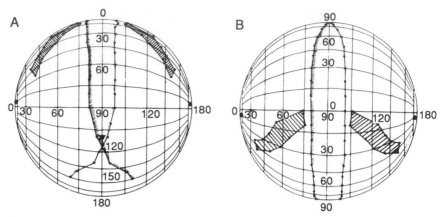

Figure 2.2 The retinal fields of the mallard duck showing the limited degree of binocular overlap. ▧▧▧ , margin of the left eye retinal field; ●●● , margin of the right eye retinal field. The data are presented as if projected onto a sphere around the bird's head. The coordinate system is conventional latitude and longitude, but placed with the poles opposite each eye and the equator vertical. The equator falls on the bird's medial saggital plane. In (A) the bird is facing the reader from the center of the projection. ▼ indicates the bill tip. In (B) the reader is looking down on the bird's head with the bill pointing towards the bottom of the page. The numerical values are in degrees. The hatched areas are the projections of the pecten onto the retinal field. From Martin (1986).

one hemisphere taking a dominant role in the control of given behavior, conflict between the hemispheres may be avoided.

In other words, visual constraints on brain functioning may have been a prime reason for brain lateralization to evolve. This evolutionary step is likely to have predated the birds, instead making its first appearance in reptiles or fish with their laterally placed eyes. Once brain function lateralization evolved for visual purposes, it may have been elaborated on for other neural processing, such as bird song, as we discuss later.

Of course, the resulting asymmetry of responses to visual stimuli may cause the animal problems when a stimulus is seen by one eye only (i.e., it is in the lateral, monocular field of vision). What if the bird responds only to some types of stimuli seen in its right monocular field and to others seen in its left? That does not appear to be a viable system for survival. Clearly, the dominance of one hemisphere over the other cannot be absolute for all situations, or, alternatively, species with laterally placed eyes may turn the head to ensure that the stimulus is seen with both eyes. Perhaps for some aspects of stimulus processing (e.g., for predators) no laterality exists, whereas for other stimuli the avian brain deals with inputs from the left and right eyes quite differently.

Acuity, detection, elicitation of escape (or prey catching) might be performed equally well in response to stimuli perceived by either eye. Indeed, Rogers and Anson (1979) have shown that in chicks there is no lateralization of the habituation of the orientation response to a large, novel stimulus viewed by either of the monocular fields. Nevertheless, there is a large amount of evidence, primarily for the young chicken, that birds do process the information received from the left and right eyes quite differently.

Functional Lateralization of Visual Behaviors

Monocular testing of young chickens (*Gallus gallus domesticus*) has revealed a large number of functional lateralities. Tested in the same task, a chick's performance varies according to whether it is using the left or right eye. The same stimulus elicits different responses according to whether it is presented in the left or right monocular visual field.

In the chicken, neural systems which receive input from the right eye (and are primarily on the left side of the brain) appear to be particularly concerned with the specialized use of conspicuous cues to assign stimuli to categories, whereas those fed by the left eye (and primarily in the right side of the brain) are more involved with all the properties of the stimulus itself, including its position in space (Andrew 1988). There is lateralized specialization for two major modes of information analysis, which must be carried out simultaneously by any animal capable of complex behavior (Andrew 1991). The systems fed by the left eye analyze the detailed and specific properties of any given stimulus, assessing its novelty and its spatial position. This eye system is used to build up a map of the environment, monitor its organization, and respond to changes in it. The systems fed by the right eye, by contrast, are able to process information necessary for classifying stimuli into categories. The right-eye system is less responsive to the unique properties of each stimulus, including its position in space, and better able to assign stimuli to categories accommodating a range of differing exemplars (Andrew 1991). Such categorization is important when rapid decisions must be made about the appropriate response to perform. It is an ability that would serve well for classifying types of food objects.

In fact, in visual tasks requiring pecking for food objects scattered among nonfood objects, chicks using the right eye shift to pecking at the food objects sooner than do those using the left eye. For example, when chicks are tested monocularly on a task requiring them to discriminate food grains scattered randomly on a background of small pebbles adhering to the floor, and differing from the grains in texture and hue but not in size, color, or shape (Figure 2.3), those using the right eye direct their pecks at food grains and avoid pecking pebbles faster than those using the left (Andrew, Mench, and Rainey 1982; Mench and Andrew 1986; Zappia and Rogers 1987). At first the chicks peck randomly at grain and pebbles, but within 60 pecks those using the right eye shift to pecking predominantly at grain. Within the 60 pecks allowed no significant shift away from pecking pebbles to pecking grain occurs in chicks using the left eye. Although we may say that chicks using the right eye learn faster than those using the left, it could also be that those using the left eye prefer to peck at pebbles and so show no shift away from pecking them. If so, left–eye/right–eye differences may depend on lateralized neural circuits which do not involve learning per se. For example, neural circuits fed by the left eye may be more concerned with novelty rather than learning to feed. Chicks trained binocularly learn not to peck at pebbles or shift away from pecking pebbles to pecking grain at the same rate as those using the right eye, which suggests dominance of the right eye for this task.

In female chicks, lateralization in monocular tests on the pebble–grain discrimination task is less marked, and less easily revealed, compared to

Figure 2.3 A chick pecking for grain scattered on a background of small pebbles adhered to the floor. The grains differ from the pebbles in texture and hue but not in size, color, or shape. Males tested monocularly using the right eye learn faster than those using the left eye.

males. Zappia and Rogers (1987) found no left–right eye difference in the performance of females aged 14 to 16 days tested on this visual discrimination task. Mench and Andrew (1986) did, however, report a right eye advantage in younger females tested on the same task.

Other evidence indicates that the right-eye dominance for control of pecking in the pebble–grain discrimination task may be a manifestation of specialization of the left hemisphere for controlling the responses required by this task. If development of the left hemisphere is disrupted by injecting the protein synthesis inhibitor cycloheximide or the neuro-transmitter glutamate into the left hemisphere on day 2 posthatching, chicks tested binocularly in the second week of life show no shift away from pecking at random at pebbles and grains (Rogers and Anson 1979; Howard, Rogers, and Boura 1980; Rogers and Hambley 1982). The chicks appear to show permanent retardation of visual discrimination learning. Similar treatment of the right hemisphere has no effect on this perfor-mance.

In contrast to the sex difference apparent in monocular testing of males and females, the laterality revealed by treating the left or right hemisphere with cycloheximide or glutamate is just as marked in females as in males (Rogers 1986). The less easily revealed lateralization of un-treated females tested monocularly on this task (days 14 to 16; see pre-viously) may therefore indicate that there is a sex difference in the percep-tual input pathways that carry information from the eyes to the forebrain circuits which process the information and control the pecking response (Rogers 1986). In other words, females and males appear to have equiva-lent degrees of lateralization for the cognitive processes required to per-form this task; but because either eye of the female learns as well as the right eye of the male, both eyes of the female may have access to these cognitive areas, whereas in males it is only the right eye which has this access (see later for evidence in support of this).

Thus far, we have been cautious not to conclude that shifting away from pebbles to peck at grain necessarily implies learning in the peb-ble–grain discrimination task, since chicks using the left eye may be able to discriminate grain from pebbles but prefer to continue pecking at both pebbles and grain as they respond to each stimulus for its own unique characteristics rather than as a category or type. Nevertheless, other evidence supports the interpretation that, in fact, chicks using the right eye learn, and remember, visual discrimination tasks better than those using the left. Gaston (1984) trained chicks aged 3 to 25 days to discrimi-nate between cross and triangle patterns in an operant chamber using food as reward. The chicks using the right eye acquired the pattern discrimination with fewer trials than those using the left, and they showed better transfer of the learning to the other side of the brain when tested for recall monocularly using the other eye. Also, Gaston and Gaston (1984) trained chicks binocularly on a visual pattern discrimination task and then tested them for recall monocularly with either the left or right eye open. Recall of the task occurred only in the chicks using the right eye.

This latter test in which the chicks were trained binocularly (Gaston and Gaston 1984) generates competition between the two sides of the brain and so reveals dominance of the right-eye system. It is possible that only the right eye (of males) has access to the circuits that can learn and consolidate memory for this task, which would mean that although the chicks were trained with both eyes open, they were essentially performing monocularly. As Andrew (1991) has suggested, either eye system may take charge of the behavior depending on the nature of the task, and in this case it is the right-eye system. It is only in tasks such as this, in which the animals are trained binocularly, that dominance can be demonstrated. Monocular training, by itself, may reveal lateralization or specialization of the systems fed by each eye, but demonstration of dominance requires competition between the two eye systems.

Right-eye advantage in a visual discrimination learning task and for control of pecking rate has also been reported for the adult pigeon (Güntürkün 1985; Güntürkün and Kesch 1987). Pigeons trained to discriminate grains from pebbles are more accurate and have a faster pecking rate when they use the right eye (Güntürkün and Kesch 1987). The task chosen by these authors was similar to the pebble–grain task used to test chicks. Safflower seeds were mixed with pebbles resembling the grains in their range of colors and shapes. The birds were allowed to feed from a trough containing the mixture and, over several trials, the number of grains ingested was calculated and the number of pecks performed was scored. The pigeons tested using the right eye consumed significantly more seeds than those using the left. This higher success was due to better discrimination when using the right eye. Similarly, when pigeons were trained for pattern discrimination (using a number of patterns) by operant conditioning, the pecking rate was found to be faster for those using the right eye (Güntürkün and Hoferichter 1985). Like the young chicken, adult pigeons also have right-eye dominance for retention performance of a visual discrimination task (von Fersen and Güntürkün 1990). In this task the pigeons were trained to discriminate 100 different visual patterns from 625 similar patterns and then they were tested for retention either binocularly or monocularly (Figure 2.4). Retention performance was significantly higher when the right eye was used compared to the left eye. The retention was, however, even higher when the pigeons were tested binocularly, which indicated that the two eye

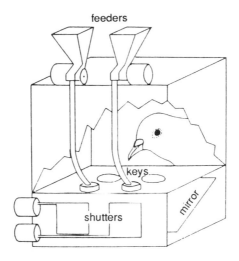

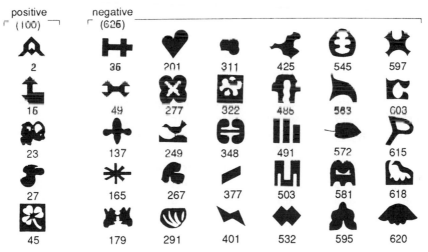

Figure 2.4 Pigeons were tested using a simultaneous instrumental discrimination procedure. They were trained to discriminate 100 different visual patterns from 625 similar patterns, examples of which are illustrated. Positive means rewarded stimuli; negative means nonrewarded stimuli. When tested monocularly, the pigeons performed better with the right eye. From von Fersen and Güntürkün (1990).

systems must cooperate in this case. The two eye systems similarly cooperate in visual discrimination learning (Watanabe et al. 1984).

It should be noted that although young chicks show the same lateralization for visual discrimination performance as adult pigeons, adult chickens may not be comparable to the adult pigeons because the left–right eye difference in visual discrimination learning in the chick may no longer be present in adulthood. Male chicks tested at 3 weeks of age show equal ability to learn with the left and right eyes (Rogers 1991), which suggests that in chickens the lateralization for this task may be present only during during the early stages of development. However, adult chickens have not yet been tested on any of these visual discrimination tasks and no conclusions can be drawn until this has been done. Other evidence suggests that the visual discrimination performance of adult chickens may differ from that of adult pigeons in terms of lateralization because Schulte (1970) has reported that monocularly trained hens learn to discriminate faster than binocularly trained ones, and monocular hens show better retention than binocular ones. This contrasts with the finding with pigeons and suggests that in chickens interference occurs when the two eye systems are used together. It seems rather difficult to accept that two eyes would be worse than one. However, the report by Schulte (1970) was made prior to the discovery of lateralization in the avian brain, and no mention was made of which eye was being used in the monocular condition.

A left–right eye difference is also evident in chicks tested in an avoidance task involving pecking at a bead coated with methylanthranilate (Andrew 1988; Andrew and Brennan 1985). Methylanthranilate makes the bead taste bitter, and the chick shows a disgust response of bill shaking and wiping after it pecks the bead. When tested for memory retention of the task, those chicks tested using the right eye avoided pecking the bead on which they had been trained and differentiated it from other beads of differing color. Chicks tested using the left eye generalized to avoid all beads irrespective of their color. Chicks using the left eye do, however, pay attention to color when they are tested for habituation to pecking a colored bead. Pecking at the bead dishabituates when its color is changed. Thus, it seems that chicks using either the left or right eye show differences in attendance to the perceptual properties of stimuli according to the task on which they are being tested.

The differing abilities of the left and right eye systems have been used to explain some recent, interesting observations made by Vallortigara and Andrew (1991). Chicks were raised in isolation in cages with a table tennis ball, painted red and with a white bar on one face, suspended at their eye level. They were later tested by being placed in the middle of a runway with the familiar ball suspended at one end and an unfamiliar one, with various rotated positions of the white bar, suspended at the other end. Their choice of approach and staying near each object was scored. Males raised with a ball with a white vertical bar and tested using the left eye chose to associate with the familiar test-object when the unfamiliar object had a horizontal bar and with the unfamiliar object when it had an oblique bar (45° to the vertical). Males using the right eye chose at random in both of these choice tests. To summarize, those using the right eye did not differentiate their responses between objects presented in the test unless the unfamiliar object differed markedly from the test-object (e.g., by having no bar at all or having a horizontal white stripe around the entire circumference of the ball). Only then did they choose to associate preferentially with the familiar object. Thus, chicks using the left eye are responsive to small changes, whereas those using the right eye ignore small changes. Such choices were also evident in response to other chicks. Male chicks raised with a companion chick and then tested in the runway with a choice of their companion and a strange chick avoided the stranger when they were tested using the left eye, and chose at random when they were tested using the right eye. Presumably the systems fed by the right eye were interested in chicks as a category and not in the small differences that designate individuals. This result is consistent with faster learning to peck at grain in the pebble–floor task, as discussed already: Chicks using the right eye also categorize food versus nonfood objects and direct their pecks accordingly.

Female chicks also show left eye–right eye differences in the imprinting choice test. Under all conditions of testing with the balls, females chose the familiar model; but when given a choice between familiar and strange chicks, the females using the right eye, like males using the right eye, showed no preference.

The left-eye system is more responsive to the spatial position of stimuli than is the right. For example, habituation of pecking at a violet, illuminated bead following repeated presentations of the bead to the left eye is dishabituated when the position of the stimulus is changed, but similar dishabituation does not occur when the position of the stimulus presented to the right eye is changed (Andrew 1983). In this test it was actually the position of the entry of the bead into the chick's cage that was changed. The chick was first habituated to the bead being introduced into the cage repeatedly from above and then the bead was introduced from below. The same involvement of the left eye system with spatial or topographical cues, and the right with categorization of stimuli, has been demonstrated in a very different testing situation. Vallortigara, Zandforlin, and Cailotto (1988) trained chicks to locate the position of a food source using spatial cues provided by the positioning of a small box. The chicks were tested binocularly. Their learning performance was superior when the box was placed on the chick's left side. In a similar task which used color discrimination as the cue for the food source instead of spatial cues, the chicks learned better when the color cue was placed on their right side (Vallortigara 1989). This result may be consistent with right-eye superiority for categorization of nontopographical stimuli.

Left-eye superiority for spatial orientation has also been demonstrated in a task requiring chicks to search for food using cues from large, relatively distant objects (Andrew 1988; Rashid and Andrew 1989). The chicks were trained binocularly to find a small patch of food buried in sawdust in one corner of an arena. The position of the food had to be determined by using features of the testing room and features of the walls of the arena together with cues placed near the food source. At test, no food was present and the arena and internal cues were rotated through 180 degrees so that two areas of the tray were indicated as food sources—one being specified by the unchanged cues of the room external to the arena (distant cues), and the other by the shifted features of the arena walls and the cues inside the arena (near cues). The chicks were tested monocularly. From day 9 of testing on those using the left eye showed a superior searching strategy. They spent most of their time searching in the two areas, indicating that they used both the distant and the near cues to specify the food source. The chicks using the right eye searched inefficiently over the entire arena. The left-eye system (and right hemisphere) is thus the one that attends to and encodes information about the spatial position of stimuli. Andrew (1988) suggests that the left-eye system is better suited to building up a description of the spatial organization of the environment because it examines the relationships between stimuli, encoding their unique characteristics rather than assigning them to categories, as the right-eye system does.

This specialization of the left eye should make it more responsive to novel stimuli, and indeed earlier reports had shown a left eye–right hemisphere advantage for responding to novelty. In males, responses to novel stimuli are more marked to stimuli that are present in the left monocular visual field than compared to the right, although this varies at some ages (Rogers and Anson 1979; Andrew and Brennan 1983). Also, the chick is more likely to produce distress vocalizations (peeps) when novel stimuli are presented to the left eye (Andrew et al. 1982). The lateralized response to novelty was first discovered by introducing small red beads simultaneously into the chick's left and right monocular visual fields and scoring to which bead the chick oriented, usually to peck at it (Rogers and Anson 1979). The beads were moved toward the chick from behind when the chick was feeding from a dish. The chick was twice as likely to orient to the bead on its left side, which was interpreted to mean that novelty detection was more likely to occur in the left monocular field of vision.

Lateralization of Fear Responses, Attack, and Copulation

Consistent with the lateralization for response to novelty, the left and right hemispheres of the chick are lateralized for control of fear. Fear responses depend to some extent on lateralized control processes located in the archistriatal regions of the forebrain, the right archistriatum being more involved than the left. This was delineated by injecting kainic acid, which causes a localized lesion of the brain tissue surrounding the injection site, into the left or right archistriatum of 5-day-old chicks (Phillips and Youngren 1986). Fear behavior in an unfamiliar environment was scored 12 days later. The kainic acid treated chicks made distress calls (peeped) less often than the controls; the chicks that received kainic acid injections into the right archistriatum had lower peeping scores than those injected in the left archistriatum. As fear responses in the open field require novelty detection primarily based on topographical cues, the results of Phillips and Youngren could stem from the left eye–right hemisphere's specialization for processing topographical information and novelty detection (see earlier).

The right hemisphere appears to be dominant in controlling attack and copulation behaviors (Rogers 1982; Zappia and Rogers 1983; Bullock and Rogers 1986). A marked laterality in controlling attack and copulation was first discovered by injecting cycloheximide or glutamate into the left or right hemisphere of the chick on day 2 posthatching (Rogers 1980a; Howard et al. 1980). Attack and copulation levels were measured in the second week of life using the standard hand-thrust tests developed by Andrew (1966). The hand is used to simulate a chick either attacking (fingers arched and thrust at the chick's head height) or copulating (hand held flat, palm facing floor and below chest height of the chick) and the responses of the chick to the hand are ranked (Figure 2.5). For attack, a score of 0 is given if the chick turns away from the hand, and a maximum score of 10 is given if the chick pecks at the hand and performs attack-leaping at the hand. For copulation, a maximum score of 10 is given if the chick mounts the out-stretched hand and couches fully with pecking, treading, and pelvic thrusting. Following injection of the left, but not the right, hemisphere with glutamate or cycloheximide, attack and copulation levels were elevated. Similar treatment of older chicks had the same effect (Bullock and Rogers 1986). In another set of experiments chicks were treated with the hormone testosterone, which also elevates attack and copulation (Andrew 1966), and then were tested monocularly for these behaviors. When the chicks were tested using the left eye, attack and copulation levels were elevated as expected following the testosterone treatment, but when they were tested using the right eye these

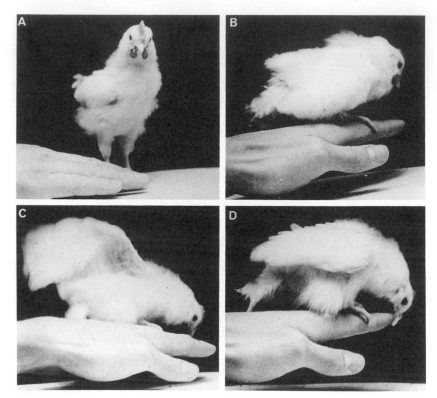

Figure 2.5 A chick performing copulation in the standard hand-thrust test. The photographs are taken from a series of videotaped stills. The bird mounts the hand, shows treading, grasps the hand, and performs pelvic thrusting. From Zappia and Rogers (1983).

behaviors remained at the low levels characteristic of untreated controls (Rogers Zappia, and Bullock 1985). Taken together, these data may indicate that the left hemisphere system suppresses attack and copulation, and that the right hemisphere is specialized to activate attack and copulation.

Asymmetry for courtship behavior has been reported in the zebra finch, *Taeniopygia guttata*. During courtship the male approaches the female so that he views her with his right eye (Workman and Andrew 1986). It should be noted that this direction of asymmetry (right eye–left hemisphere) is the opposite to that of chick copulation, and one might query whether it is, rather, the female who positions herself to view the male with her left eye–right hemisphere. Workman and Andrew (1986) measured the preferred approach orientation of the male to the female as the male moved along a narrow perch. Their result has been disputed by ten Cate et al. (1990), who measured the orientation of male zebra finches as they moved around a circular corridor surrounded either on the inside or the outside by cages containing females, and found no evidence for asymmetry in the males' courtship behavior. These differing results may stem from the two very different testing situations or from the fact that in the experiments of Workman and Andrew, but not those of ten Cate et al., the males were tested only with their female partner (Workman and Andrew 1991). The latter reason is, however, disputed by ten Cate (1991). Monocular testing of zebra finches may help to solve this debate.

The forebrain of the bird is involved in the control of attack and copulation, but lower brain regions are also involved in these responses (Barfield 1969). There are receptors for the hormone testosterone in the septal, archistriatal, and preoptic regions of the brain (Barfield, Ronay, and Pfaff 1978) among other regions (Harding, Walters, and Parsons 1984; Vockel, Pröve, and Balthazart 1990). Given the well-documented involvement of the preoptic region of the hypothalamus in the copulation behavior of birds, there may be lateralization at this level of brain organization too. In fact, Hutchison, Steimer, and Hutchison (1986) have shown an asymmetry in the levels of an important enzyme in the left and right preoptic regions of the ring dove. The enzyme is aromatase, which converts testosterone into estrogen, a step which must occur if testosterone is to activate behavior. The level of aromatase is higher in the right preoptic area, indicating that testosterone has a greater effect on the right side. Therefore, the right preoptic area appears to be more involved in the elevation of copulation that occurs when testosterone levels rise.

Taken together, these data indicate that all of the regions on the right side of the brain which are involved in the control of copulation behavior, including higher and lower levels of neural organization, may have a dominant role over their equivalents on the left side of the brain.

Sex Differences and Changes in Functional Lateralization with Age

It is important to state that in all of the tasks which have revealed lateralization in the chick, both age and sex are important variables. Left–right laterality for fear responses and for avoiding pecking at pebbles in the pebble–floor test varies with age, and in some monocular tests females show weaker asymmetry than do males (Andrew and Brennan 1983, 1984; Vallortigara 1989; Zappia and Rogers 1987).

We have already mentioned the sex difference in lateralization in chicks tested monocularly in the pebble–grain discrimination task. At an age of 5 days, both females and males show better learning with the right eye than the left (Mench and Andrew 1986), but in the second week of life females show equally fast learning with either eye, whereas males still learn better with the right eye (Zappia and Rogers 1987). These data suggest that in chickens there may be a sex difference in the time courses of lateralized development. Incidentally, adult male and female pigeons show no difference in lateralization of visual discrimination (Güntürkün, Emmerton, and Delius 1989).

Developmental changes in lateralization are clearly demonstrated in chicks tested for age-related fear responses to a novel, violet colored bead. Andrew and Brennan (1983) have shown that on day 5 posthatching, male chickens presented with a violet bead perform more fear responses (backing away from the bead, jumping and peeping) when they are tested monocularly using the right eye than when they use the left eye. At this age the amount of fear displayed by chicks tested binocularly is equivalent to that of those tested using the right eye. On day 8 this reverses, as greater fear is shown by male chicks using the left eye. At this age the fear scores of chicks tested binocularly is equivalent to that of those using the left eye, suggesting that there has been a reversal of dominance for control of this behavior. By day 10 low levels of fear are elicited from both of the monocularly tested groups (i.e., there is no longer any detectable lateralization of control for this behavior), but very high levels of fear are scored for the binocularly tested group. Andrew and Brennan (1983) suggest that coupling between the two sides of the brain (hemispheres) has occurred by day 10.

Females tested with the violet bead show no left eye–right eye differences at any age (Andrew and Brennan 1984). Both left- and right-eyed females show a time-course similar to that of right-eyed males. These developmental changes in lateralization and dominance are correlated with changes in cellular function in the forebrain of the chick (Rogers 1986, 1991; Rogers and Erhlich 1983). If the drug cycloheximide is injected into the left forebrain hemisphere of male chicks at any one time between day 2 and day 5 posthatching, learning to discriminate grains from pebbles tested on day 14 is slowed (as discussed previously). Similar treatment of the right hemisphere of males in this time period is without effect. On day 7 neither hemisphere is affected by treatment with cycloheximide, and so too on day 9. A very brief period of left hemisphere susceptibility occurs on day 8, and of right hemisphere susceptibility on days 10 and 11. These data indicate that the developing male chick brain passes through developmental changes that occur first in the left and then in the right hemisphere. Females have a different time course of susceptibility to cycloheximide treatment of the left or right hemispheres (Rogers 1991). On day 2, as for males, treatment on the left hemisphere and not on the right leads to retarded learning of the pebble discrimination task, but this is the only age at which females show any effect from the treatment. These biochemical phases may correlate with shifts in dominance between the left and right hemispheres during early development. On, and probably up to, day 8 the left hemisphere appears to be dominant, and there is a shift to right hemisphere control on days 10 and 11.

Consistent with changes in hemispheric neurochemistry, the behavioral development of the chick undergoes a series of rather precisely timed phases which might well be explained by shifts in dominance. For example, it may not be coincidental that young chicks raised with the hen under seminatural conditions first begin to move around their farmyard environment independently of their mother on day 10 posthatching (Workman and Andrew 1989). On day 10 they show a sharp increase in running ahead of the hen instead of simply following her and a sharp increase in the amount of time they are out of view of the hen. At this age the right hemisphere may become dominant and, as discussed earlier, this hemisphere is concerned with the spatial relations of stimuli and novelty. The shift to dominance of the right hemisphere on day 10 may therefore generate in the young chick increased curiosity and the ability to assess topographical relationships of its environment.

The changing hemispheric dominance for control of behavior during development is rather well demonstrated in a task in which two identical stimuli are introduced simultaneously into the left and right monocular fields of vision and a note is made of at which one the chick pecks (data by M. Dharmaretnam, presented in Workman, Kent, and Andrew 1991). Two dead *Tenebrio* beetles were adhered to each prong of a "pitch fork" and this was introduced from behind the chick's head. Chicks aged 8 days turned to peck at the beetle in their right visual field, indicating left hemisphere dominance, whereas those aged 10 days and more pecked the beetle on the left, which indicates right hemisphere dominance.

A similar shift of dominance has been demonstrated in chicks allowed to view a hen for the first time (Workman et al. 1991). The chick had to put its head through a hole to view the hen placed at a distance of 50 cm, but it was otherwise unrestrained. On day 8 of life the chicks preferred to view the hen with the right monocular field (left hemisphere), whereas on day 10 they preferred to use the left field (right hemisphere). This form of lateralization was present in both males and females; indeed, the shift from right eye on day 8 to left eye on day 10 was even stronger in females.

Given these marked changes in the lateralization of the developing chick, it is reasonable to assume that not all of the lateralities present in young chicks will be retained in adulthood. Unfortunately, because there has been very little investigation of lateralization in adult chickens, we cannot draw conclusions. Lateralization is not, however, completely lost in adulthood. Hens wearing a monocular occluder of the right binocular field of vision score higher numbers of threat displays than do those with similar occlusion of the left field (Rogers 1991). A recent study has shown that hens look up to scan with the left eye when they are played a recording of a chicken's alarm call for an aerial predator (Marler, personal communication, 1991, and Evans and Marler, 1992). This result is most interesting because it suggests that the vocalization carries meaning which elicits motor responses associated with the left eye (notably, the one more interested in novelty) and the neural circuits connected to it.

Visual lateralization does not appear to be sex-dependent in the adult pigeon (Güntürkün and Kischel 1992). Male and female pigeons (12 animals in each group) were trained binocularly on an operant visual discrimination task and tested monocularly, as described previously. Pecking responses were higher in the pigeons tested using the right eye, and there was no significant difference between the sexes.

Factors Influencing the Development of Functional Lateralization for Visual Behaviors

Functional lateralization in the chicken is influenced by light stimulation of the embryo (Rogers 1982; Rogers 1990; Zappia and Rogers 1983). During the later stages of incubation, the embryo is oriented in the egg with its head turned against the left side of the body so that the left eye is occluded and the right eye remains exposed next to the air sac and able to receive light stimulation entering through the egg shell and membranes (Figure 2.6). This asymmetrical stimulation of the eyes plays a significant role in determining the direction of lateralization in the brain.

Throughout the early stages of embryonic development the chick embryo lies on the yolk sac on its left side, with its right eye uppermost, next to the egg membranes and shell (Freeman and Vince 1974). Although this orientation of the developing embryo may have some, as yet unknown, consequences on lateralization, it has no influence on the development of the neural pathways which respond to light because the retina is not yet functional. As the embryo develops it adopts a position in the egg in which both eyes are occluded by the yolk sac and then, over the last 5 or so days of incubation, as the yolk sac shrinks, the head tilts to the left side of the body and becomes positioned so that only the left eye is occluded (Freeman and Vince 1974). Coincidentally with the adoption of this posture, the first visually evoked responses in the optic tectum of the chick occur on day 17 of incubation. On day 18, the first electroretinogram (of embryonic form) can be detected and the first behavioral responses to light occur, and on day 19 the electroretinogram reaches its posthatching form and light-evoked potentials can be detected in the contralateral forebrain hemisphere for the first time (Freeman and Vince 1974; Sedláček 1972, Thanos and Bonhoeffer 1987). Thus, during the last days of incubation the visual pathways are developing functional connections, and it is during this period when the lateralized light stimulation can affect lateralization of brain function.

Chicks hatched from eggs incubated in darkness show no population bias in lateralization for performance of the pebble–grain pecking task or for control of attack and copulation behaviors (Rogers 1982; Zappia and Rogers 1983). Although each individual chick may retain lateralization of these brain functions, there is no consistency in the direction of this lateralization and so no population bias. In other words, some individuals appear to be lateralized in one direction and some in the other. If the eggs are incubated in darkness up until day 17 and then placed in the light, the population of chicks that hatches will be lateralized (Zappia and Rogers 1983). Conversely, if the eggs are incubated in the light up until day 17 and then kept in darkness until after they hatch, no lateralization will be present in the chicks which hatch. Hence, it is only during the last 5 days of incubation, when functional visual connections are developing, that light stimulation establishes a lateralized bias of brain function.

Strikingly, as little as 2 hours of light stimulation given on day 19 of incubation is sufficient to establish lateralization in the population (Rogers 1982). It may not be necessary for this 2 hours of light exposure to occur in one interval. Possibly, shorter periods of exposure spread over the last 5 days of incubation are sufficient, but this has yet to be tested. Nevertheless, it seems highly likely that eggs being incubated by the hen receive sufficient light exposure for lateralization to be established within the population. So far, there has been no investigation of lateralization in birds, such as wild fowl, incubated and hatched in the natural environment, but Workman and Andrew (1989) have reported that lateralization is present in domestic chicks hatched from eggs incubated by hens in

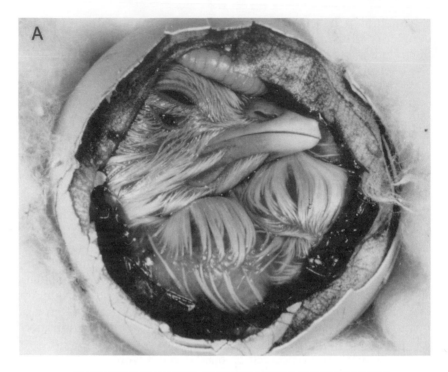

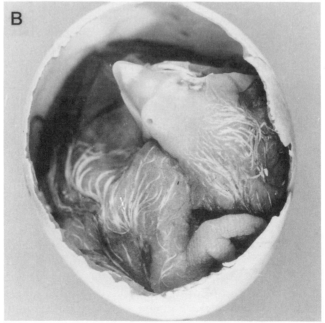

the farmyard. At an age of 8 days, these chicks showed a right-eye preference for viewing the human observer.

The presence or absence of lateralization in groups of chicks hatched from eggs incubated in the light or in darkness, respectively, is not simply a nonspecific response to the presence or absence of stimulation, but rather a specific effect of light in determining lateralization of brain development. Quite dramatically, the direction of lateralization in the population can be completely reversed in the laboratory by manipulating the embryo so that the left eye is exposed to light but the right is not (Rogers 1990). This is achieved by withdrawing the embryo's head from the egg on day 19/20 of incubation, occluding the right eye and exposing the left eye to light. Groups of chicks treated in this way have been shown to have reversed lateralization for controlling attack and copulation and for pecking in the pebble–grain task (Figure 2.7). Males and females are affected equally. When glutamate is injected into the right hemisphere of these chicks, performance on the pebble–grain test is impaired and copulation scores are elevated, whereas glutamate treatment of the left hemisphere has no effect (Rogers 1990). This is the reverse of the usual situation in chicks hatched from eggs incubated in the light or in the control chicks for this experiment, each of which had the head removed from the egg and an eye-patch applied to the left eye.

These experiments demonstrate a rather remarkable ontogenetic process for the differentiation of brain lateralization in the chick. It is possible that the lateralized light stimulation induces lateralized developmental events which lay the basis for further lateralized development. In other words, the light may trigger a chain of lateralization steps which is ultimately manifested in a wide range of functions. Initially, however, the light stimulation of the right eye stimulates development of regions of the brain that receive direct input from this eye, and these are primarily on the left side of the midbrain, the left tectum and left side of the thalamus, which, in turn, has input to the forebrain (see next section). The development of lateralization in the chick is therefore an elegant model of the interaction of genetic and environmental influences in the

---

Figure 2.6   A, Orientation of the chick embryo in the egg during the last 3 days before hatching. The angle of viewing is from the air-sac end of the egg. Note that the left eye is occluded by the body, whereas the right eye remains exposed to light input which passes through the shell and membranes. Even when the eye is closed, light can pass through the transparent eyelid. At this stage of development the eyelid opens and closes frequently. From Rogers (1990). B, Parrot embryo (*Agapornis roseicollis*) just prior to hatching. Here the view is from the side. Note that the embryo adopts the same posture as the chick embryo, with the right eye exposed to light stimulation. From Rogers (1986).

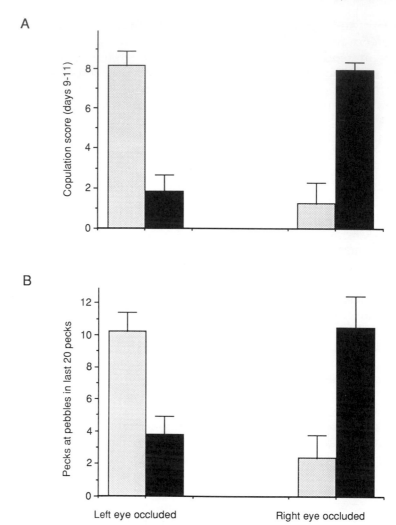

Figure 2.7 Reversed lateralization for (A) copulation and (B) visual discrimination performance caused by stimulating the left eye of the embryo with light. On day 19/20 of incubation, the embryo's head was withdrawn from the egg and an eye patch was applied for 24 hours to either the left eye (mimicking the normal situation; see Figure 2.6) or the right eye. After hatching, the chicks were tested for lateralization by injecting glutamate into either the left hemisphere (open bars) or the right hemisphere (black bars) followed by testing for copulation and visual discrimination learning in the pebble–grain task. The controls, which had occlusion of the left eye, show elevated levels of copulation and slower learning when injected in the left hemisphere compared to the right. In the test animals this was reversed. Group means for 8 to 10 subjects and standard errors are plotted. For copulation, data from days 9 to 11 are given. For visual discrimination, the number of pecks at pebbles in the last 20 pecks of the task indicates learning rate (higher scores mean slower learning).

course of development. The orientation of the embryo in the egg is most likely to be genetically determined, and this is the key to lateralized environmental stimulation, which produces brain function lateralization.

It is not known whether the same development processes occur in other avian species; yet most species tilt the head to the left, leaving only the right eye exposed to light stimulation (Rogers 1986). Incidentally, even if the right eye is closed, the eyelid in most avian species is transparent and so allows light to reach the eye. In chicks it is known that the eye opens and closes frequently during the last stages of embryonic development (Freeman and Vince 1974) and more so in response to auditory and tactile stimulation (Vince and Toosey 1980). Hence, sensory input in modalities other than vision may enhance the effect of the light stimulation in determining lateralization.

There are some, although few, avian species that do not tilt the head in the embryo stages of development (Baltin, 1969). One of these is *Alectura lathami*, the Australian brush turkey, the eggs of which are incubated in dark mounds of litter and rotting organic material. Because the embryos receive no light stimulation before hatching, one might hypothesize that this species has no lateralization of brain functions.

Lateralization in the chicken brain is also influenced by the circulating levels of the hormone testosterone. If male chicks are injected on day 2 posthatching with a slow-release form of testosterone and then tested monocularly on pebble grain discrimination in the second week of life, they show reversed lateralization (Zappia and Rogers 1987). When using the left eye, they make the shift to peck for grain and to avoid the pebbles, but not when using the right eye. Thus, the hormonal condition also interacts with the genetic and environmental determinants of functional lateralization for control of visual behaviors by the chick brain.

Lateralization of Memory Storage

Extensive studies of imprinting in chicks have led to the discovery that there is lateralization for forming and recalling memories in the chicken forebrain. Although both hemispheres are involved in laying down the imprinting memory, each hemisphere is differentially involved (see Horn 1985; McCabe 1991). Interestingly, the very first report of neurochemical asymmetry in the chick brain was made, and argued away as not being important, in a study by Horn, Rose, and Bateson (1973) of the incorporation of radioactive uracil into protein in the brains of "split-brain" chicks (with the supraoptic commissure sectioned). The chicks were injected with radioactive uracil before being trained on an imprinting stimulus, a flashing red or yellow light. Each was trained monocularly, with either the left or right eye occluded. After 50 minutes exposure of the chicks to the imprinting stimulus, biochemical analysis was made of the incorporation of the radioactive uracil into protein, the molecules thought to store long-term memory. The authors reported that, in addition to finding increased incorporation in the roof of the forebrain contralateral to the open eye, they also found that incorporation in the left and right sides was asymmetrical. Irrespective of the eye used in training, the incorporation of uracil was significantly higher in the left midbrain region. In the roof of the forebrain, the incorporation was higher on the right side, but this difference was not quite significant. The asymmetry was clearly unexpected, and the authors explained it away by saying that, as no such differences were found in their unoperated (control) chicks, the asymmetry must have resulted from sectioning the decussation. Perhaps they were right. On the other hand, perhaps they had overlooked the first evidence for lateralization in the chick brain. Later research by the same group has clearly demonstrated that the memory processes for imprinting differ between the left and right hemispheres. These studies came some 10 years later after lateralization had been demonstrated in the chick brain by injecting cycloheximide into the left and right hemispheres (Rogers and Anson 1979; see earlier).

Lesioning studies have revealed that the intermediate medial hyperstriatum ventrale (IMHV) region of the left hemisphere (present in the roof of the forebrain; Figure 2.8) is involved in both the early and later storage of imprinting memory, but the IMHV region of the right hemisphere is involved only in the early events of memory storage (Cipolla-Neto, Horn, and McCabe 1982). If a chick is imprinted and then the right IMHV is lesioned 3 hours later, followed by lesioning of the left IMHV some 26 hours later, no memory is present on retest. By contrast, if the left IMHV is lesioned first and then the right, the chicks can recall the memory on retest (see Horn 1985, 1990).

Because this was an important discovery of differential use of each hemisphere in memory formation, it is worth explaining in more detail. The chicks were imprinted on day 1 posthatching by exposing them to either a rotating red box or a rotating stuffed jungle fowl for 2 hours (equal numbers were trained on each of these stimuli). Approximately 3 hours after the imprinting training, the chicks were anaesthetized and assigned to different groups. In one group, the right IMHV region was lesioned, and the chicks were retested for imprinting preference some 18 hours later (Test 1). The preference test involved placing the chick in a running wheel with each of the imprinting stimuli placed on either side (see Horn 1985 and Figure 2.9). The approach of the chick to each stimulus was measured in terms of the number of revolutions of the running

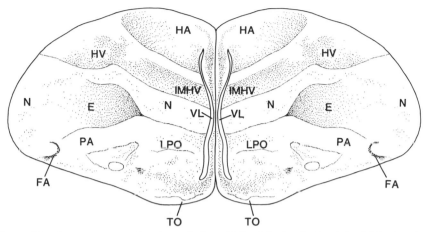

Figure 2.8 A transverse section through the middle of the forebrain of the chick, showing the position of the intermediate medial hyperstriatum ventrale (IMHV) region used for storage of the imprinting memory, and also memory of the passive avoidance task. HA, hyperstriatum accessorium; HV, hyperstriatum ventrale; E, ectostriatum; N, neostriatum; PA, paleostriatum augmentatum; FA, fronto-archistriatal tract; LPO, lobus parolfactorius; TO, tuberculum olfactorium; VL, lateral ventricle. Adapted from Kuenzel and Masson (1988).

wheel made in either direction, and from this a preference score was calculated in terms of approach toward the imprinting stimulus as a percentage of the total number of revolutions of the wheel. A high percentage (usually in the region of 70–80%) indicated a preference for approaching the training stimulus and hence memory recall. The chicks

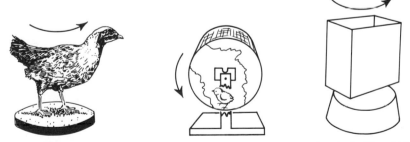

Figure 2.9 The apparatus used to test imprinting percentage in the chick. The chick is placed in a wheel that turns as the chick walks or runs. Two stimuli, the imprinting stimulus and another, are placed on either side of the wheel. These stimuli are rotated. The number of revolutions of the wheel in each direction is counted and a imprinting percentage score is calculated from the number of revolutions made towards the imprinting stimulus divided by the total number of revolutions. Adapted from Horn (1985), with permission.

given Test 1 after lesioning of the right IMHV showed memory recall. Five hours later they were reanaesthetized and the left IMHV was lesioned. A second preference test (Test 2) was then given on day 3. In Test 2 these chicks showed no memory recall (Figure 2.10). A second group of chicks had the left IMHV lesioned first, after which they showed memory recall in Test 1, and then the right IMHV was lesioned. In Test 2 these chicks showed memory recall (Figure 2.10). That is, memory is present following the left—right sequence of lesioning but not after the right—left sequence. The controls were given sequential lesions in the visual Wulst regions of the left and right forebrain hemispheres, and they had memory recall at both Test 1 and 2 irrespective of the sequence of lesioning. Thus, the left IMHV is involved throughout both the long-term and short-term processes of memory formation, whereas the right IMHV has a transient role in the initial (short-term) processes only. In the right hemisphere the memory may be shunted away from the IMHV to a store somewhere else in the hemisphere. This site is not yet known, although there are suggestions that it might be the lobus parolfactorius (LPO) or the archistriatum (Rose and Csillag 1985; Salzen, Parker, and Williamson 1975).

The right IMHV appears to act as a buffer store for memory, a role now assigned to the mammalian hippocampus (McNaughton et al. 1986). Horn and Johnson (1989) have suggested that, as is the case for the mammalian hippocampus, the right IMHV may add to the "depth" of processing by contributing contextual information during learning. Incidentally, there is asymmetry of the mammalian hippocampus (see Chapter 3). The memory trace in the right hemisphere may then move to another region that has a larger capacity for storage and perhaps an ability to modify and extend the memory through subsequent experience, giving it more flexibility for use in a variety of contexts (Horn 1990).

Lateralized involvement of the IMHV regions in imprinting learning

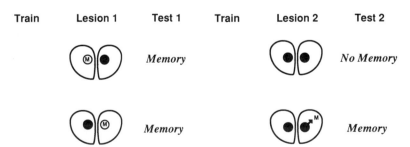

Figure 2.10 This is a schematic representation of the sequential lesioning experiment of Cipolla-Neto *et al.* (1982), which demonstrated lateralization of memory storage for imprinting (see text). Adapted from Horn (1985).

is also apparent at the subcellular and neurochemical levels. Imprinting learning leads to changes in the synapses of the left IMHV, but no significant changes occur in the right IMHV (Bradley, Horn, and Bateson 1981). In untrained control chicks, the length of the postsynaptic density of the synaptic apposition zones is greater in the right IMHV than in the left. Training removes this difference by increasing synaptic apposition length in the left IMHV. Imprinting has also been shown to correlate with a lateralized change in the number of receptors for glutamate (NMDA receptors, to be precise) in the IMHV region. Eight hours after imprinting, there is a significant increase in the number of glutamate receptors in the left IMHV, and not in the right (McCabe and Horn 1988; also recently confirmed by Johnston, Rogers, and Johnston, in press). The elevation in glutamate receptor numbers may be a manifestation of long-term memory formation in the left IMHV, as it does not occur prior to 8 hours after training (Horn and McCabe 1990; McCabe and Horn 1991). Also, no increase in glutamate receptor number occurs in the right IMHV, and, as demonstrated by the lesioning studies, the right IMHV apparently plays no part in long-term storage of imprinting memory.

There is an increasing amount of evidence that glutamate (or NMDA) receptors are involved in certain forms of plasticity in the nervous system (e.g., Rauschecker and Hahn 1987). It would seem, therefore, that the lateralized change in glutamate receptor numbers may reflect learning-dependent plasticity changes in the left hemisphere. Changes in glutamate receptor numbers may also occur in other regions apart from IMHV and also elsewhere in the right hemisphere, but this awaits further study. A pharmacological study indicates that the latter may be correct: Injection of glutamate into the left, but not the right, IMHV has been found to cause amnesia of the passive avoidance task, whereas placement of the injection into the right, but not the left, neostriatum causes amnesia (Patterson et al. 1986). Clearly, different regions of each forebrain are used to store memory of the same task, and most likely therefore different types of memory are simultaneously laid down in each hemisphere.

The forebrain hemispheres are also differentially involved in long-term memory retrieval or storage of the passive avoidance task of pecking at a bead coated with methylanthranilate (Rose 1989). As described earlier, the chick is trained by pecking at a bitter-tasting bead coated with methylanthranilate. This evokes a disgust response from the chick and, provided that it remembers, it will avoid pecking a similar bead when it is presented at retest. Performance of this task results in different patterns of neural activity, as detected by an uptake of radioactive 2-deoxyglucose, in the left and right hemispheres (Rose and Csillag 1985). The radioactive sugar accumulates in those neurones which have been active. The chicks were given a pulse of radioactive 2-deoxyglucose lasting for 30 minutes just after training with the bead, and the pattern of neural activity was determined from autoradiograms of the forebrain. Chicks trained with the methylanthranilate-coated bead were compared to controls which pecked at a bead coated with water. Compared to the controls, the trained chicks showed increased accumulation of 2-deoxyglucose in the left IMHV and the left lobus parolfactorius (LPO). Thus, as for imprinting, the left IMHV appears to be more involved than the right.

Lesioning studies have also demonstrated the differential role of the right and left IMHV regions in memory retention of this task (Patterson, Gilbert, and Rose 1990). Chicks, which have had the left IMHV lesioned on day 1 posthatching and which are trained on day 2, are amnesic when tested for recall on day 3. Lesions of the right IMHV do not impair the memory. Furthermore, as for imprinting, passive avoidance learning causes neurochemical changes and changes in subcellular structure in the left, but not the right, IMHV. In the left IMHV the mean number of neurotransmitter-containing vesicles per synapse is increased following training, and the density and size of dendritic spines on large, multipolar projection neurones is increased (Stewart et al. 1984; Stewart 1991), and the spine head diameter is also increased (Patel and Stewart 1988). In the left IMHV, the concentration of the enzyme, protein kinase C, which phosphorylates a certain protein located at the synapse, increases and amnesia is produced by injecting an inhibitor of this enzyme into the left hemisphere (Burchuladze, Potter, and Rose 1990). Glutamate receptor binding is also increased in the left IMHV (Steel and Rose, in preparation). None of these changes occur in the right IMHV.

Although lesioning the right IMHV does not result in amnesia in binocularly trained and tested chicks, the right IMHV is capable of taking over from the left in chicks trained and tested monocularly. Sandi, Patterson, and Rose (1992) have found that in chicks wearing an eyepatch on the right eye (using the left eye), lesions of the right IMHV disrupt learning and memory acquisition, and lesions of the left IMHV are ineffective. These rather anomalous results, in comparison to the binocular situation at least, may occur as a consequence of the monocular experience causing shifts in brain organization. In addition, the same researchers have demonstrated that interocular transfer of the passive avoidance task occurs more readily from the right eye to the left (i.e., when training is with the right eye and testing with the left) than vice versa. That is, interocular transfer is lateralized and the left eye has better access to information acquired by the right eye than the other way around. This result is consistent with that of Gaston (1984) discussed previously: Chicks trained in a visual discrimination task using the right eye not only learn faster but have more of the learning transferred to the other eye than vice versa.

In addition to IMHV, the LPO and possibly the paleostriatum augmentatum (PA) regions are involved in memory formation of the passive avoidance task. In the LPO region, the postsynaptic thickening length is larger on the right side in controls and on the left side in chicks trained in the passive avoidance task (Stewart, Csillag, and Rose 1987). In the trained chicks the volume of the synaptic boutons increases in the left LPO only, this being a likely correlate of memory formation. Stewart et al. (1987) also examined the subcellular structure of the PA region of the forebrain and found no asymmetry or changes with training in the postsynaptic density thickening. They did, however, find that the density of synapses and the number of vesicles per synaptic bouton to be higher in the right PA, and that this asymmetry was removed by training. Apparently, in control chicks, neurotransmission in the right PA is greater than in the left, and passive avoidance training facilitates transmission on the left side thereby removing the asymmetry.

These data indicate that in more than one region of the control chicken brain there are a number of subcellular asymmetries which reflect asymmetries in neurotransmission. Training imposes new asymmetries in some regions and removes them in others, and these shifts are clearly correlates of memory formation. The subcellular asymmetries at any one time are the result of interaction between developmental events and associated learning and memory formation.

It also appears that at least some different neurochemical events may occur in each hemisphere in correlation with memory storage. Barber and Rose (1991) report that injection of 2-deoxy-D-galactose, which inhibits brain glycoprotein fucosylation, into the right hemisphere causes amnesia of the passive avoidance task, but injection of this drug into the left hemisphere has no effect on memory recall. There is considerable evidence that long-term memory formation is associated with enhanced incorporation of precursor sugars into glyoproteins, these being destined for the synaptic membrane where they alter neurotransmission (Rose and Jork, 1987). According to this experiment, this process occurs in the right hemisphere only.

By elucidating the sites of these lateralized neurochemical events of memory formation within each hemisphere, the complex events underlying memory storage of various tasks may be pieced together. It is clear that several brain sites are involved in memory storage and that these vary with time and differ between the hemispheres. Discovery of lateralization in the chicken brain, the species now accepted by many as the best model system for studying memory, provided the basis for a rather major step forward in understanding the biological correlates of memory formation.

Based on the extensive data collected by his research group, Rose (1991) has suggested a model in which the memory "flows" from left IMHV to right IMHV and then on to the left and right LPO regions. If the left IMHV is lesioned before training on the passive avoidance task, amnesia results, but if bilateral lesions of LPO are made before training, no amnesia results. Bilateral lesions of LPO made after learning do, however, cause amnesia (Gilbert, Paterson, and Rose 1991). Thus, memory must "flow" on to LPO for long-term storage. It is possible that other regions of the forebrain (such as PA) are also involved, but here we simply want to emphasize that the memory process of this task and imprinting are clearly lateralized in the forebrain. Lateralized use of the forebrain is also most likely for other memory processes. This lateralized storage may, in turn, give rise to the lateralized processing of new information and responses. It may be manifest in the head movements used by the chick during memory recall in retest, although that has not yet been investigated.

Lateralized Control of Singing and Auditory Processing

The first convincing example of functional asymmetry in a nonhuman nervous system was that of vocal control in a number of species of songbird. We have chosen not to discuss this research first, as might be appropriate in a historical context, but rather have followed the probable evolutionary sequence of lateralization in the avian brain, adhering to the hypothesis that it was primarily the constraints of visual processing which brought about the evolution of lateralization. Indeed, the numerous examples of the lateralization of visual processing in birds which do not sing and which are considered to be more primitive than songbirds, suggest that lateralization for control of song was a later development.

In the early 1970s, Nottebohm (1970, 1971, 1972) reported that sectioning the left hypoglossus nerve of the adult male chaffinch, *Fringilla coelebs*, resulted in the loss of most of the components of song, sub-song, and calling, whereas sectioning the right hypoglossus resulted in loss of few or none of these components.

Avian vocalizations are produced by expelling air past the elastic membranes of the syrinx, and in songbirds the syrinx consists of two parts, one in each bronchus (Nottebohm 1972; Figure 2.11). Each is an independent sound source. The musculature of each part is innervated separately by a branch of the left or right hypoglossus, the tracheosyringealis nerve,

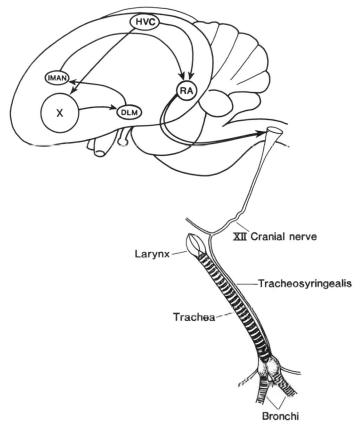

Figure 2.11 The song nuclei involved in the song system of a songbird. The projections from HVC to RA to the nucleus of the XIIth cranial nerve (the hypoglossus) are the primary ones involved in controlling the motor aspects of song. The auditory input to the system enters HVC via the nucleus ovoidalis and field L, not shown here. In HVC there is interaction between sensory and motor activity. HVC also projects to Area X, and that via DLM to lMAN and back to RA. This recursive loop is important for song learning. HVC, high vocal center; RA, robust nucleus of the archistriatum; XII, hypoglossus nerve which gives rise to the tracheosyringealis nerve which innervates the syrinx; DLM, medial portion of the dorsolateral thalamic nucleus; lMAN, lateral part of the magnocellular nucleus of the anterior neostriatum. Adapted from Nottebohm (1989).

and the musculature on the left side of the syrinx is larger than on the right (Nottebohm 1971, 1977). There has been some debate about whether each part of the syrinx of songbirds operates independently, an issue relevant to the interpretation of the results of sectioning the left or right hypoglossus nerves. Nowicki and Capranica (1986) have shown that for producing one of the vocalizations of the black-capped chickadee (*Parus atricapillus*) the two parts of the syrinx are coupled in a nonlinear fashion. But, although there can be interaction between the two parts, evidence points to each part being functionally decoupled for most vocalizations, which means that most components (syllables) of song are produced by the left side of the syrinx alone. Recently, Hartley and Suthers (1990) have conducted experiments to further test this theory using the canary (*Serinus canarius*). They plugged the bronchus unilaterally and followed this by denervating the ipsilateral syringeal muscles. Male canaries which had had the right bronchus plugged, and therefore were singing with the left side of the syrinx alone, produced almost normal songs. Those with the left bronchus plugged sang poorly. Sectioning the right nerve had little effect on the songs of birds with the right bronchus plugged, but sectioning the left nerve of birds with the left bronchus plugged impaired song production. The researchers concluded that in male canaries most syllables are sung by the left side of the syrinx alone, although some are produced by the right syrinx acting either alone or coupled to the left. Furthermore, one can say that irrespective of the side of the syrinx used, it is the left nerve that controls song production.

When the left hypoglossus of the adult chaffinch is sectioned, the loss of song components is irreversible, but when it is cut before the bird has started to develop its sub-song and song (i.e., before the bird's first spring) the right hypoglossus assumes control and normal song develops (Nottebohm 1971). Thus, in the young bird there is either no hypoglossal dominance or the system has plasticity, in that after sectioning the left hypoglossus, the right is still able to take over. If the left hypoglossal nerve is sectioned slightly later, during the spring development of sub-song and plastic song, the right hypoglossus can partly, but not fully, take over control of singing. Left hypoglossal dominance develops along with the development of song and with the increasing complexity of the vocal template.

Left hypoglossal dominance was later found to be present in the white-throated sparrow, *Zonotrichia albicollis* (Lemon 1973), white-crowned sparrow, *Zonotrichia leucophrys* (Nottebohm and Nottebohm 1976), and the canary, *Serinus canarius* (Nottebohm 1977; Nottebohm and Nottebohm 1976). After sectioning of the right tracheosyringealis nerve in the adult male canary, an average of only one-tenth of the song syllables disappeared; there were brief silent gaps in the song, noticeable only on sound spectrographs and not directly to the human ear. Sectioning the left tracheosyringealis had a dramatic effect. Nottebohm (1977) described these birds as performing like actors on the silent cinema screen. The birds performed all the beak, throat, wing, and breathing motor acts usually associated with singing; yet for most of the time, it was almost silent except for very faint clicks made by uncontrolled vibration of the syringeal membranes due to the passage of air across them. Then unexpectedly the few remaining syllables of song issued forth loudly and clearly, to be followed again by breathy silence.

Nottebohm did not state how long the left-sectioned birds continued this behavior unperturbed by auditory feedback, but he did report that unilateral hypoglossectomy had the same effects on both deaf and hearing canaries (Nottebohm and Nottebohm 1976) and chaffinches (Nottebohm 1971). It therefore seems that the template for adult song is used with little, if any, reference to auditory feedback, even though auditory feedback is required for these species to be able to learn to sing, since young chaffinches deafened before the beginning of sub-song practice sing less well as adults than birds deafened after sub-song practice (Nottebohm 1972).

As in the chaffinch, left hypoglossal dominance develops in the canary along with the development of stable adult song. While left hypoglossal section during the first 2 weeks after hatching causes the right hypoglossus to assume complete control of song, performing this operation during the third and fourth weeks causes incomplete control by the right hypoglossus. Later, when song is developing or has become stable, left hypoglossal section leads to poor singing since the right is much less able to assume control (Nottebohm, Manning, and Nottebohm 1979). At least this is true for singing in the first year after the operation, but unlike chaffinches, the reduced singing of left hypoglossal sectioned canaries is not permanent. In the next year song improves even though it is under right hypoglossal control. Nottebohm et al. (1979) suggest that this species difference may occur because canaries develop new song repertoires each year whereas chaffinches do not.

The next step was to find the higher brain centers involved in song production, and to see if lateralization existed at this level of neural organization. Nottebohm (1977) and Nottebohm, Stokes, and Leonard (1976) were able to show that lesioning of the "high vocal center" of the neostriatum (HVC; originally incorrectly called hyperstriatum ventrale, pars caudalis) in the left forebrain hemisphere of the adult canary disrupts song, whereas a similar lesion in the right hemisphere has a much smaller effect on song (see Figure 2.11). It should be mentioned here that the brain centers involved in the control of song largely connect to motor functions on the ipsilateral side, and not the contralateral side, as is more commonly the case for vertebrate motor functions. Following lesioning of the left HVC, the canaries were able to produce no more than one of the syllables of song that they had produced preoperatively. Their song was a monotonous, simple succession of notes of rising and falling pitch, with apparently one syllable remaining. Following lesioning of the right HVC, the canaries were able to sing one-third to three-fifths or more of the syllables that they had produced preoperatively. Some of these birds were totally unaffected by the lesion.

As seen previously for the peripheral innervation of the syrinx, there is some plasticity within the adult canary central nervous system; 7 months after lesioning the left HVC of adult male canaries, in the next season of song, there was some degree of song recovery. A new repertoire developed and, although it was not as accomplished as that of the intact bird, it did possess most of the features of normal song. Subsequent sectioning of the right tracheosyringealis nerve in these birds confirmed that the right hemisphere had assumed control of singing, although with less perfection than the original control by the left side (Nottebohm 1977).

Degenerating fibers have been traced from HVC lesions to two other nuclei in the forebrain, one called area X in the parolfactory lobe and the other being the nucleus robustus archistriatalis (RA) in the archistriatum (Nottebohm et al. 1976). Unilateral lesions of area X had no effect on song, but lesions of the left RA did disrupt song. RA has direct connections to the motor nucleus innervating the syrinx (Figure 2.11). As for the HVC, lesions of the RA on the left side were found to cause greater loss of song than lesions of the RA on the right side. Lesions of either side led to a marked reduction in the number of syllables in the song, but only lesions of the left RA resulted in a reduced frequency range of the song fundamental.

Thus, control of singing is primarily located in specific areas of the left forebrain hemisphere. In canaries there are sex differences in the sizes of area X, HVC, RA, and the hypoglossal nucleus. They are all smaller in females. Because females do not sing, the size of these nuclei may be related to the singing function. However, no left–right asymmetry in the sizes of these nuclei has been found, except in the hypoglossal nuclei. Thus, while the hypoglossal asymmetry in structure may well relate to left hypoglossal dominance for song, higher level asymmetry is functional and not structural.

Autoradiographic studies using zebra finches have shown an accumulation of testosterone in the HVC, the hypoglossal nuclei, and another area connected to RA, the nucleus intercollicularis (Arnold, Nottebohm, and Pfaff 1976). Higher concentrations of testosterone-metabolizing enzymes, aromatase and 5α-reductase, are also present in the forebrain nuclei involved in song (Vockel et al. 1990). This may reflect the androgen dependence of song in this species. Nevertheless, no asymmetry of testosterone receptors or metabolic enzymes has been reported. An asymmetrical interaction with testosterone does, however, exist more peripherally in the neuromuscular complex of the syrinx. In the presence of testosterone, the levels of acetylcholine esterase and choline acetylase are higher on the left side of the syrinx than on the right (Luine et al. 1980). Coincident structural, biochemical, and functional asymmetries are therefore apparent peripherally in the motor neurons and muscles producing singing, but at the higher levels of song control functional asymmetry has not been matched by asymmetries in structure or biochemistry.

The function of the right HVC in an intact songbird brain is not known. Perhaps it is used in analyzing and comprehending the songs of other birds, or in storing a memory of the individual's own song. Although the HVC is a forebrain nucleus in the motor pathway for control of song (Margoliash and Konishi 1985), neurons in the HVC display auditory responses which reach it from the auditory region of the forebrain, field L (Kelly and Nottebohm 1979). In fact, a subset of auditory neurons in the white-crowned sparrow has been shown to be highly selective for the individual's own song (Margoliash and Konishi 1985). Each bird was reared in the laboratory in a sound-attenuated chamber and exposed to a recorded song. This species learns to sing, but its own song will be somewhat different from the "tutor" song. It was found that the neurons in the HVC preferred the bird's own song to the tutor song, which suggests that these neurons must be specified at the time when the song crystallizes or after. Since song is learned by auditory feedback, the bird listens to its own performance, HVC auditory neurons may be involved in the development of the motor program for song.

McCasland and Konishi (1981) were able to distinguish between the sensory activity of the neurons and the motor activity involved in singing. Careful comparison of the time courses of these two patterns of neural activity allowed them to show that many spikes of activity, which occurred in response to hearing a song, occurred at times when there were no spikes seen in the motor records. That is, the peaks in auditory activity and motor activity occur at different times. Once the bird has learned its song, the temporal pattern of sensory activity is no longer influenced by auditory feedback, as it occurs even in deafened birds. Thus, once the program for the individual's song has been established, the neurones display the temporal patterning of the learned song even in the absence of auditory feedback. By having neurons specified for its own song, the bird may be able to recognize the slightly different songs of conspecifics by reference to its own song. Thus, HVC appears to be the brain site where interaction between sensory and motor activity occurs for the purpose of learning and maintaining the individual's own song and also of recognising conspecific songs.

Surprisingly, although these studies of HVC in the white-crowned sparrow were conducted after Nottebohm and Nottebohm (1976) had shown left hypoglossal nerve dominance for control of song in this species, no attempt was made to compare the auditory responses of the left and right HVC. In the texts of these papers (McCasland and Konishi 1981; Margoliash and Konishi 1985) the authors refer to the HVC as if it were one nucleus in the forebrain, but HVC nuclei occur in the left and right hemispheres. The papers do not report in which HVC the electrodes were placed, although all the footnotes to the figures in McCasland and Konishi (1981) refer to the left HVC. Consequently, from these particular studies it is impossible to deduce anything of the relative involvement of left and right HVC in these functions. However, a later study by McCasland (1987) used bilateral recording from the left and right HVC, using four species of songbirds, including the canary, and found that both the left and right HVC make equal electrophysiological contributions to the production of song. This result appears to be incompatible with the lesioning studies demonstrating clear lateralization of singing control in the canary. McCasland (1987) concluded that the lateralization for song control was not central, but present only at the level of neuro-muscular interaction in the syrinx. However, Nottebohm et al. (1990) have recently shown asymmetry in the auditory perceptual mechanisms for song.

The nuclei ovoidali in the thalamus, left and right, receive auditory input. Each nucleus ovoidalis projects to field L in the neostriatal region of its ipsilateral forebrain hemisphere (Kelly and Nottebohm 1979), and field L, in turn, projects to HVC in the same hemisphere. Thus removal of, say, the left nucleus ovoidalis removes auditory input to the left HVC. These nuclei were lesioned unilaterally in zebra finches and the ability of the lesioned birds to detect alterations in recorded song was assessed. The birds that had the right nucleus ovoidalis lesioned took longer to recognize a missing harmonic in a syllable of a song compared to those with the left nucleus lesioned, but they were better at discriminating between two familiar songs. This demonstrates that lateralization of song perception does indeed occur at the central level and, in fact, on the input–perceptual side. It is likely therefore that the left and right HVC are both receiving auditory input, but either each receives a different kind of input or each HVC processes the input differently.

Nottebohm et al. (1990) suggest that in the zebra finch, the left hemisphere may be better at discriminating between stimuli which differ in a variety of ways, and so is involved with "holistic" perception, and the right hemisphere is better at making discriminations which require more "analytical" processing of input. One is always tempted to draw such parallels to the human brain; but, in our opinion, such speculation is premature given the limited information on laterality of song perception. Also it should be noted that involvement of the left hemisphere in holistic and the right hemisphere in the analytical perception does not appear to fit with the more extensive data on lateralized perception in the chicken and pigeon (see earlier). On face value, the left hemisphere of the chick and pigeon brain appears to perform analytical processing, whereas the right may be more "holistic." The direction of lateralization for songbirds suggested by Nottebohm would also be opposite to that known for lateralization in rats and humans (see later). The direction of lateralization may be of very little consequence for a given species on its own, but songbirds evolved after birds without song, such as the chick or pigeon, and so one might expect the organization of their hemispheres to in some way reflect their evolutionary origins. If there are indeed visual perceptual requirements for lateralization in birds, then song learning and perception may be an elaboration of this lateralization. Once a given hemisphere has become specialized for certain kinds of processing and for making certain kinds of decisions, this may be extended to a wide range of functions. To test this we need to look for correspondences across a wide range of avian species in terms of hemispheric special-ization.

Many intriguing questions remain about the neural pathways involved in song. For example, why does the right HVC remain suppressed for the rest of the singing season when the left is lesioned? In other words, when the left HVC has been lesioned, why is there a delay until after the sex steroid hormone levels have subsided and are then reelevated before the right HVC can take over and control singing? Elevated testos-terone levels permit neuroplasticity in the adult canary brain (Nottebohm 1987, 1989), and this neural plasticity is clearly necessary for the right HVC to assume control of singing after lesioning the left. Possibly, rising levels of testosterone at the beginning of the reproductive season are essential to trigger the combined processes of song production and neural plasticity. It should be noted that in chaffinches, which unlike canaries do not embellish their song repertoire each season, the right HVC does not take over song control after the left HVC is lesioned (Nottebohm 1987). Canaries are, according to Nottebohm, "open-ended learners" that retain neural and functional plasticity in adulthood, whereas chaf-finches are "critical-period learners" that lose and never regain the ability to add to their repertoire.

In canaries, several of the song nuclei, including HVC, increase in volume in the spring when new syllables are added to the vocal repertoire, and decrease again in the autumn when syllables are lost and the song becomes as unstable as that of juvenile birds (Nottebohm 1989; Nordeen and Nordeen 1990). Canaries learn a new song repertoire each year (Nottebohm and Nottebohm 1978) and neurogenesis in the song nuclei may be associated with the new repertoire and with song perception (Nottebohm et al. 1990). It is remarkable that new neurons can form in adulthood. They grow long axons which project to RA and thus become part of the efferent pathway for song control (Kirn, Alvarez-Buylla, and Nottebohm 1991). Most of these new projection neurons survive for at least 8 months, suggesting that they remain part of the vocal control circuit long enough to participate in the annual renewal of the song repertoire. As yet no comparison has been made of new neuron formation in the left and right HVCs, and, surprisingly, even the latest study by Nottebohm's group (Kirn et al. 1991) simply lumped together the data for the left and right sides, claiming that they did not differ significantly on the basis of three or four subjects per group. It is likely that there is no asymmetry in the formation of new neurons, but one might expect studies to collect sufficient data to make comparisons of the left and right HVCs.

Neurogenesis can be induced in female canaries by treating them with testosterone (Goldman and Nottebohm 1983) and the size of the vocal repertoire acquired correlates with the volumes of the HVC and RA. Interestingly, the increase in volume of HVC in females treated with testosterone also depends on auditory feedback; the growth of HVC following testosterone treatment is greatly attenuated in deafened canaries (Bottjer, Schoonmaker, and Arnold 1986). Again, we know nothing of any left–right asymmetries in females. It is not known whether the interaction of testosterone treatment and auditory feedback induces lateralization for control of singing in females.

Finally, it should be mentioned that, although left hemisphere dominance has been shown in a number of species of songbirds, this dominance may not be characteristic of all songbirds, or for control of vocalization in birds other than songbirds. The zebra finch, *Taeniopygia guttata*, for example, has little hemispheric asymmetry in song control (Nottebohm, cited as unpublished observations in Nottebohm et al. 1990) and, unlike the canary, no asymmetry in the size of the left and right hypogossal nuclei (Nottebohm and Arnold 1976). Nottebohm (1976) has also studied control of vocalizations in the orange-winged Amazon parrot, *Amazona amazonica*, finding no lateralization in this species. Sectioning either the left or right tracheosyringealis nerve (branches of the hypoglossus) leads to a minor and temporary change in the structure of the vocalizations, lasting for only 1 week. Hence there is no evidence for left or right dominance in terms of hypoglossal control of the parrot syrinx. It should be mentioned that, unlike songbirds, in parrots each side of the syrinx is innervated by each tracheosyringealis nerve. No study has been made of hypoglossal control of vocalizations in nonsinging birds such as the domestic fowl, but there is asymmetry in the organization of the hypoglossal innervation of the syrinx (Youngren, Peek, and Phillips 1974). The left hypoglossus innervates the musculature of both the right and left sides of the syrinx, whereas the right hypoglossus innervates only the right side. This organization may, in itself, suggest that there is lateralized control of syringeal function.

It is worth speculating that, as in songbirds, the left hemisphere of the forebrain of the chicken is dominant for control of auditory processing and possibly vocalizations because cycloheximide or glutamate treatment of the left hemisphere leads to a slowing of the rate of habituation to an auditory stimulus, whereas treatment of the right hemisphere has no effect (Rogers and Anson 1979; Howard et al. 1980). The effect of the drugs occurs during a sensitive period from hatching to day 3 (Rogers, Drennen, and Mark 1974). Presumably, neurons which are differentiating in the left hemisphere and which will later be used for processing auditory information are susceptible to the drugs during this period. No other study of lateralized auditory processing in chickens has yet been made.

Lateralization of Proprioceptive Responses and Olfaction

So far, there is little evidence for lateralization of perceptual processes other than those of vision and audition, apart from the following reports. Güntürkün and Hoferichter (1985) have reported lateralized somatosensory neglect in pigeons which have had unilateral lesions of the occipitomesencephalic tract (OM). Those with lesions of the left OM showed a marked lack of attention to tactile stimulation of the right side of the body (i.e., to a hand touching the body). Those with lesions of the right OM tract or the OM tracts on both sides showed no such effect.

Hemispheric specialization for responses to olfactory stimuli has recently been demonstrated in young chicks (Andrew and Vallortigara, personal communication, 1991). Chicks reared with a ball having a persistent scent, or with a scentless ball, tend to choose a ball which smells like the one to which they have attached, when presented with a scented and an unscented ball. When one or another external naris is temporarily blocked, chicks using their right nostril (so having direct olfactory input to the right hemisphere) choose correctly, but chicks using their left nostril do not, instead associating with the balls at random.

Structural Asymmetries in the Avian Brain

Although there is lateralization in the avian brain for control of song in some species of songbird, this functional asymmetry is not correlated with asymmetry in the structure of the forebrain nuclei involved in song control (HVC, area X, and lMAN). All of the forebrain nuclei and connections are present in both hemispheres. It is possible that those in the right hemisphere are used for some aspects of song, possibly in the perception and analysis of conspecific song (see earlier).

Nevertheless, other structural asymmetries do exist in the avian brain. The first report of which was for asymmetry in the size of midbrain auditory nuclei in two species of birds (Cobb 1964). In the oilbird *Steatornis caripensis*, the left auditory nucleus (nucleus mesencephalicus lateralis, pars dorsalis) is 20% larger than the right, and in the saw-whet owl, *Aegolius acadicus*, the left nucleus is 10% larger than the right. It should be stated, however, that only small sample sizes were used in this study. Both of these species have large numbers of auditory projections adapted for echo location which is used for navigation in the dark and for capturing prey. The asymmetry in the neural structures of the owl correlates with the asymmetry in the construction of the ears (see later) and is likely to reflect differential neural processing according to the incoming perceptual information.

Asymmetry of nuclei size in the avian brain also exists for the habenular nuclei. In 2-day-old male chicks the volume of the medial habenular nuclei was found to be significantly larger on the right side of the brain compared to the left (a difference of 6% in volume). In females there was no significant asymmetry at the group level, but individual females were found to be lateralized. In females treated postnatally with testosterone, however, a group bias developed similar to that of untreated males (Gurusinghe and Ehrlich 1985a,b).

Asymmetry of the habenular nuclei in the chicken appears to have an evolutionary origin as there is an obvious asymmetry of these nuclei in the frog, the left side being larger than the right (Kemali 1977) and in the lizard, the left habenula has an extra nucleus not present in the right (Kemali and Agrelli 1972).

The significance of this structural asymmetry is not known because the function of the habenular nuclei is not yet understood. They are connected to the pineal gland and receive input from the archistriatum of the forebrain, the corticoseptal, and the preoptic regions. It is unlikely that asymmetry of the habenulae has any direct involvement in the lateralized functions so far discussed, unless the connections to the archistratium are linked to the lateralized role of this region in fear responses (see earlier) or, in connection with the preoptic regions, in lateralization of copulation responses.

Structural asymmetry is known to be present in the visual pathways of pigeons and, in particular, of chickens. This is most likely to relate to the functional lateralities in visually guided behavior.

There are two main visual pathways in birds, the tectofugal and the thalamofugal (Webster 1974; and see Figure 2.12). In the tectofugal pathway, projections from the ganglion cells in the retina go to the tectum on the contralateral side of the brain. Cells from this region project forward to the ipsilateral nucleus rotundus, and cells from here project to the ectostriatal region of the forebrain (Karten and Hodos 1970). In the thalamofugal visual pathway, ganglion cells from the retina project to the contralateral side of the thalamus and the cells receiving this input project from the thalamus to the hyperstriatal regions in each side of the forebrain (Karten et al. 1973). Although there are more projections from each side of the thalamus to its ipsilateral hyperstriatum, the contralateral hyperstriatum receives a substantial number of projections, ranging up to 50% of the ipsilateral projections (Boxer and Stanford 1985; Rogers and Bolden 1991).

One form of asymmetry in the structure of the tectofugal pathway has so far been reported for the pigeon, and a substantial asymmetry in the thalamofugal pathways has been investigated and found in the chicken. Structural asymmetries in the visual pathways must mean that there are different visual inputs to each side of the brain, and this is likely to contribute substantially to lateralization of function for a range of visually guided behaviors.

The thalamofugal visual system of the pigeon has been well studied (Miceli and Repérant 1982, 1985; Miceli et al. 1980), but not in terms of its potential laterality. Studies of the latter have so far been limited to the chicken. Asymmetry in thalamofugal visual projections to the forebrain of the chick was first reported by Boxer and Stanford (1985). Using horseradish peroxidase to label neurons, these authors reported finding a greater number of projections from the left side of the thalamus (which receives input from the right eye) to the right hyperstriatum than conversely from the right side of the thalamus (which receives input from the left eye) to the left hyperstriatum (Figure 2.13). Subsequently, Rogers and Sink (1988) and Rogers and Bolden (1991) confirmed this report using

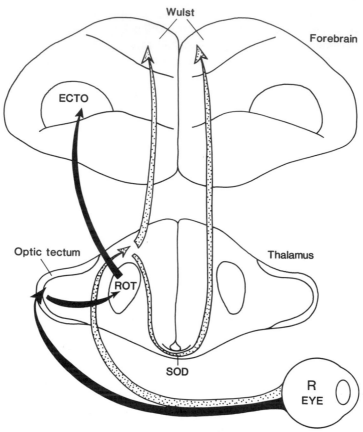

Figure 2.12   The two main visual pathways in the avian brain. Axons of the ganglion cells of the retina project to either the contralateral tectum or the contralateral thalamus. The tectofugal pathway (black arrows) projects from the tectum to the ipsilateral nucleus rotundus (ROT) and finally to the ipsilateral ectostriatal region (ECTO) of the forebrain. The thalamofugal pathway (dotted arrows) projects from the thalamus to both the ipsilateral and the contralateral hyperstriatal regions of the forebrain (the visual Wulst). The contralateral projections in this pathway cross the midline in the supraoptic decussation (SOD). The tectal commissure, part of the tectofugal pathway, is not represented here (see Figure 2.1). Transverse sections of the forebrain and the thalamus are illustrated. The representation is schematic. Projections from the right eye only are illustrated.

a relatively new technique of injecting fluorescent dyes (True Blue and Fluorogold), which are transported retrogradely along the nerve cell axons to label the cell bodies of neurons which project to the region of the brain injected with each particular dye (Bentivoglio et al. 1979). True Blue was injected into the hyperstriatum on one side of the forebrain

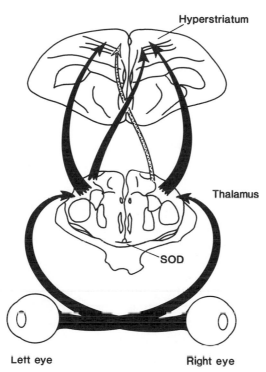

Figure 2.13   Asymmetry in the thalamofugal visual projections of the male chick. There are more projections from the left side of the thalamus (which receives input from the right eye) to its contralateral hyperstriatum than from the right side of the thalamus (input from the left eye) to its contralateral hyperstriatum. Note that the contralateral projections cross the midline in the supraoptic decussation (see Figure 2.12), but for clarity this has not been represented here.

and Fluorogold into the hyperstriatum on the other side. The cell bodies labeled with these dyes on each side of the thalamus were counted. To control for variations in the amount of dye injected into the forebrain absolute counts of the labeled cells were not used, but instead the C/I ratio was calculated of the number of cell bodies labeled in the thalamus contralateral to the injection site to the number labeled on the side of the thalamus ipsilateral to the injection site. The C/I ratios determined following injections into the right hyperstriata were larger than those determined by injecting the left hyperstriata. The asymmetry was highly significant. This asymmetry has been reported for males of two domestic strains of chickens (Boxer and Stanford 1985; Rogers and Sink 1988) and also for a strain of feral chickens from North West Island, Queensland, Australia (Adret and Rogers 1989).

From the C/I ratio it is not possible to determine whether the asymmetry is located in the contralateral or ipsilateral projections from thalamus to forebrain, or both. However, the earlier study by Boxer and Stanford (1985) demonstrated that the asymmetry is present in the contralateral projections. More cells on the left side of the thalamus project to the contralateral (right) hyperstriatum than vice versa.

The asymmetry was found to be present in young male chicks aged up to just before 3 weeks of age, by which time the asymmetry disappears. Interestingly, and as predicted for the data obtained with monocular testing and cycloheximide injection (see earlier), females were found to have symmetry of these visual projections (Adret and Rogers 1989). Subsequent to the latter report, females injected with dyes on day 2 have been shown to have asymmetry in the thalamofugal visual projections (and in the same direction as in males) but to a far lesser degree compared to males (Rajendra and Rogers, submitted). A large sample size with the possibility of double labeling with the dyes was necessary to reveal this lesser degree of asymmetry in females.

In young males it may be the better development of visual projections from the left thalamus (right eye) to the forebrain which affords the right eye with superior learning ability in visual discrimination tests. If so, this would imply that the tasks may require processing by both hemispheres since the left thalamus has well-developed projections to the ipsilateral and contralateral hyperstriata. Yet, cycloheximide or glutamate treatment of the left hemisphere only slows the learning, and treatment of the right hemisphere is without effect. As already discussed, there are two main sets of visual projections to the avian forebrain, the other travels from the tectum to the ectostriatal region of the forebrain. The relative contributions of these two systems to performance in visual discrimination tasks is unknown; right-eye superiority in males may rely on cooperation between the ectostriatal system in the left hemisphere and the well-developed bilateral projections to the hyperstriata in both hemispheres (Figure 2.12 and 2.13).

In the pigeon, Güntürkün et al. (1989) found that there is asymmetry in the tectofugal visual system. This study found that there are significantly larger cells with more dendrites in the deeper layers (10 to 15) of the right tectum than in the left tectum. The average surface area of neurones in these deep tectal layers was calculated to be 10% greater on the right side. A later study (Melsbach et al. 1991) has shown that the larger cell size on the right side is confined to layer 13 of the tectum. In layers 2 to 12 of the tectum the cells are larger on the left side.

The cells in layers 2 to 12 receive input from the retina or from other regions of the tectum. The cells in layer 13 of the tectum receive inputs from the tectal commissure, which connects the left and right tecta, as well as from two regions of the forebrain, the Wulst or hyperstriatum, and the archistriatum. Güntürkün and Böhringer (1987) argued that the tectal commissure has an asymmetrical influence whereby the left tectum inhibits the right. This was deduced from an experiment in which they sectioned the tectal commissure of the pigeon and found that lateralization for pecking in a visual discrimination task was reversed. In controls pecking was faster when the pigeon used the right eye, whereas in the group with posterior and tectal commissures sectioned (these two commissures make up a single complex, see Figure 2.1) it was faster when using the left eye. The structural asymmetry of cell size in layer 13 of the tecta would seem to reflect this lateralized function of the tectal commissure. The greater surface area of tectal neurons on the right side may be matched by receiving more input from neurons originating in the left tectum (Güntürkün et al. 1989). In other words, the neurones from the left tectum which send their axons in the tectal commissure and synapse on cells in layers 10 to 15 of the right tectum, might be either greater in number than their counterparts going from right tectum to left tectum, or they may have more arborization and more end terminals. If so, the left tectum would have more influence on the right tectum than vice versa. This asymmetry stems from the right tectum being visually deprived relative to the left during incubation (see later).

Unlike the asymmetry in the thalamofugal projections of the chicken, the morphological asymmetries in the tecta of the pigeon are identical in females and males (Melsbach et al. 1991). As yet, there has been no investigation of structural asymmetry in the tectofugal system of the chicken, but there has been investigation of the role of the tectal commissure in functional lateralization in the chicken (Parsons and Rogers, submitted). Chicks with sectioned tectal and posterior commissures were found to have lateralization in their pecking responses at a red bead. Lesioned chicks using the right eye pecked more at the bead each time it was presented over eight trials. The trend for increased pecking was much less in lesioned chicks using the left eye. In sham-operated controls, pecking at the bead showed no significant change, or a slight decline in pecking across trials, with use of either eye. Thus, for this function, the intact commissures appear to suppress a form of lateralization present at midbrain level. In chicks the commissure appears to suppress novelty pecking driven by the right eye and left tectum. The behavioral functions for which this may be important must depend on the visual functions subserved by the tectofugal system, but these are not yet well delineated. Lesions of the tectofugal pathway are known to impair visual discrimination learning in pigeons (Bessette and Hodos

1989). The tecto-ectostriatal system is said to parallel the geniculate visual system in the mammal (Dubbledam 1991), but more neurophysiological studies are necessary to delineate the roles of the two visual systems in birds. For some functions, such as species recognition, the two visual systems cooperate, or at least either system can assume the function of species recognition when the other had been lesioned (Watanbe 1991; and personal communication, 1991).

Factors Influencing the Development of Structural Asymmetry of the Thalamofugal Visual Projections

We have already discussed the manner in which the orientation of the chick embryo in the egg and the consequent lateralization of light input to the eyes during the last stages of incubation determines the direction of functional lateralization. The same effect also occurs for the development of structural asymmetry in the thalamofugal visual projections (Rogers and Sink 1988; Rogers and Bolden 1991).

Embryos were exposed to various lighting conditions during the last stages of incubation, and the organization of the thalamofugal visual projections in males was examined following injection of the fluorescent tracers on day 2 posthatching. As for functional lateralization, the direction of asymmetry in the visual projections was reversed by withdrawing the embryo's head from the egg on day 19/20 of incubation, applying a patch of black tape to the right eye, and exposing the left eye to light for 24 hours before removing the eyepatch (Rogers and Sink 1988). Controls, which had their heads withdrawn from the egg and a patch placed on the left eye, retained the usual direction of asymmetry. Thus, lateralized light input to the thalamus determines the asymmetrical development of the visual projections from this region of the brain to the forebrain. This asymmetry persists after hatching, despite the fact that the chicks receive light input to both eyes after hatching, suggesting that the period of sensitivity to light stimulation is over by the time of hatching.

If male embryos are incubated in darkness and not exposed to light until after hatching, so that both eyes are simultaneously stimulated with light, they develop a symmetrical organization of the thalamofugal visual projections (Rogers and Bolden 1991). Thus, the period of sensitivity to light affecting the development of these visual projections can extend into the immediate posthatching period if the eggs receive no light exposure prior to hatching. Since male embryos which have been exposed to lateralized light input prior to hatching retain asymmetry of their visual projections despite binocular light stimulation of the eyes after hatching, the light exposure prior to hatching may end the sensitive period before hatching occurs. This closure of the sensitive period for establishing the direction of lateralization is similar to the closure of the sensitive period for imprinting. The latter is closed by exposure to the imprinting stimulus, but remains open for an extended period if no imprinting occurs (i.e., if the chick is held in darkness; see Horn 1985).

Female embryos are oriented in the egg in the same way as males and so receive lateralized light stimulation prior to hatching. Nevertheless, they have a lesser degree of asymmetry of their thalamofugal visual projections. The development of these visual projections in females must therefore be less influenced by light, or the sensitive period for females may extend into the posthatching period so that an initial asymmetry in the projections is removed by exposure of both eyes to light after hatching. In contrast to the organization of these visual input projections, lateralization of higher levels of neural processing in the female *is* dependent on lateralized light stimulation of the developing embryo. As we have already discussed, equally in males and females, light exposure during incubation determines the direction of functional lateralization revealed by injecting glutamate into the left or right hemisphere.

Some recent studies have demonstrated that the levels of the sex steroid hormones during the sensitive period of development just prior to hatching influences the response of the developing thalamofugal projections to light stimulation (Schwarz and Rogers, in press; Rogers and Rajendra, submitted). If either testosterone or estrogen is injected into the egg on day 16 of incubation, so that the circulating levels of the hormone present in the embryo are elevated over the last days of incubation, the asymmetry in the thalamofugal projections (determined after hatching) is lessened. The levels of the sex hormones modulate the responsiveness of the developing visual projections to the lateralized light stimulation. Different levels of sex hormones may therefore explain the sex difference in the asymmetry of the thalamofugal visual projections. Indeed, in male embryos, estrogen levels are much lower than in females and, interestingly, testosterone levels decline over the sensitive period during the last few days before hatching (Woods, Simpson, and Moore 1975; Tanabe et al. 1979). In males, it is during this temporary period of lowered levels of testosterone, against a background of low levels of estrogen, that the lateralized light stimulation is able to affect asymmetrical development of the visual projections.

In female embryos, estrogen levels rise slowly throughout incubation and are high during the period of sensitivity to light stimulation. This may well explain the much lesser degree of asymmetry in the thalamofugal visual projections of female chicks. Each side of the thalamus of the females has well developed projections to each side of the forebrain.

To summarize, two factors have been shown to influence the growth of visual projections from thalamus to forebrain: light stimulation and sex hormones. When the levels of estrogen and testosterone are raised, presumably above a threshold level, the growth of projections from both sides of the thalamus is promoted and no, or little, asymmetry results. When the levels of these hormones are low (as in males), light alone stimulates the growth of these projections and, in the normal condition, that stimulation occurs only in the left thalamus, thereby generating asymmetry. A similar stimulation of neuronal growth by testosterone, in this case acting synergistically with auditory stimulation, has been demonstrated for growth of the song nuclei in canaries (Bottjer et al. 1986), although this is not an asymmetrical effect.

By manipulating light exposure and hormonal condition of the embryos it is possible to produce chicks with and without asymmetry in the thalamofugal visual projections, or with reversed asymmetry in these projections. Consistently, the various forms of asymmetry in these projections correlates with performance lateralization in the pebble–grain discrimination task. The shift from pecking at random to pecking for grain and avoiding the pebbles occurs in chicks using the eye which connects to the side of the thalamus that has well-developed projections to the forebrain, and in chicks with symmetry of the thalamofugal visual projections no lateralization of function occurs. From these correspondences we may infer that the thalamofugal visual pathway has a major, if not exclusive, role in the performance of the pebble–grain discrimination task.

The fact that control (untreated) females tested monocularly on pebble–grain discrimination show less performance lateralization than males (see earlier) may depend on the lesser degree of asymmetry in the thalamofugal visual projections in females. Hence, a structural asymmetry may be the basis of the sex difference. Nevertheless, performance on this task obviously requires information processing at higher levels of neural organization in the forebrain, than in just the visual input pathways. At this level of neural organization, females and males appear to have equal degrees of lateralization; their performance is disrupted to the same extent by unihemispheric treatment with glutamate or cyclohex-imide (see earlier). It would therefore be imprecise to make a global statement that the brain of the female chicken is less lateralized than the male, without specifying the limitations of this statement by referring to the level of neural processing to which one is referring. This point has implications for the way in which we discuss lateralization in other species.

Based on the demonstrated role of light in determining the direction of asymmetry in the chicken, Güntürkün and Böhringer (1987) suggested that asymmetry in the pigeon may result from lateralized light stimulation of the embryo's eyes, since the pigeon embryo is oriented in the egg in the same way as the chick. As Güntürkün and Böhringer point out, prolonged monocular deprivation of kittens that have had the optic chiasma sectioned (having visual input restricted to the ipsilateral projections from the retinae) results in asymmetries in the corpus callosum. The hemisphere which has received visual inputs has a much greater influence over the visually deprived hemisphere than vice versa (Cynader, Lepore, and Guillemont 1981). During embryonic development of the pigeon, the right tectum is visually deprived compared to the left, and so the left may assume a greater influence over the right via the tectal commissure. This would explain both the lateralized effects of sectioning the tectal and posterior commissures and the structural asymmetries in cell sizes in the tecta reported by Güntürkün's group (see earlier). Indeed, very recently this group has demonstrated that light exposure of the pigeon embryo does indeed determine the direction of lateralization in pigeons: dark-reared pigeons have no lateralization of visual behavior and no structural asymmetries in the tecta, whereas light-reared pigeons have the usual pattern of asymmetry (Güntürkün 1990).

Extra-Brain Structural Asymmetries in Birds

In many of their somatic organs birds have marked asymmetry. Asymmetry is an essential feature of avian embryogenesis and it persists into adulthood. In females, for example, only the reproductive organs on one side of the body develop to maturity. Asymmetry in these organs is of little concern to us here, but at least some of the somatic asymmetries may have bearing on lateralities in brain function. Structural asymmetries which influence perception or motor behavior place contraints on both the development and function of the central nervous system.

The example most relevant to the lateralities in visual function discussed already is that of asymmetry in the refraction of the left and right eyes in chicks. The left eye is more hypermetropic (farsighted) than the right, as measured at 4 weeks of age (Noller 1984). Thus the left eye is focused for stimuli more distant from the animal than the right eye. Setting aside the added complexity that the chick has a ramped cornea which allows simultaneous focus for short focal distances in the frontal field and longer focal distances in the peripheral field, in general, the relatively hypermetropic left eye may be better at detecting novel stimuli and in assessing topographical cues from the more distant visual space, whereas the right eye is focussed for near vision and so is better able to discriminate, for example, grain from pebbles. Therefore, the refraction asymmetry may be consistent with right-eye specialization for visual discriminations made at pecking distance, and left-eye specialization for novelty responses. If so, there may be a continuous asymmetry from eye structure, through visual projections to neural circuits within the forebrain.

It is conceivable that the eye asymmetry in refraction, similar to the development of the neural pathways, stems from the lateralization of light stimulation of the eyes during embryonic development. The growth of the vertebrate eye is dependent on visual experience, and "form vision" in particular. If the eyelid of the young chick is sutured or if a translucent eyepatch is applied, the eye grows larger and myopia (shortsightedness) develops (Schaeffel and Howland 1989). Form deprivation, by applying translucent occluders, in pigeon squabs (aged 1 to 40 days) has also been shown to affect eye growth, the refractive state being hypermetropic in this case (MacFadden and Salmelainen 1990). Whether myopia or hypermetropia develops as a result of preventing form vision may depend on the age of the animal and the duration of wearing the patches. While neither eye of the chick embryo would receive form vision, as this is precluded by the egg shell, other aspects of visual stimulation (e.g., light flicker, intensity, and wavelength) may conceivably effect eye growth and thus lead to the differential refraction of the eyes.

If light stimulation during embryonic development does establish this difference in refraction between the eyes, it may do so during the same sensitive period during which the asymmetry in visual projections develops, or even during the earlier stages of embryonic development when the embryo is lying on its left side and the left eye is next to the yolk sac. While the effects of light stimulation on the physical construction of the eye and on the growth of visual projections may possibly occur at different stages of development, the end result would be the same: lateralization of the visual system so that each side of the brain receives different information. This means that the lateralization for visual processing may be imposed on the avian brain by lateralized perceptual inputs.

A similar situation occurs for auditory processing in owls. We have discussed previously that the oilbird and the saw-whet owl have a larger number of auditory projections on the left side of the brain. Asymmetry is also present in the structure of the external ears of the owl, and is used in locating a sound source. The asymmetry involves the external ear only; it does not extend to the middle or inner ear. The size and form of the asymmetry varies between owl species. In most species, the asymmetry is confined to the soft structures of the ears, but in some species there is also asymmetry of temporal parts of the skull (Norberg 1977).

In the barn owl, *Tyto alba*, the right ear is directed slightly upward and the left slightly downward (Norberg 1977). This species has a complete facial ruff of feathers, forming a parabolic dish, which is transparent to sound (the barn owl is a nocturnal hunter), and removal of part of this ruff reveals the preaural flaps which are asymmetrical. The left preaural flap is positioned somewhat higher than the right. The right flap is directed about 15 degrees upward compared to the left. The right ear is more sensitive to high frequency sounds located above the mid-horizontal plane of the owl's head, and the left is more sensitive to high frequency sounds located below this plane. Thus, as a sound source moves, say, upward the high-frequency components of the sound become louder in the right ear. This information can be used for accurate location of the elevation of a sound source without the need to tilt the head. Position of the sound in the horizontal plane is determined from the difference of the arrival time of the sound at the two ears, and has nothing to do with the structural asymmetry of the ears (Knusden 1981). In species of owl with symmetrical ears, the location of a sound source in the vertical plane is achieved by tilting or cocking the head 90 degrees or more in the vertical plane. According to Norberg (cited in Neville 1976), young *Aegolius funereus* with asymmetrical ears must learn to interpret the asymmetrical information from both ears by cocking the head alternatively to the left and right. Thus the owl learns that, for example, when higher frequencies are louder in the right ear and lower frequencies are equally loud in both ears, the sound source is located above the owl's mid-horizontal plane and not to its right.

In addition to the asymmetry of the external ears, in *Aegolius funereus*, the skull is also asymmetrical (Norberg 1977, 1978). The left ear opening is about 6.5 mm lower than the right (Figure 2.14). This means that a number of the skull bones are asymmetrical. The middle ear and cochlea are symmetrical. However, there are three pairs of air spaces communicating with the middle ear and one of these pairs is asymmetrical. As Norberg (1978) states, the asymmetry is obviously designed to bring about differential reception of sound between the two ears. Quite clearly, neural analysis of audition must be binaural, with comparisons being made of the information from the two ears to take advantage of the asymmetries.

Since the asymmetries in the external ears and the openings in the skull are not matched by anatomical asymmetries in the middle and inner ears (Norberg 1978), there may be symmetry in the neural receptors and in the organization of the eighth nerve. Nevertheless, at any one moment

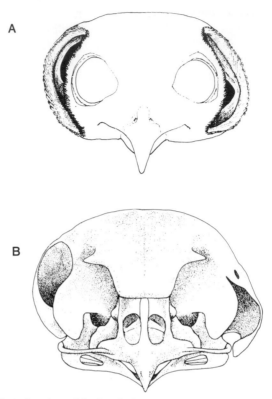

Figure 2.14 A. Anterior view of the head of Tengmalm's owl, *Ageolius funereus*, showing the asymmetry of the external ear cavities. B. View of the skull showing that in this species it is markedly asymmetrical. From Norberg (1978).

in time the auditory information being transmitted to the brain will be different for each ear. This in itself is a form of lateralization imposed on the brain by the structural asymmetry of the receptor organs, much the same as that resulting from the differential refraction ability of the eyes. Perceptual input asymmetries are manifest as lateralized brain function; but the lateralization of information processing may well be more radical than just this. If, as found for one species of owl so far (see earlier), the left nucleus mesencephalicus lateralis, pars dorsalis is larger on the left side, it may play a greater role in information processing. This nucleus is the avian analogue of the inferior colliculus (Camhi 1984), and it is known to be involved in auditory localization and in prey catching behavior by owls (Knusden and Konishi 1979). It has a spatial auditory map and, perhaps, the map in the nucleus on the left side is more detailed.

So far, there has been no attempt to look at which stage of development the asymmetry in either the external ears or the auditory nuclei first develops, apart from Norberg's report that there is asymmetry in the ear openings of fledgeling *Aegolius funereus*, about 30 days old (Norberg 1977). It appears very likely that asymmetry during embryonic development leads to asymmetry of the ears, and therefore throughout development each ear may be functioning differentially. Furthermore, we should consider that the leftward tilt of the head during embryonic development leads to occlusion of not only the left eye but the left ear. Although this occlusion of the left ear would simply attenuate reception of extraneous sound, it would add to any differential perceptual input between the ears. The left ear of the embryo is in fact placed directly over the thoracic cavity and so it receives greater input of sound from the embryo's own heart beat, adding yet another factor of lateralized sound perception. If these factors do indeed play a role during development, lateralization of auditory information processing may be the end result of a complex interactive process of development between structural asymmetry (of the ears, and as imposed by the orientation of the body) and the developing neural pathways for audition.

We have discussed the possible influences on brain lateralization of asymmetry of the perceptual organs and in perception. Asymmetry in the structure of organs used to perform motor behavior may also determine lateralization. The most notable example of the latter is the New Zealand wry-billed plover, *Anarhyncus frontalis*, which has its beak curved toward the right. This species feeds by dislodging stones in search of prey (Neville 1976). The curvature of the beak means that prey objects are more likely to first appear in the right visual field and, parallel to the chick and the pigeon, this eye is likely to be the one specialized to discriminate between food and nonfood objects. Again, there appears to be functional consistency between asymmetry in an extra-brain structure and lateralization of perception and other brain processes.

Lateralization of Foot Use in Birds

Parrots and cockatoos show a preference for using one foot for feeding (footedness) at a population level. Most species are left-footed, but there are exceptions. In a comprehensive review of footedness in parrots, Harris (1989) traces the first reference to foot preference back to 1646, when unknown species were referred to as being left-footed. The issue of footedness in parrots received revived interest in the late nineteenth century, spurred on by Broca's discovery of lateralization for control of speech in humans. At this time, according to Harris (1989), a British neurologist William Ogle observed 86 parrots of unknown species at the London Zoological Gardens and found 63 to be left-footed. Other reports of left-footedness were made at about the same time, some extrapolating to claim that all animals are left-footed or rather "left-handed." There were critics of the time also, some claiming that laterality would be disadvantageous and that symmetry was more adaptive. Today, we still hear the same skeptics state that the lateralization of perception in birds cannot exist because it would be disadvantageous!

In 1938, Friedman and Davis reported left-footedness for manipulating food objects in several species of African parrots. These researchers observed zoo animals and scored the foot used during feeding on 20 occasions for each bird. Reanalysis of their data revealed that, of the 15 species which they listed, 7 had significant footedness. Six species were significantly left-footed, and only 1 species, *Tanygnathus megalorhynochos*, was significantly right-footed (Rogers 1980b). The sample sizes in this study were, however, very small.

Australian cockatoos and parrots also have footedness for manipulating food objects (Rogers 1981; 1989). Both caged and wild parrots were scored. A strong bias for left-footedness was found in eight of the nine species scored. The exception in this study was the crimson rosella, *Platycercus elegans*, which showed right-footedness. Cannon (1983) has reported right-footedness in two other species of this genus, the eastern rosella, *Platycercus eximius*, and the pale-headed rosella, *Platycercus adscitus*. The reason why some species of parrot are right-footed while the majority are left-footed is not known. Possibly it relates to the type of food eaten. When the parrot feeds by holding the food object in one foot, it also uses the beak to manipulate the food object and it turns the head to view the food with the eye that is on the same side as the foot holding the food. If parrots have lateralization of visual perception, maybe visual requirements are associated with the limb used for holding food.

In parrots, only those species which feed while perched in trees use the prehensile-footed style, not those that feed predominantly on the ground. Nos and Camerino (1984; cited in Harris 1989) studied five species of South American parrots housed in the Barcelona Zoo and found that footedness was stronger when the parrots were perched above ground than when they fed on the ground. These researchers observed very few individuals from each species, but most were left-footed.

Footedness for manipulation of food objects also occurs in other avian species. Vince (1964) made a detailed study of foot use of the Great Tit, *Parus major*, in feeding. She gave captive Great Tits a variety of food objects and noted which foot they used to stabilize themselves against

the perch for feeding purposes. Individuals showed significant foot preference, but there was no population bias. Of the nine birds tested,
three were strongly right-footed, two less strongly right-footed, and four
strongly left-footed. The birds were also tested with objects of food tied
to a length of string hanging in a glass cylinder. The Great Tits retrieved
these objects by pulling up a loop of string with the beak, holding it with
one foot, and pulling up again. A foot preference was also found for
each individual in this task, but as before there was no population bias.

A population bias of footedness was, however, found for goldfinches,
Carduelis carduelis, tested on a task requiring them to manipulate doors
and catches using the beak and a foot in order to obtain a food reward
(Ducker, Luscher, and Schultz 1986). All the birds preferred to use the
right foot.

Walker (1980) has stated a general hypothesis that limb use preferences
occur only in those species that use their limbs for manipulative tasks.
Footedness may perhaps occur only in those avian species that use their
feet in feeding. Pigeons, for example, do not manipulate food objects
with their feet, and Güntürkün, Kesch, and Delius (1988) have reported
the absence of footedness in pigeons tested by sticking a piece of tape
on the tip of the beak and scoring the foot used in the first attempt to
remove it. They found no bias in foot use at either the population or
individual level. The lack of footedness in pigeons tested on this task is
species rather than task specific. They tested a small number of parrots
on the same task and found the preferred foot used to remove the tape
was consistent with their footedness for manipulation of food objects.

Some individual pigeons do, however, display footedness in other
situations. Davies and Green (1991) investigated foot use in pigeons
during taking off for flight and landing. They tested the birds in a flight
tunnel and scored foot use using frame-by-frame analysis of videotapes
of these behaviors. No population bias of footedness was found for
either taking off or landing, but some individuals showed footedness
for landing, although not for taking off. The authors proposed that
footedness may occur in landing only as it is a more stressful maneuver
and requires fine visuomotor control. This, they suggest, strongly contrasts to removing a piece of sticking tape from the beak, which uses
grooming movements and may be entirely under proprioceptive control.

Although this result illustrates that footedness, at the level of the
individual, can occur in a species which does not use the feet to manipulate objects, the original hypothesis may still hold for footedness present
at the population level. That is, a population bias of footedness may
occur only in those species which use the feet to manipulate objects.

Budgerigars have also been tested for foot use in removing a piece of
sticky tape from the beak (Rogers and Workman, in press). They were

chosen for study because they are parrots, but they do not use their feet when feeding. Nine individuals were scored for a mean of 20 trials each. Like the pigeons tested on this task, the budgerigars showed no footedness either at the population level or as individuals. This supports the hypothesis that laterality occurs only when the feet are used to manipulate objects.

Yet, contrary to this hypothesis there is footedness, at the population level, in a species which does not use its feet to directly manipulate objects. Chickens (*Gallus gallus*) do not use their feet to pick up and manipulate food or other objects, but they frequently scratch the ground when searching for food. Rogers and Workman (submitted) scored the foot used to initiate a bout of scratching the ground in search of food. Six chicks of a feral strain were scored for 40 scratching bouts each. Though both feet are used in this behavior, there was a significant tendency (68%) to initiate a bout of ground scratching by using the right foot (Rogers 1989). When 10-day-old chicks were tested on the task requiring removal of sticky tape from the beak, a stronger right-foot bias was found (84%). Apparently, it is not manipulative ability alone which confers a population bais of footedness in avian species, but rather active use of the feet in any manner for feeding or searching for food.

The fact that chickens show right-footedness in searching for food correlates with the specialization of the right eye in tasks requiring the chick to search for food and to perform visual discrimination. It makes logical sense that chickens have right-footedness for initiating scratching of the ground to expose grains of food or insects, as the right eye better discriminates food from nonfood objects.

The pigeon has the same lateralization of eye use in visual discrimination learning as does the chicken (see earlier), but it does not have footedness at the population level and it does not use its feet to scratch the ground while feeding. This may suggest that a population bias for footedness (in both feeding and nonfeeding tasks) in avian species may have developed secondarily to lateralization of visual functions at the perceptual level, and therefore occurs only in species which must coordinate eye and foot use in feeding, either to manipulate the food or to uncover it by scratching the ground. In other words, if the feet are used in feeding, laterality of foot use may occur as a result of the constraints placed upon it by lateralization in perceptual or cognitive processes linked to either eye.

Unfortunately, we know nothing as yet of functional lateralization in the forebrain of parrots or cockatoos, except that *Amazona amazonica*, which is 75% left-footed (Friedman and Davis 1938), does not have vocalization control lateralization (Nottebohm 1976). Unlike songbirds, the syrinx of parrots does not have two separate sound sources and there can be no lateralization of vocalization control at this level of organization. It is highly likely that parrots and cockatoos do have laterality of other forebrain functions. As we have discussed, lateralization can occur at a number of different levels of neural organization, and those levels at which it occurs may vary from species to species. Zebra finches, for example, have no, or possibly only slight, laterality for control of their vocalization, but they do show functional laterality for copulation responses (Workman and Andrew 1986; and see earlier). The right-footedness of the crimson rosella, *Platycerus elegans*, may indicate a different, if not inverted, laterality at higher levels of central processing in this species.

Summary and Conclusions

In this chapter we have discussed the specialized features of the avian visual system which may have provided an impetus to the evolution of brain lateralization. This lateralization is clearly evident in a range of visually guided behaviors, and it can be revealed simply by testing the birds monocularly. For example, it has been shown for both pigeons and chickens that visual discrimination learning is faster when the right eye is used. The right eye system is specialized for categorizing objects, such as food versus nonfood objects. This eye system pays less attention to the unique properties of each stimulus than does the left eye system. The left-eye system is specialized to respond to the unique properties of each stimulus, and thus to respond to novelty. In particular, the left-eye system is specialized for processing topographical information; it responds to the changed position of a stimulus and is used to build up a map of the environment and respond to changes in it. In addition, the left-eye system of the chick has been shown to control copulation behavior.

Because, in general, information received by one eye is processed by the contralateral hemisphere of the forebrain, it can be deduced that the left hemisphere is specialized to categorize objects, and the right hemisphere is specialized for processing topographical information, for responding to novelty, and for controlling fear and copulatory responses. Unilateral administration of certain drugs into the forebrain hemispheres confirms these deductions. However, we must take into account that this division is not absolute, as there are left side to right side connections, and vice versa. Indeed, structural asymmetry has been discovered in one set of visual projections, the thalamofugal visual projections, in the young male chick. Each side of the thalamus sends visual projections to the Wulst, or hyperstriatal, regions of each of the forebrain hemispheres.

There are more contralateral projections from the left side of the thalamus to the right hyperstriatum than vice versa. Other anatomical asymmetries are also known in the tectofugal visual system of the pigeon.

At the subcellular level of organization, a number of asymmetries are known to occur in the chicken brain in connection with memory formation. The left and right hemispheres are differentially involved in memory formation. Different synaptic and neurochemical changes occur in each hemisphere and in different anatomical regions.

In young chicks there are sex differences in both the structural and functional asymmetries, and a role of the sex steroid hormones has been demonstrated. Furthermore, asymmetrical stimulation of the eyes by light influences the development of both the structural and functional asymmetry, as shown in the chick and very recently in the pigeon. The embryo is oriented in the egg in such a way that its right eye can be stimulated by light, whereas the left eye is occluded. This promotes better development of the projections from the left side of the thalamus (fed by the right eye) to the forebrain, and organizes the direction of functional asymmetry so that a population bias is established. Thus, the development of lateralization in the avian brain results from the interaction of genetic, hormonal, and environmental (light) factors.

Birds also display laterality of auditory perception and in control of song production. In owls there are differences in the sizes of auditory nuclei on the left and right sides of the brain and, in some species, the external ears and the skull are quite asymmetrical. The latter allows the owl to locate a sound source in the vertical plane without the need to cock its head. A number of species of songbird have lateralized control of singing by the left hemisphere. This functional asymmetry is not matched by structural asymmetry, apart from the enlarged musculature on the left side of the syrinx. Some evidence also exists for lateralization of proprioception and olfaction in birds, but the study of these is only just beginning.

Finally, we have discussed foot use in birds. Many species have been shown to have footedness similar to handedness in humans. Those avian species showing footedness use the feet either to manipulate food objects (e.g., parrots and cockatoos), or they use the feet to reveal food objects (e.g., chickens scratching the ground).

Given the growing list of both functional and structural asymmetries in birds, one begins to wonder whether it is not asymmetry but symmetry which, when it occurs, needs to be explained. Birds have provided numerous examples of lateralization at the perceptual level of neural processing (particularly for visual processing), at the level of higher cognition and memory formation, and on the output side of motor control (song and footedness). Some functions appear to involve lateralization at all of these levels of neural integration, from perceptual input through processing to motor output, whereas others may involve lateralization at one level and not another.

Most of the examples discussed in this chapter are of lateralization present at the population level, all individuals in the population having laterality in the same direction. At this point it may be worth speculating whether population lateralization has any role in organization of social structure. We have discussed data showing that a bird is more likely to attack, or copulate with, a stimulus perceived with the left eye (in the left monocular visual field). Considering this the other way around, a conspecific may avoid being attacked, or copulated with, if it approaches another individual from the latter's right side. If all individuals in the population are lateralized in the same direction, laterality could be used as a basis for constructing social organization. There is some evidence that this may be so. The structure of the social hierarchy was compared for groups of chicks hatched from eggs which had received exposure to light (and therefore were all lateralized in the same direction), and groups of chicks hatched from eggs incubated in darkness (and therefore without lateralization at the population level). The social hierarchy was more stable in the groups hatched from eggs exposed to light (Rogers and Workman 1989). Although there may be other explanations for this result, it is not inconsistent with the prediction that social organization may be influenced by the presence of lateralization at the population level.

In some species of birds laterality occurs in individuals and not at the population level. Foot preference in the Great Tit is an example of this, half the individuals preferring the left foot to hold food and the other half preferring the right foot. There has been a tendency to see lateralization at the population level as being more important than that present only at the individual level. Yet, for some brain functions, lateralization may be advantageous irrespective of the direction which it takes. And for some functions, population diversity may be desirable. We need more information across a range of behaviors and species before carrying this issue further.

We might also wish to consider lateralities which occur transiently as different from, but just as important as, those which persist over longer periods of time. For example, this would appear to be the case for some of the neurochemical changes associated with memory formation. Unihemispheric sleep is another example. Birds spend a certain percentage of sleeping time with only one hemisphere asleep (Ball et al. 1988). The latter researchers showed correlation between unilateral EEG sleep activity and closure of the contralateral eye. The other hemisphere stays awake, presumably monitoring the external environment. Thus, unilateral sleep may possibly be used during migration, although this is unknown. Indeed, unihemispheric sleep also occurs in the Cetacea, another group of animals which navigates over long distances (Mukhametov, Lyamin, and Polyakova 1985). When one hemisphere is asleep, the hemispheres of the brain are functioning asymmetrically with most processing lateralized to the "awake" hemisphere. This is therefore a transient asymmetry which has potential ramifications for the behavior of the whole animal at any given time depending on which hemisphere is asleep. That is, when the left hemisphere is asleep, the functions of the right will gain temporary dominance and vice versa. So far, there has been no investigation of whether one hemisphere tends to sleep more than the other, but it is worth considering that lateralized perceptual inputs (auditory, visual, or even olfactory) may lead to lateralized changes in arousal. It is known, for example, that imprinting causes a transient change in the firing activity of neurons in the IMHV regions of the chick forebrain (Davey and Horn 1991). The changed activity is characteristic of the waking pattern; it is as if imprinting "wakes up" the forebrain. Although this change is not lateralized, the rate of spontaneous firing of neurons in the IMHV regions does change in a lateralized manner (it increases on the left side). Further study of changes in neuronal firing patterns as a consequence of learning other tasks which involve lateralized perceptual inputs to the forebrain hemispheres may possibly reveal awake–sleep lateralities.

There are two main lessons to be learned from study of lateralization in birds. First, lateralization may take many different forms in addition to dominance of one hemisphere, or side of the brain, over the other. It can involve differential use of each side of the brain for a given function. This differential use may be a matter of degree, one side being used to a greater extent than the other, or it may involve equal but different use of each side of the brain. Differential use may take the form of different neural sites, different neurochemical changes, or different pathway integration. Birds have provided us with examples of all of these.

Secondly, lateralization develops as a result of the interaction between genetic, hormonal, and environmental factors. Here studies of the developing chicken embryo have been most enlightening, and they have important implications for our understanding of asymmetry in other species. Study of the role of lateralized light input of the chick embryo in determining the direction of asymmetry in the visual pathways and in visually guided behaviors has illustrated that any hypothesis that ties the determination of lateralization to a unitary cause is oversimplified.

Environmental factors interact intimately with genetic and hormonal factors as the developmental events leading to asymmetry of brain structure and function unfold.

This does not imply that selective pressures of evolution have not been operating. On the contrary, it is most unlikely that the large number of lateralities present in birds at the population level are simply the result of chance developments. Evolutionary processes must have been operating, and we have discussed how constraints of visual perception may have provided a basis for the evolution of brain lateralization.

Further Reading

Adret, P., and Rogers, L.J. Sex difference in the visual projections of young chicks: A quantitative study of the thalamofugal pathway. *Brain Research*, 1989, 478:59–73.

Andrew, R.J. The development of visual lateralization in the domestic chick. *Behavioural Brain Research*, 1988, 29:201–209.

Andrew, R.J. The nature of behavioral lateralization. In R.J. Andrew (ed.), Neural and Behavioural Plasticity: The Use of the Domestic Chick as a Model. Oxford: Oxford University Press, 1991, pp. 536–554.

Cipolla-Neto, J.; Horn, G.; and McCabe, B.J. Hemispheric asymmetry and imprinting: The effect of sequential lesions of the hyperstriatum ventrale. *Experimental Brain Research*, 1982, 48.22–27.

Feisen, L. von, and Güntürkün, O. Visual memory lateralization in pigeons. *Neuropsychologia*, 1990, 28:1–7.

Harris, L.J. Footedness in parrots: Three centuries of research, theory, and mere surmise. *Canadian Journal of Psychology*, 1989, 43:369–396.

Horn, G. Neural basis of recognition memory investigated through an analysis of imprinting. *Philosophical Transactions of the Royal Society Lond. B.*, 1990, 329:133–142.

Nottebohm, F. Asymmetries in neural control of vocalization in the canary. In S. Harnard, R.W. Doty, L. Goldstein, J. Jaynes, and G. Krauthamer (eds.), Lateralization in the Nervous System. New York, Academic Press, 1977.

Rogers, L.J. Development of lateralization. In R.J. Andrew (ed.), Neural and Behavioural Plasticity: The Use of the Domestic Chick as a Model. Oxford: Oxford University Press, 1991, pp. 507–535.

Rose, S.P.R. How chicks make memories, the cellular cascade from c-fos to dendritic modelling. *Trends in Neurosciences*, 1991, 14:390–397.

Zappia, J.V., and Rogers, L.J. Light experience during development affects asymmetry of forebrain function in chickens. *Developmental Brain Research*, 1983, 11:93–106.

Asymmetries in Rats and Mice   3

Introduction

There are numerous examples of asymmetry in rats and mice. Some of
these asymmetrics are present in individuals but without a consistent
bias in the population (e.g., preference of forelimb use in feeding, or
"pawedness," and direction preference for turning or rotating). Other
asymmetries occur in individuals and, because the majority of individuals show the same direction of asymmetry, there is also asymmetry in
the population as a whole. Asymmetries present at the population level
have been described for a range of behaviors and for neurochemical and
structural parameters as well.

Apart from the need to document these asymmetries, perhaps the
most notable finding to arise from the studies on rats and mice is that
the many forms of asymmetry depend on early experience and hormonal
condition. The important studies of Denenberg and his co-workers (see
Denenberg and Yutzey 1985) have shown that early experiences can
generate laterality in the brain or enhance laterality that is already present, and that the effect of early experience can be modified by the hormonal condition of the pups. Indeed, in this respect the development of
asymmetry or lateralization in rodents clearly corresponds to that of birds
(see Chapter 2). The form that it takes in any individual or any population
is determined by the interaction of genetic, environmental, and hormonal
factors. As also shown in birds, the developing brain is in continuous
interaction with its environment and, in the context of laterality, it displays a remarkable degree of neural plasticity and behavioral adaptation.
In fact, it is studies using rats that have shown that brain plasticity is
not entirely lost in adulthood. Even adult rats show changes in the
thickness of the cortex and changes in laterality in response to being
raised either in enriched and or impoverished environments (Diamond
1985). Taken together with the similar studies of the avian brain, these
findings have major implications for our understanding of brain functions and the development of lateralization in the human brain.

In Chapter 2 we hypothesized that having laterally placed eyes and
relatively small commissural systems may have set a requirement for

lateralization of the brain. Birds have a relative lack of left side–right side interconnections at the forebrain level; the main commissural systems (the tectal and posterior commissures) in the avian brain being at a lower level of neural organization in the midbrain. Because birds consequently process information from each of the two visual fields largely separately, hemispheric dominance may prevent conflicting responses to simultaneous inputs from the two visual fields. In contrast to birds, rodents have well-developed connections between the left and right hemispheres, mainly via the corpus callosum, but also including the hippocampal, anterior, and posterior commissures (see Figure 3.1). As we will discuss, the corpus callosum plays a role in the lateralization of the hemispheres. It is also worth noting here that the size of the corpus callosum, and the connectivity of the neurons which travel in it, is influenced by early experience and hormonal condition. Much of the research on asymmetry in the rodent brain has focused on this aspect.

Given the well-developed corpus callosum in mammals, perceptual information is easily conveyed to both hemispheres. We might therefore expect to find in rodents less requirement for lateralization of visual information, as the coupling of the hemispheres might eliminate or re-

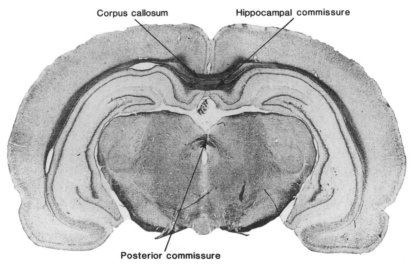

Figure 3.1   A transverse section through the rat brain at the level of the cortex. The neural tracts which connect the left and right sides of the brain have been labeled: the corpus callosum, the hippocampal commissure, and the posterior commissure. From Pellegrino, L. J., Pellegrino, A. S., and Cushman, A. J. (1967).

duce potential problems of conflict for control of responses to stimuli perceived simultaneously in the two monocular visual fields, each demanding a different response (see Introduction to Chapter 2). Additionally, mammals have both ipsilateral (noncrossing) and contralateral (crossing) projections from each eye to the brain, which might be seen as further reducing the need for any form of hemispheric dominance. However, rats and mice have laterally placed eyes, and they have a relatively small number of ipsilateral projections from the retina to the brain. In rats, generally only 5% of the optic fibers do not cross over at the optic chiasma (Hayhow, Sefton, and Webb 1962). The ganglion cells of the retina which give rise to these noncrossing fibers are in the extreme temporal retina and they have a narrow visual field in the binocular area. Therefore, the majority of the visual field uses the fibers that cross in the optic chiasma. The albino rat has even fewer noncrossing fibers than does the Norwegian hooded rat, a pigmented strain (Lund 1965).

Therefore, in rats and mice, visual input to the first relay station in the brain, the lateral geniculate nucleus, is largely from the contralateral eye and, in this sense, they are similar to birds. From each lateral geniculate nucleus the information is fed forward to the ipsilateral visual cortex; it may then cross to the other hemisphere via the corpus callosum. Information that reaches a hemisphere via the corpus callosum is qualitatively different from that which reaches it directly. Consequently, we may expect to find in rodents at least some of the same requirements for lateralization of information processing as occur in birds, although these requirements may operate to a lesser degree. In fact, as for birds, monocular testing of albino rats has recently been found to reveal lateralized behavioral functions, and these lateralities are present at the population level.

On the evolutionary tree, birds branched off before mammals evolved, and so both birds and mammals share a common reptilian ancestor. As it is unlikely that brain lateralization arose *de novo* in both birds and mammals, a strong argument can be made that it was already present in the reptilian ancestor, which also would have had laterally placed eyes and very small commissural systems. This is supported by the fact that there are at least some asymmetries in the brains of presently existing reptiles (see Chapter 1). Yet, while rodents (and, for that matter, other mammals) may show remnants of an evolutionary need for brain lateralization, they appear to have elaborated the asymmetrical organization of the brain to include a far wider range of functions and structures. In an article on morphological asymmetries in the rat brain, Diamond (1985) aptly questioned, "Is any part of the nervous system really symmetrical?"

She was referring to structural asymmetries only: Add the functional asymmetries and the question becomes more pertinent.

Postural and Motor Asymmetries in Rats and Mice

Laterality of Forelimb Use

Rats and mice show a preference to use one forelimb over the other to reach for food. This preference is referred to as "pawedness." In general, about half of the individuals are right-pawed and half left-pawed, so that there is no population bias equivalent to handedness in humans or footedness in parrots.

Paw preference was first discovered in rats by Peterson nearly six decades ago, using a task that required the animals to reach into a tube to obtain a food reward (Peterson 1934). Most rats were found to exhibit strong paw preference, and the population consisted of equal numbers of left- and right-preferent individuals.

Collins (1975) tested mice in a similar task, requiring them to reach into a cylindrical tube attached to the front wall of the cage to obtain a food reward (Figure 3.2). For each individual mouse the number of reaches with the right and left paws was scored. Individuals of an inbred strain were tested for 50 reaches each. The percentage of right paw entries into the tube was calculated. Most mice were found to be either strongly left-pawed or strongly right-pawed, and few used both paws relatively equally. Thus, half the population was found to be lateralized in one direction and half in the other (Figure 3.3). The females were more strongly lateralized than the males. The pawedness for each individual was consistent over weeks and even months (Collins 1985). Recently, Signore et al. (1991) have used Collins' procedure to test over 300 male and female mice of a variety of inbred strains and their results collectively confirm the original findings. Other strains, tested by Betancur, Neveu, and Le Moal (1991), have been found to be only weakly lateralized. Nevertheless, even in these strains, the females were more strongly lateralized than the males.

These tests were conducted in an "unbiased world," in that the tube was placed in the center of the cage front so that each forelimb could be used with equal ease. Additional mice were tested in biased worlds which made use of one or the other of the forelimbs more difficult. For example, a right-biased world was created by locating the feeding tube flush with the right side wall of the cage and the reverse arrangement created a left-biased world. The distribution of paw preferences in the population was influenced by the biased worlds (Figure 3.3). In the right-

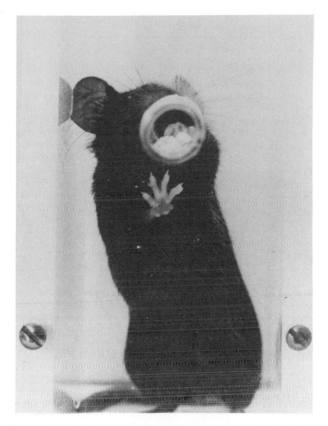

Figure 3.2 A mouse being tested for paw preference in reaching for food placed inside a tube attached to the front of a cage. The tube containing the food is placed equidistantly from each of the side walls. Thus, this animal is being tested in an "unbiased world." From Collins (1985).

biased world, more of the mice exhibited right-pawedness, but a residual of approximately 10% were left-pawed. The reverse was the case in the left-biased world. These effects of the biased worlds illustrate that both genetic and environmental factors interact to determine paw preference.

It appears that genetic and environmental factors may determine different aspects of paw preference. It is possible to genetically select lines of mice that have stronger or weaker lateralization, but not for more right-pawedness or left-pawedness per se (Collins 1969, 1977, 1991). That is, genes may specify the degree or strength of lateralization, but not its direction. The latter definitely can be influenced by environmental factors. It should also be mentioned that being able to select for lines of

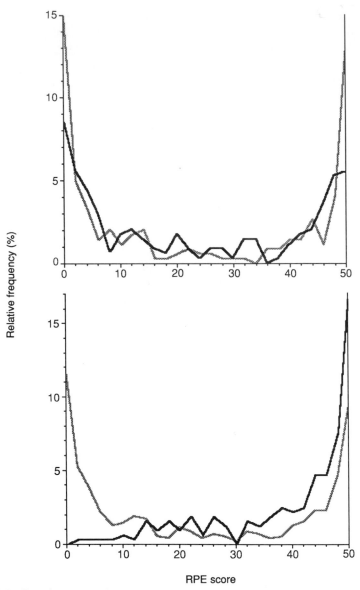

Figure 3.3   Data for paw preference in mice obtained using an apparatus the same as, or similar to, the one illustrated in Figure 3.2. The distribution of "right paw entry" (RPE) scores is plotted for over 300 mice of each sex. Each mouse was tested for 50 reaches. The graphs at the top of the figure are data for males (black line) and (grey line) females tested in the unbiased world (i.e., with no physical restrictions on use of either the left or right paw). Note that the majority of mice are either strongly left pawed (clustered at the 0 RPE score mark) or strongly right pawed (clustered at the 50 RPE score mark). Also note that

mice with different degrees of lateralization does not necessarily imply that the genetic influence is a direct one: It is possible that the selection was for, say, different maternal–offspring interaction, and that the differences in rearing conditions produced by this indirectly determine the degree of lateralization of paw preference.

Selection for a line of mice that is strongly lateralized ("high" line) is carried out by testing for paw preference, selecting animals that show strong paw preferences, and then inbreeding them. This procedure is repeated over generations. Selection for a weakly lateralized line ("low" line) involves choosing mice that are ambidextrous and inbreeding them. Significant line differences may emerge by as early as the third generation (Collins 1991). These differences continue to increase with each subsequent generation. After eleven generations, Collins (1991) relaxed the selection and allowed each line to continue to breed at random within the line. He found that significant differences between the lines remained for at least a further sixteen generations. Hence, the determinants of the degree of lateralization were remarkably stable within each of the selected lines.

Males and females were found to respond differently in the biased worlds. The males became more lateralized than the females (Collins 1977). To Collins, this result was expected; if the genetic complement is associated with weaker lateralization in males, males should be more influenced by the biased environment.

Furthermore, hormonal factors influence the distribution of pawedness in the population. In a population of mice carrying a mutation which makes them insensitive to testosterone, males were found to be just as lateralized as females and more strongly lateralized than their male siblings not carrying the mutation (Nosten, Roubertoux, Degrelle, and Leoyer 1989).

There is a relationship between paw preference and development of the corpus callosum. Schmidt, Manhães, and de Moraes (1991) have studied paw preference in a strain of mice in which 20% of the pups are born with a very small corpus callosum. This group exhibited a significant left-paw bias (78%) at the population level. Hence, population lateraliza-

---

females have more asymmetrical paw use than the males. The bottom graphs illustrate the effect of testing the mice in a biased world. The distributions are for males and females combined. Data for mice tested in an unbiased world (grey line) is plotted for comparison to data for mice tested in a right-biased world (black line) (i.e., with the tube located next to the left-hand wall of the cage, relative to the mouse, so that reaching with the left paw was more difficult). Note the shift toward right-pawedness in the latter situation. Approximately 90% of the mice exhibited paw preference consistent with the world bias, whereas 10% resisted this influence. Adapted from Collins (1985).

tion emerges when there is a defect in the development of the corpus callosum. In other strains of mice, defects of the corpus callosum have been reported to have a variety of different effects. For example, Ward, Tremblay, and Lassonde (1987) reported that a reduced corpus callosum size is associated with weaker paw preference in mice of the strain 129/ J, but not in mice of the strain BALB/cCF. Also, lines of mice selected to be either strongly or weakly lateralized for paw preference have a different organization of the corpus callosum (Cassells, Collins, and Wahlsten 1990). In the weakly lateralized line, the dorsal commissure of the fornix (part of the corpus callosum) is composed of a greater number of discrete bundles compared to that of the strongly lateralized line.

There is one mouse strain that completely lacks the corpus callosum. Gruber et al. (1992) have tested this strain for paw preference in reaching for food and compared it to a strain with normal development of the corpus callosum. Two experiments were conducted in different laboratories. In one laboratory, males of the acallosal strain were found to have reduced paw preference, but this did not occur in the females. In the other laboratory, all of the acallosal mice, males and females, were found to be ambilateral. Each laboratory tested animals of a different age and the researchers consider this as a likely explanation for the difference in results. Nevertheless, the overall finding is less lateralization in the acallosal animals, which lends support to Denenberg's proposal (Denenberg 1981) that the corpus callosum mediates inhibition of one hemisphere by the other and so serves to generate lateralization.

The corpus callosum clearly has a role in paw preference, but it is not a simple one. Other regions of the brain may also be involved. In fact, several regions of the brain do differ in size asymmetry according to the degree of pawedness. Lipp, Collins, and Nauta (1984) made measurements of asymmetry in cortical thickness in several regions of the cortex in two lines of mice: one selected over 11 generations to show a strong degree of paw preference and the other to show a weak degree of paw preference. The latter were largely ambilateral. Individuals of the strongly lateralized line had larger asymmetries in four regions of the cortex. In addition, the two lines could be differentiated by their asymmetries in the volume of the hippocampus. These regional asymmetries are not, however, reflected in overall weight of the left and right hemispheres: There is no significant difference between the weights of the left and right halves of the brain in either the strongly or weakly lateralized lines of mice (Cassells et al. 1990).

Nevertheless, Ward and Collins (1985) have found that mice of a strongly lateralized line have heavier brains than do mice of a weakly lateralized line, and both lines diverged from the control, a randomly bred line. Interestingly, the mean size of the corpus callosum of both the strongly and weakly lateralized lines was found to be smaller than that of the control group. A more recent study (Cassells et al. 1990) of the brains of mice from the strongly and weakly lateralized lines has confirmed that the more strongly lateralized line is characterized by a heavier brain than the weakly lateralized line. The more strongly lateralized line was also found to have a larger cross-sectional area of the corpus callosum, but, when this was expressed in relation to brain weight, the more strongly lateralized line actually had a relatively smaller corpus callosum. The size of the corpus callosum relative to brain size was not related to strength of pawedness, whereas brain size was positively correlated with strength of pawedness. The conclusion reached in this study is that the difference in brain size between the strongly and weakly lateralized lines is an important correlate of the difference in strength of pawedness, but not of the difference in the size of the corpus callosum per se.

The reasons for the difference in brain weight are, as yet, unknown. But, as the authors point out, it cannot be assumed that the causes are entirely genetic because the effect of selecting for strongly and weakly lateralized lines may have produced two groups which manipulate objects in their environment to different degrees. For example, the strongly lateralized line may interact more with the environment and so experience "environmental enrichment," which is known to alter cortical thickness and asymmetry. Hormonal condition may also differ between the two lines or, along with selection for different degrees of limb-use laterality, other behaviors may have been selected, either as a consequence of the changed degree of laterality, or in parallel with it but not directly related to it. Thus, for example, females of the strongly lateralized line have been found to be more aggressive in a standard laboratory test (Scott, Bradt, and Collins 1985). This result is, incidentally, most interesting in comparison to the known laterality for control of attack behavior in the chicken (see Chapter 2) and the rat (see later).

Lateralized behavior may also be passed on from generation to generation by a process of learning. Collins (1988) has demonstrated that mice can learn lateralized behavior by observation. A "teacher" mouse was trained to open a door either on the right or the left side of a cage to obtain a food reward. A "class" of five mice watched the teacher from cubicles behind a barrier. Male observers that had watched the teacher open a door on the left side performed sinistrally, whereas those that had watched the teacher open a door on the right side performed dextrally. Females observers did not acquire the teacher's laterality. Therefore, in males at least, observation learning may contribute to lateralized motor preferences. We have seen earlier that males are also more responsive to the environmental influence of a biased world.

Pawedness does in some way develop in response to reward. If weakly lateralized (ambilateral) rats are rewarded by intracranial self-stimulation in the lateral hypothalamus for reaching into a feeder tube, they shift their paw preference to use the paw contralateral to the side of the hypothalamus that has been stimulated (Hernandez-Mesa and Bureš 1985). Paw preference of strongly lateralized individuals is not affected by this procedure. The authors conclude that because the reward affects integration of sensory and motor elements, motor asymmetries occur as a consequence of asymmetries in motivational and attentional mechanisms. Whatever the exact explanation, these results do provide a basis for concluding that Collins' mice, tested in the biased worlds, may shift paw preference as a consequence of gaining reward (food reward in this case) by reaching with a particular paw.

Sidedness and Circling

Rats turn in circles, or rotate, spontaneously at night—the time of the circadian cycle when they are most active—or in the daytime after they have been given an injection of amphetamine (Glick and Cox 1978; Glick and Shapiro 1984, 1985). In both situations, the direction of circling is the same for each individual. Like pawedness, half the population of rats prefer to circle clockwise and half counterclockwise.

The circling direction of an individual is related to asymmetry of the dopamine levels in the nigrostriatal basal ganglia system (Jerussi and Glick 1976). Unilateral damage to the nigrostriatal region causes the rat to circle to the side of the lesion. Administration of a dopaminergic drug, such as amphetamine, potentiates the circling behavior. This led Glick and co-workers to conclude that the direction of circling in normal, untreated rats must be determined by lower levels of dopaminergic activity on the side ipsilateral to the direction of circling. The asymmetry in nigrostriatal function is present in 90–95% of normal rats (Glick, Jerussi, and Zimmerberg 1977; Jerussi, Glick, and Johnson 1977). In fact, in the majority of rats there is asymmetry of striatal dopamine content (Zimmerberg, Glick, and Jerussi 1974), striatal dopamine metabolism, and dopamine-stimulated adenylate cyclase activity (Jerussi et al. 1977), all of which are associated with dopaminergic neurotransmission. The dopamine asymmetry was later confirmed by studying the effects of unilateral lesions of the dopamine system using the drug 6-hydroxy-dopamine, which selectively destroys dopamine neurons (Robinson and

Becker 1981). The results were the same as for placing lesions in the nigrostriatal region, resulting in circling to the side of the lesion.

This asymmetry is rather persistent. Although rats can be trained to circle in a direction opposite to their spontaneous preference, within 2 to 3 weeks of training, they revert to the original, preferred direction (Zimmerberg, Stumpf, and Glick 1978). This appears to occur because the training has no detectable effect on the direction of asymmetry of the dopamine activity in the nigrostriatal regions on each side.

Just as strength of pawedness is genetically determined, so too may genetic determinants contribute to the strength of rotation or circling. Rats of different strains, or even those of the same strain but from different colonies, show different degrees of rotation (Carlson and Glick 1989). Furthermore, Glick (1985) was able to genetically select lines of rats in which the females are either strongly or weakly rotating. Yet, as Glick et al. (1986) recognized, environmental factors (such as diet, housing conditions, and stress) may also contribute to the degree of rotational behavior.

Strongly and weakly rotating rats show different behavioral responses to treatment with amphetamine (Glick et al. 1986). Rats derived from a Sprague-Dawley strain were treated twice with the same dose of d-amphetamine. Those that rotated weakly (less frequently in a fixed time interval) in response to the first dose of amphetamine, rotated more frequently following the second dose, whereas those that rotated strongly in response to the first dose rotated less frequently after being given the second dose. The preference percentage for rotation in one direction or the other showed a similar pattern of changes: Rats that had a rotation preference of 70–85% (calculated as the ratio of the number of rotations in the individual's preferred direction to the total number of rotations) in response to the first injection of amphetamine displayed increased percentage preference in response to the second injection, whereas those with a percentage preference of 95% and above showed decreased preference in response to the second injection. The authors explained their results on the basis that dopamine levels in the nigrostriatum differ in strongly and weakly rotating individuals and so the sensitivity to amphetamine differs accordingly, and in a lateralized manner. The preferred direction of rotation is not, however, affected in these cases, and therefore the direction of asymmetry in dopamine levels cannot be affected.

Sex differences have also been reported in the effects of dopaminergic drugs on circling behavior. Females show a higher rate of circling and more prolonged circling in response to treatment with amphetamine (Robinson et al. 1983; Robinson 1985) or cocaine (Glick, Hinds, and

Shapiro 1983). Although males metabolize amphetamine more rapidly than do females, this is not the explanation for their lower response to the treatment because males injected with higher doses of amphetamine also rotate less than the females treated with a lower dose (Brass and Glick 1981). There is evidence that the circulating levels of the sex hormones affect the amphetamine-induced circling. Becker, Robinson, and Lorenz (1982) have shown that the amphetamine-induced circling rate varies across the estrus cycle. On the day of estrus, females rotate more than they do 24 hours later at diestrus. Brain levels of amphetamine were measured and the doses of amphetamine administered were adjusted to give equivalent brain levels. Even with this adjustment, males rotated less than females at proestrus, estrus, and day 2 diestrus. It would appear that activity of the nigrostriatal dopamine system is modulated by the levels of the sex hormones.

Thus far, we have discussed the influence of the sex hormones on the strength of rotation. In addition, sex hormones have a significant, but small, influence on the direction of rotation. In larger samples, females of more than one strain of rat have been found to have a slight population bias to circle to the right (in the region of 57%; Robinson 1985). By contrast, males of two strains have now been shown to have a preference (60%) for circling towards the left (Robinson 1985). Although the sizes of these population biases is small, it is interesting that the direction preference for females is opposite to that of males. Thus, apparently the sex hormone levels also have some limited ability to modify the direction of asymmetry in dopamine levels in the nigrostriatal system.

We have mentioned that it is possible to select for strength of pawedness in mice. Collins (1985) tested his "high" and "low" lines of mice (those with strong and weak lateralization, respectively) for circling behavior in a swimming tank by scoring clockwise and counterclockwise rotation. He found that the high line had a more strongly preferred direction of circling than did the low line. Thus, there may be some association between strength of pawedness and strength of rotation in the population. This needs further investigation at the level of the individual. The first obvious question is whether the paw preference of an individual correlates with that individual's circling preference.

Nevertheless, circling preference is associated with a range of other behaviors. Indeed, the direction of asymmetry of dopamine activity appears to determine a range of spatial preferences. The preference to press the lever on the left or right side of an operant box correlates with the direction of circling determined following amphetamine treatment, and side preferences in T-mazes are similarly correlated (Zimmerberg et al. 1974). Consequently, Glick and Shapiro (1985) concluded that circling is a stereotyped form of spatial behavior and that spatial tendencies derive from nigrostriatal asymmetry. Rats with stronger circling preferences were indeed found to be better at discriminating left from right (Zimmerberg et al. 1978). In fact, rats that did not have clear circling or side preferences were found to be hyperactive and to have difficulty learning in a number of tasks (see Glick and Shapiro 1985). For example, rats were first identified as "rotators" or "nonrotators" and then they were trained in a spatial-escape task. They were placed in one arm of a T-maze and foot shock was administered. Escape could be gained only by entering one of the arms, either the one ipsilateral to the individual's rotational direction or the one contralateral to it. The rats were trained and then tested for retention. The rotators scored significantly better retention. The amount of retention in rotators was equal whether they had been trained to avoid the shock by entering the arm ipsilateral or contralateral to their side preference. Nevertheless, relearning was faster if the ipsilateral arm was used, indicating learning is better when the inherent preference is reinforced.

While Glick and Shapiro (1985) claimed that dopamine asymmetry was related to spatial behavior in general, it must be recognized that the spatial abilities which they tested were rather limited and highly related to side preferences. Studies by Denenberg (1981), to be discussed later, have demonstrated that the right hemisphere is specialized for spatial performance and that this is a form of lateralization that is present at the population level. That is, there is no strong correlation between the form of lateralization that Denenberg tested and circling performance, because the latter is either not present at the population level or, in some strains, present to only a very small degree. It is, therefore, incorrect to view circling preference as an indication of some general brain asymmetry causally related to all spatial abilities.

Notwithstanding this, the asymmetry in dopamine neurotransmission may modify the population (or group biased) laterality present in the hemispheres (see later), and it may therefore explain some of the individual differences in behavior (Carlson and Glick 1989). In those individuals which have a right-side preference, based on lower dopamine in the right nigrostriatal region, there should be enhancement of the right hemisphere bias, whereas in individuals with a left-side preference there should be some degree of modification of the influence of the right hemisphere. Indeed, in a population of Sprague-Dawley derived rats slightly more rats had a right circling preference (the difference was significant; 55% right bias), and these right-preferent rats had stronger preferences than the left-preferent ones (Glick and Ross 1981). This might be explained if the population bias for spatial ability present in the right cortex is modified by the distribution of dopamine asymmetry in the nigrostriatal region. As a follow-up, bilateral lesions of the frontal cortex, which would remove the bias on spatial behavior at the cortical level leaving only the nigrostriatal system, were found to remove any population bias for circling (Ross and Glick 1981). Thus, a form of lateralization present at the population level interacts with another form of lateralization present only at the individual level.

Behavioral differences are known to occur in left- and right-circling rats. For example, when cocaine is administered to female Sprague-Dawley derived rats, the right-circling ones rotate more than the left-circling ones (Glick, Hinds, and Shapiro 1983). For males, the reverse occurs. Another example emerges in rats tested in a shuttlebox shock-escape task; in right-circling males a positive correlation is found between the strength of rotational bias and performance deficit on the task, whereas a negative correlation occurs for left-circling males (Carlson, Glick, and Hinds 1987). Differing circling preferences also correlate with reaction to foot-shock stress (Carlson and Glick 1989). Left-circling rats are more reactive than right-circling ones. The right-circling rats failed to learn to escape the footshock (they became "helpless"), whereas the left-circling rats escaped it. Instead of escaping, the right-circling rats learned to cope with the foot shock by improving their performance in a bar-pressing task. The direction of circling is therefore a predictor of a number of other behaviors.

Incidentally, humans show circling behavior similar to that seen in rats (see Chapter 5). By attaching an electronic device to a waist belt it is possible to measure circling in humans over a day's activities, and preferential rotation to the left or right has been revealed (Bracha, Seitz, Otemaa, and Glick 1987). In humans, therefore, this also may be determined by asymmetry in the dopamine levels in the nigrostriatal region.

Asymmetry of Tail Posture

The newborn rat displays asymmetry of tail posture. Ross, Glick, and Meiback (1981) held rat pups in a symmetrical position with the head and tail in line for 5 seconds and then released them. The asymmetrical position into which they moved was scored. The females had a significant bias to move the tail to the right (62%), whereas the males had a nonsignificant bias to the left (57%). As adults, the rats rotate in the same direction as the tail posture which they adopt. Thus there is correspondence between the population biases for the direction of asymmetry in tail posture and the population rotational biases in males and females of the same strain. In the Sprague-Dawley strain used by Glick and co-workers, females have a right preference on both measures, and males prefer the left. Strain differences exist for both of the measures, but the correlation between tail posture and the rotational direction persists across strains. The Purdue-Wistar strain has an overall (males plus females) left bias for both tail posture and direction of rotation (Denenberg 1984; Denenberg, Rosen, Hofmann, Gall, Stockler, and Yutzey 1982). Females of the Purdue-Wistar strain are also more strongly lateralized than the males. Thus it seems that the tests for tail posture in neonates and for rotation in adulthood may both reflect the same asymmetry in dopamine levels in the nigrostriatal regions of the brain.

Tail posture asymmetry is influenced by sex hormone levels during uterine development. Rosen et al. (1983) administered testosterone, dihydrotestosterone, or the vehicle, sesame oil, to pregnant females. The pups from these females were assessed for tail posture on the first day of life. The control group showed the usual left bias characteristic of the strain which they used. The dihydrotestosterone had no effect on either females or males, nor did the testosterone affect the males. The testosterone-treated females, however, had the tail posture shifted to the right.

It is also possible to modify the tail posture bias of pups by exposing the mother to alcohol. Zimmerberg and Reuter (1989) treated pregnant females with alcohol given in the food, and then they tested tail posture in the pups born to these mothers. In the strain of rat that they used, the female controls had a left bias of tail posture, and male controls had no strong bias. The females of mothers treated with alcohol had a weaker left bias than the control females, and the exposed males had a strong left bias, resembling that of control females. This shift in tail posture in males may be due to the action of testosterone, because alcohol exposure reduces testicular development. It also reduces circulating testosterone level in neonates.

Behavioral Lateralization Present at the Population Level

While paw preference and preference for circling in either a clockwise or counterclockwise direction usually have only a small, if any, bias away from a 50–50 distribution in the population (half of the individuals being lateralized in one direction and half in the other), a large number of other behaviors in rats are controlled by processes that are clearly lateralized at the population level (Denenberg 1981). That is, not only is there lateralization for these behaviors in each individual, as for pawedness and circling preference, but all, or most, individuals in the population have the same direction of lateralization. Evidence for lateralization in a population of animals implies that evolutionary processes may have been operating, although not in the absence of environmental influence, as we will discuss in this chapter.

To summarize the findings thus far, the rodent brain has a potential for being lateralized and ontogenetic factors operate to unmask or suppress this potential. Although it is commonly accepted that population lateralization reflects evolutionary processes and, by implication, that this is not so for lateralization present only at the individual level, this may not necessarily be the case. Evolutionary processes may well have selected for a lateralized brain, but not for the direction of its lateralization—precisely as Collins demonstrated in the selection of strongly versus weakly lateralized genetic lines of mice. Nevertheless, for the thesis that we are developing in this book, lateralization clearly present at the population level is of more importance because this is the nature of the lateralization that is present for tool use and language in humans.

Lateralization of Activity Levels

In 1978 Denenberg et al. published a paper which demonstrated that population lateralization occurs in rats tested in the open field (a large open arena). At the same time, these researchers reported the influence of early experience on the presence of this lateralization (Denenberg, Garbanati, Sherman, Yutzey, and Kaplan 1978). Rat pups were either "handled" or "nonhandled." Handling involves removing the pups from their mother and placing them in isolation for 3 minutes in a small container before returning them to their mother. This procedure is repeated daily for the first 21 days of life, that is, until weaning. The nonhandled controls are left undisturbed throughout this time. In the original experiment, half of each group (i.e., handled and nonhandled) was raised in an enriched environment consisting of a large cage with other rats and "playthings," whereas the other half was raised in normal laboratory housing. Thus, four groups were formed. From 50 days of age until adulthood all received the same treatment in standard laboratory cages. As adults they were treated surgically, receiving either ablation of the right hemisphere, ablation of the left hemisphere, a sham operation, or no surgery. Thirty days later, they were tested in the open field. Hemispheric specialization for control of activity was found in the handled, but not in the nonhandled, rats. In the nonhandled animals, irrespective of the postweaning housing conditions, ablation of either the left or right hemisphere (the neocortical region, to be specific) led to increased activity. However, handled rats that were raised only in the standard laboratory cages showed markedly increased activity scores when they had only the left hemisphere intact (right hemisphere ablated), and they were significantly more active than their counterparts with only the right hemisphere intact. The asymmetry for this behavior was present at the population level. The handled rats raised in the enriched environment also displayed asymmetry, but the other way around. Those rats that had only the left hemisphere intact had very low levels of activity, whereas those with only the right hemisphere intact had significantly higher levels. Therefore, in this paradigm, asymmetry develops only in the rats that receive early experience, and the direction of the asymmetry varies according to postweaning stimulation.

Asymmetry for control of activity has also been demonstrated by ligating the cerebral artery on either the left or right side, and by lesioning the frontocortical region of either hemisphere. In male rats, ligation of the right middle cerebral artery or, alternatively, lesioning of the right frontocortical region of the hemisphere produces hyperactivity, whereas identical treatment of the left side does not (Robinson 1979; Pearlson and Robinson 1981). This result is consistent with that of Denenberg et al. (1978), assuming that the rats received the equivalent of handling in neonatal life, although the authors made no mention of this. More specific lesions of the right transcortical fibers, which are thought to use noradrenalin as their neurotransmitter, also increase activity, whereas similar lesions of the left do not (Kubos and Robinson 1984). In females, the unilateral frontocortical lesions fail to reveal asymmetry for activity (Lipsey and Robinson 1986). This sex difference in asymmetry appears to depend at least to some extent on an effect of the hormone testosterone on the developing brain. For example, asymmetry is not present in prepubertal males, and it does not develop in castrated males. Starkstein et al. (1989) showed the asymmetry for activity levels by placing suction lesions in the left or right frontal cortex of rats of various ages up to 90 days. It was not until 90 days of age that lesioning the right side produced hyperactivity. Also, testosterone treatment was found to reinstate the asymmetry in the castrated males. This suggested a maturational effect, influenced by testosterone.

Considered together, these data on activity in the open field clearly show that asymmetry for control of this behavior is present in the rat brain, and that the development of this asymmetry depends on both early experience and hormonal condition. In the course of normal development, early experience and hormonal condition may be interactive factors influencing the development of lateralization. For example, the amount of testosterone circulating in the pup determines the behavior which the mother directs toward it: The mother directs a greater amount of anogenital licking behavior to intact male pups and females treated with testosterone, as compared to untreated females or castrated males

(Moore 1982, 1984). It is not yet known whether a different amount of anogenital licking affects lateralization per se, but it is known to influence the development of sexually dimorphic behavior: More licking leads to increased male-type behavior, irrespective of genetic sex.

The exact nature of the effect of handling is as yet unclear, although the procedure does change the maternal behavior directed toward the pups. The mother licks the pups more when they are returned to her after the 3-minute separation (Smotherman, Brown, and Levine 1977). Coupled with the effect of circulating testosterone on maternal licking, handled males may receive enhanced amounts of anogenital licking, which in turn may produce greater lateralization of the brain. This is a plausible hypothesis which remains to be tested. That is, when changes in brain structure (such as changed size of the corpus callosum; see later) are found to result from changing sex hormone levels in neonatal life, the possibility that these may be caused indirectly by changes in mother–pup interaction must always be considered as an alternative to the direct effects of the hormone on the developing brain, particularly when handling has an additive or synergistic effect with the testosterone treatment (see later).

Lateralization of Spatial Abilities

The direction rats take to move off from a starting position in one corner of an open field is also lateralized at the population level and influenced by handling (Sherman et al. 1980). Nonhandled male rats with only the left hemisphere intact showed a preference to move off toward their right side, and those with only the right hemisphere intact showed a preference to move off toward their left side. That is, the rats moved off toward the ablated side. However, the directional preference of those with only the right hemisphere intact was stronger than that of those with only the left hemisphere intact (see Figure 3.4). Thus there is asymmetry for this behavior at least in terms of degree, the right hemisphere having a greater influence. Intact nonhandled males had no directional preference, whereas intact handled males showed a significant bias to move off toward the left. Handling appears to unmask dominance of the right hemisphere.

Females also show asymmetry for control of this behavior, but the pattern of its occurrence differs from that of males. Nonhandled intact females were found to have a preference for moving off to the left, and handling removed this preference in females (Sherman et al. 1983; Robinson, Becker, and Camp 1983). Thus the population bias, when it occurs in intact males or females, is for moving off to the left and, therefore, a

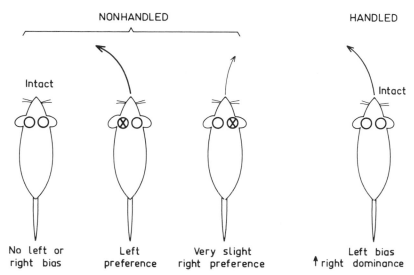

Figure 3.4 A summary of the results obtained by Sherman et al. (1980). The choice to move off in a left or right direction has been scored in the open field for handled and nonhandled male rats. Intact, nonhandled animals show no directional bias. Ablation of the left hemisphere (indicated by the cross) causes the rat to show a strong bias to move off toward its left. Ablation of the right hemisphere causes the rat to more off, with a less strong bias, toward its right. Intact animals which have received handling in early life show a preference to move off toward their left. These data indicate that handling unmasks dominance of the right hemisphere, as it directs moving off to the left. From Rogers (1986).

right hemisphere dominance. The right hemisphere, therefore, controls this form of spatial ability, although the reader will recall that in each individual, spatial bias can be modified by asymmetries in dopamine levels in the nigrostriatal system (see earlier). For control of spatial ability, the rats brain shows interesting differences at different levels (or sites) of neural organization: Individual lateralities exist at one level of organization (the striata) and a population bias exists at another level (in the cortex).

Although making unilateral lesions is a successful method for revealing lateralization in the brain, it is also possible to show lateralization in albino strains of rats simply by testing the animals monocularly, as was done for birds (Chapter 2). We have already mentioned in some detail that only some 5% of the optic fibers are uncrossed in albino rats. Also, when rats are tested in a Swim Maze and are required to locate a hidden platform by using distal visual cues, it is probable that they rely mostly on the extreme peripheral field of vision and therefore they may use only the crossed optic fibers. That is, the visual information from one eye goes to the contralateral side of the brain, although it may subsequently reach the ipsilateral hemisphere via the corpus callosum. Researchers in Denenberg's laboratory (Cowell and Waters, personal communication, 1990) completed testing Purdue-Wistar rats monocularly in the Swim Maze. Asymmetry was revealed in handled, but not in nonhandled, males. Handled males tested using the left eye were found to learn the task in fewer trials than those using the right eye, and they performed at the same level as handled animals tested binocularly. The right hemisphere, therefore, must be dominant in binocularly trained animals. These data confirm this group's earlier report of the specialization of the right hemisphere (and left eye) for spatial performance.

But, contrary to this study, Adelstein and Crowne (1991) have found significantly better learning in female Sprague-Dawley rats tested monocularly using the right eye in the Swim Maze. When tested using the right eye, the rats had shorter latencies to reach the platform and entered fewer quadrants of the circular pool over trials than did those using the left eye. This right eye advantage occurred in both intact controls and test animals with the corpus callosum sectioned. The finding clearly indicates specialization of the right eye and left hemisphere for the spatial learning required in the Swim Maze. The explanation for the discrepancy between this result and that obtained in Denenberg's laboratory may be a sex difference in the organization of brain asymmetry, as Adelstein and Crowne tested females and Denenberg's group tested males. Alternatively, but probably less likely, there may be strain differences. The issue could be resolved by testing females and males of both strains using the same procedure.

A further, differently designed experiment indicating right hemisphere advantage for spatial ability involved testing rats in the Hebb Williams Maze (Rogers, unpublished data). On day 30 of life, male rats of the Long-Evans strain were injected with a drug which temporarily inhibits protein synthesis (cycloheximide) into either the left or right hemisphere. In adulthood, the animals were tested for problem-solving ability in the maze, their performance relying largely on spatial ability. The rats that were injected in the right hemisphere made significantly more errors (entries into blind alleys) than those injected in the left hemisphere. That is, the disruption of the development of the right hemisphere impairs spatial ability. Bianki (1981, 1988) has also reported right hemisphere specialization in rats carrying out spatial analysis using visual stimuli.

In one experiment, Bianki (1981) tested rats for analyzing the position of a spot of light according to depth localization. Inactivation of the left hemisphere with spreading depression had no effect on this ability, whereas inactivation of the right hemisphere significantly impaired the behavior. In another test, the rats had to discriminate between a vertical line and another line oriented at an angle to the vertical (various angles were used). Again, inactivation of the right hemisphere, but not the left, impaired performance. Discrimination of geometrical shapes of equal area was also more impaired by inactivation of the right hemisphere, and so too was the ability to discriminate the size of analogous geometrical figures (also see later).

Thus, a rather large number of experimental results point to specialization of the right hemisphere for spatial analysis.

Lateralization of Affective Behavior

After their initial observations of asymmetry in the rat brain, Denenberg and coworkers produced an impressive series of results demonstrating population lateralization for other behaviors (see Denenberg, 1981, 1984). Overall, these researchers have reached the conclusion that the right hemisphere of the rat brain is specialized to control spatial and affective behavior, and that the left hemisphere can suppress the right by inter-hemispheric coupling via the corpus callosum.

Specializsation of the right hemisphere of the rat for control of emotional or affective responses has been deduced from the right hemisphere's demonstrated role in the control of mouse killing (muricide) and taste aversion. Muricide is, apparently, a spontaneous, species-specific behavior of rats, and it is said to occur more frequently in more "emotional" animals. Garbanati et al. (1983) scored muricide behavior by placing a mouse in the cage of a rat for a maximum period of 5 days, noting whether any killing occurred. The other experimental procedures adopted in the same experiment were basically similar to those just described for rats tested in the open field, including handling or non-handling, enrichment or laboratory housing, and neocortical ablations in adulthood. No asymmetry was present in the nonhandled rats reared in laboratory cages: Ablation of either the left or right neocortex reduced the incidence of muricide. This suggests that the two hemispheres interact in the intact brain. Handled rats raised in laboratory cages did show asymmetry; those with only an intact right hemisphere had higher muricide scores than those with only an intact left hemisphere. Therefore, in this group of rats, muricide behavior is clearly controlled by the right hemisphere. Interestingly, rearing the rats in an enriched environment appears to annul the effects of handling. Handled rats raised in the enriched environment showed no hemispheric asymmetry.

The rats with only an intact right hemisphere had scores of muricide that were higher than those of the intact, whole-brain controls, as well as those of rats with only an intact left hemisphere. This suggests that muricide behavior is activated by the right hemisphere and inhibited by the left. Presumably, this inhibition is mediated by the corpus callosum, from left hemisphere to right hemisphere. If so, sectioning the corpus callosum should remove the inhibition and increase the level of muricide behavior. This is in fact the case (Denenberg et al. 1986).

The learned fear response of taste aversion is also controlled by the right hemisphere (Denenberg et al. 1980). The taste aversion training involved replacing the rats' water bottles with bottles containing a sweetened solution of milk on two consecutive days and, after the second occasion, giving the rats injections of lithium chloride which induces gastric disturbance. Following this treatment, the trained animals avoided the novel solution and retained the memory for this for a long time. Asymmetries for this fear-based learning response were revealed by unilateral cortical ablations. As for the tasks just described, asymmetries were found in the handled rats. Those with only an intact right hemisphere showed the greatest avoidance of the sweetened milk, those with only the left hemisphere intact consumed significantly more of the sweetened milk, and those with an intact brain consumed the most (i.e., the latter group had learned the least). The controls, which had received an injection of saline rather than lithium chloride, also showed asymmetry; in this case, those with only an intact right hemisphere avoided the milk more than either those with only an intact left hemisphere, or those with an intact brain, the latter two groups performing at the same levels. This asymmetry in controls occurred in both handled and nonhandled animals.

Several conclusions can be drawn from these results. First, in this context, handling is not required to reveal the presence of brain lateralization, at least in the control groups. Denenberg et al. (1980) interpreted this to mean that the right hemisphere is more sensitive even to the injection of saline, which acts as an unconditioned stimulus for milk consumption. Since the two counterparts to this control group, the group with only the left hemisphere intact and the group with the whole brain intact, did not differ from each other and showed less taste aversion, it may be deduced that the more fearful right hemisphere is, in the intact brain, inhibited by the left hemisphere. For the rats treated with lithium chloride, other explanations must be made. This treatment of the handled animals had a maximal taste aversion effect in the group with the right hemisphere intact, a result that is consistent with the right hemisphere being more responsive to the aversive stimulus. Yet, it might be expected that the lithium chloride treatment of the nonhandled rats would also reveal asymmetry, given that the nonhandled groups treated with saline showed it. Denenberg and coworkers hypothesize that the lithium chloride is such a powerful adversant that it eliminates all of the differences among the three nonhandled groups, because all of these nonhandled animals are very "emotional." Handling, they argue, reduces emotional reactivity and in these animals the lithium chloride can show differential effects and therefore reveal the asymmetry. Further interpretation of these results is made in Denenberg (1983, 1984).

Taste aversion and muricide are considered to be affective behaviors, activated by the right hemisphere and inhibited by the left. Asymmetry of affective behavior has also been assessed using olfactory cues, and in this case the left hemisphere is preferentially involved (Dantzer, Tazi, and Bluthé 1990). The odors chosen were associated with negative or positive social experiences; the odor of a stressed or nonstressed conspecific. Rats were given ablations of either the left or right olfactory bulb and were compared to sham operated controls and controls with bilateral bulbectomies. They were then injected with formalin into one hindpaw to induce pain and so elicit a response of licking the paw. The latency for licking the paw is delayed if the rat is stressed, such as being exposed to the odor of a stressed conspecific. The odor of a stressed conspecific delayed licking in both the sham lesioned rats and those that received right bulbectomies. It was not delayed in the rats with bilateral bulbectomies or those with left bulbectomies. That is, the left bulbectomized rats (using their right olfactory bulb) had impaired ability to react behaviorally. Given that each olfactory bulb projects to its ipsilateral hemisphere, and also that olfactory memories are laid down unilaterally in the ipsilateral hemisphere, the rats with left bulbectomies use their right hemispheres. Thus, rats using the right hemisphere do not respond to odor stressors. In other words, the left hemisphere is specialized for processing stressful olfactory stimuli. Surprisingly, the hormonal stress response to the odor was unaffected in the rats with left bulbectomies. Hence, the left hemisphere specialization for processing stressful olfactory stimuli must be confined to the control of behavioral responses and not to the processes of pituitary-adrenal control. No lateralized effects were found for olfactory recognition of conspecifics. Thus, the left hemisphere advantage is confined to the processing of stressful odors. If response to stressful odors is indeed an aspect of affective behavior, as the researchers claim, this specialization of the left hemisphere needs to be reconciled with the other studies, already discussed, which have demonstrated right hemisphere involvement with affective behaviors. The explanation cannot lie in sex difference, because the same sexes were used (males in the study of olfactory lateralization). Strain variations remain a possibility, but, in our opinion, they are an unlikely explanation.

Perhaps we need to consider more closely exactly what affective behavior is, and also the exact kind of processing that the animals are required to perform in each testing context.

Lateralization of Auditory Processing

Maternal mice have a right ear and left hemisphere specialization for processing the ultrasonic calls emitted by their pups (Ehret 1987). This left hemisphere specialization for processing species-specific vocalizations is consistent with similar specialization of the left hemisphere for auditory processing in birds (Chapter 2), for species-specific vocalizations in macaque monkeys, and for language processing in humans (see later). Ehret tested lactating house mice in a two-alternative choice test. A choice had to be made between two sounds, ultrasonic to the human ear, played using loudspeakers placed at either end of a runway. One sound resembled the ultrasonic distress calls of pups when separated from their mother and so elicited approach and retrieval behavior. The other sound was a neutral 20kHz signal not resembling pup calls. The mice were tested binaurally first, and then with either the left or right ear plugged, followed by plugging the other ear. In the binaural condition, and with the left ear plugged, the mice approached the speaker that was emitting the "distress calls." Mice with the right ear plugged were unable to discriminate between the two sounds and approached them at random. This result could mean either that the latter animals were unable to locate the speakers, or that they were unable to discriminate the nature of the two calls. A second experiment demonstrated that the right ear advantage is only for calls important in communication, which supports the second alternative explanation. When trained to discriminate between two artificial ultrasounds, mice tested monaurally were found to perform equally well with either ear. These experiments are particularly interesting to us because they indicate that left hemisphere specialization for processing species-specific communication had already evolved in mice (or their ancestors) and such a function of the left hemisphere could well provide a basis for the eventual evolution of language processing in humans.

Hemispheric Asymmetries in Cognitive Functioning

A most important series of experiments by Bianki (1983c, 1988) has demonstrated that the left hemisphere of the rat is specialized to perform sequential information processing and the right for simultaneous or parallel processing (i.e., Gestalt processing). This division of labor between the hemispheres is similar to that known for the left and right hemispheres of humans (Bradshaw and Nettleton 1981). Rats were trained to differentiate, using an operant conditioning paradigm, trinomial geometrical figure complexes presented either simultaneously or in sequence (see Figure 3.5A). Either the left or right hemisphere was then inactivated by spreading depression and the rat was tested again. Differentiation of the simultaneously presented stimuli was impaired after inactivating either of the hemispheres, but the impairment was 34% greater after inactivating the right hemisphere. By contrast, differentiation of sequentially presented stimuli was most impaired after inactivating the left hemisphere. Thus, the right hemisphere is specialized for parallel or Gestalt processing, whereas the left hemisphere is specialized for serial analysis. Specialization of the right hemisphere for parallel processing is consistent with the specialization of this hemisphere for spatial analysis, which we have already discussed.

This division of labor between the hemispheres is not absolute, as performance on either task is impaired to some extent by inactivation of either hemisphere, even though it is more impaired by inactivation of one rather than the other. Although the degree of these lateralized differences does not appear to be as great as many of the others we have already mentioned, the similarity of these results to those of humans makes them most important. It is regrettable that the reports of Bianki's work written in English do not give very comprehensive accounts of the procedures he used, and presumably this has contributed to the relative lack of acknowledgement of his research by other workers in the field.

Also, as in humans, Bianki has shown that recognition of visual stimuli by rats tested in an operant situation is faster in those using the right hemisphere, a result which he attributes to this hemisphere's ability for parallel processing (see Bianki 1988). The rats were trained for food reward to respond to a pair of geometrical, solid figures, and then they were presented with outlines of the same figures. The time to recognize the latter was scored. The time for recognition was shorter with both hemispheres functioning than when either one was inactivated by spreading depression; but rats using the right hemisphere only (with the left hemisphere inactivated) responded in significantly shorter time than those using the left hemisphere only. According to Bianki (1988), the slower recognition time of the left hemisphere is connected with its successive scanning of stimulus characteristics, compared to the simultaneous recognition made by the right hemisphere.

Mittleman, Whishaw, and Robbins (1988) have also scored reaction time to visual stimuli, but they used sequentially presented simple stimuli and found left hemisphere advantage, as Bianki's data would predict.

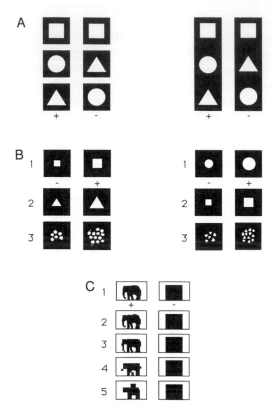

Figure 3.5 A, The stimuli presented by Bianki (1983) to test for hemispheric specialization for sequential and parallel processing in rats. The stimuli were presented on doors, one of which the rat could push open to obtain reinforcement when the positive stimulus (+) was presented. The stimuli were presented three times each for 3-second intervals interrupted by 1-second off periods. The latter intervals were used to accommodate the sequentially presented stimuli. The rats were tested with one or the other hemisphere inactivated by spreading depression. Inactivation of the right hemisphere impaired performance using the parallel stimuli to a greater extent than did inactivation of the left hemisphere, whereas the reverse was the case for the stimuli presented sequentially.

B, The stimuli presented for analysis of area. The rats were trained to respond to the larger of two solid squares or circles (1). They were then tested with two triangles or squares, respectively, each pair being of an area difference equivalent to that of the training stimuli (2). Performance was significantly more impaired by inactivating the right hemisphere. Next, they were tested for abstraction by presenting collections of unfamiliar shapes, which summated to areas equivalent to those of the training stimuli (3). This performance was significantly more impaired by inactivating the left hemisphere.

C, The stimuli used to test the rat's ability to synthesize a visual image. The rats were trained to discriminate between the image of an elephant and a square (1). They were then presented with increasingly degraded images of the elephant, (2) through (5), presented randomly. Performance of the rats with the left hemisphere inactivated by spreading depression was significantly more impaired than it was for those with inactivation of

The rats were trained to hold their noses in a central hole of a test chamber following the illumination of a light inside the hole. A light cue was then presented on either the left or right side (i.e., in the lateral field of vision) and the rat had to move its nose from the central hole to the position of the light, an action which operated a photocell and led to the delivery of a food pellet. Overall, the reaction time was shorter to the light on the right side (65% of the rats were faster at intercepting the right beam). That is, the left hemisphere responded with a faster reaction time. Although Mittleman et al. (1988) did not refer to the earlier research of Bianki, the latter's data would have predicted their result for sequentially presented stimuli.

Specialization of the left hemisphere for sequential processing is also apparent in more naturalistic behavior; namely in determining sequential choices in a radial arm maze (Bianki 1988). The sequential choices made in this task are not related to performance criteria for the task, nor are they specifically related to spatial ability. Rats using the left hemisphere adopted the most efficient sequential searching patterns significantly faster than did rats using the right hemisphere, demonstrating that the left hemisphere is better at successive analysis. Hence, the left hemisphere may be preferentially used in territorial exploration.

In addition, Bianki (1981) has shown that the left hemisphere is specialized for time analysis, and this is also consistent with its specialization for analysis of sequential stimuli. Rats were trained to move through a door on their left side if a light came on for 10 seconds, and to move through a door on the right if the light came on for 5 seconds. Inactivation of the left hemisphere caused a greater deficit in performance than inactivation of the right.

After these initial reports of left and right hemisphere specialization for cognition in rats, Bianki proceeded to report some evidence that the left hemisphere analyzes abstract characteristics of stimuli, whereas the right mainly analyzes concrete or absolute characteristics (Bianki 1982). Again, one notes the similarity to humans.

Several examples of abstract versus concrete analysis were tested. For example, rats were trained to respond to the larger of two geometrical figures as the positive stimulus (Figure 3.5B). When other geometric

the right hemisphere. The difference was already apparent for stimulus 3. The overall conclusions drawn from these results are that the left hemisphere of the rat is specialized to perform sequential processing, plus analysis, followed by synthesis. This means that it performs inductive and more abstract cognition, whereas the right hemisphere is specialized for parallel processing and making spatial analysis on the basis of concrete cues (see text). Collectively adapted from Bianki (1982, 1983a, 1983c).

figures of the same area were presented, concrete analysis could be performed to respond to the larger of the two stimuli—this ability was more impaired by inactivation of the right hemisphere. If the rat was presented with a collection of unfamiliar shapes, the total area of which was larger on the positive key (Figure 3.5B), abstract analysis had to be performed in order to make a correct response. This ability was more impaired by inactivation of the left hemisphere. Hence, the left hemisphere appears to be capable of forming concepts; it can abstract and generalize stimulus characteristics.

Thus, in rats there is a remarkable similarity to the specialization of the left and right forebrain hemispheres in the chick; the left hemisphere of the chick possessing the ability to classify objects and the right to process spatial information (see Chapter 2). As Bianki (1982) noted, the left hemisphere of the rat is specialized for abstraction, analyzing the general properties of objects and "grasping connections and relations between objects," whereas the right hemisphere stores information about the details of the object itself in the form of descriptions of each of its characteristics. This is exactly the conclusion reached by Andrew (1991) to describe the left and right hemisphere differences in the chick.

After carrying out a large number of experiments using rats trained as previously described, and using a variety of visual stimuli, Bianki (1983b) concluded that the left hemisphere of the rat is specialized for inductive processes, and the right for deductive processes. His model states that the left hemisphere analyzes the information first and then makes a synthesis; that is, it moves from dealing with single facts to dealing with the general aspects of the information (induction).

One example used by Bianki to illustrate this was the synthesis of a visual image from a fragmented, or degraded, representation of the image (Bianki 1983a). Rats were trained to discriminate between the image of an elephant (positive stimulus) and a square (negative stimulus); see Figure 3.5C. The rats were then tested using a set of fragmentary figures that resembled the positive stimulus to varying degrees (Figure 3.5C). The results showed that the left hemisphere was more capable of synthesizing a visual image from its fragments. Compared to using the left hemisphere only, the rats using the right hemisphere only began to make increasingly more errors (not recognizing the positive stimulus) as the image became more fragmented.

It is Bianki's argument that the left hemisphere of the rat has abstract and inductive cognition, and he points out that these abilities therefore did not develop either at the same time as, or after, the evolution of speech (Bianki 1983b). The extensive studies of laterality by Bianki deserve more consideration than they have received so far, but there are problems with some of the interpretations of the data. This is understandable because the experiments test complex cognitive processes in the rats, and it is frequently not possible to be absolutely sure exactly which parameters of the test are being attended to by the animal. When a rat is trained on a rather complex task requiring it to find food hidden behind a screen by following one moving cue and not another, is it primarily using spatial analysis (a specialization of the right hemisphere), or extrapolative, or inductive cognition (a specialization of the left hemisphere)? Bianki and Filippova (1985) set up such a task in an attempt to test for extrapolation ability, but, when they found that adult rats using only the right hemisphere (with spreading depression of the left hemisphere) performed better than those using only the left hemisphere, they shifted their interpretation to say that the rats must have been relying primarily on spatial information. This sort of approach serves only to confuse our understanding of laterality in the rat brain, even though they possibly may be correct. To further confound the issue, infant rats were found to perform the task better when they used the left hemisphere only (with the right hemisphere inactivated by spreading depression). It is possible that the infant rats were using different cues and different cognitive processes than the adult rats tested in the same task. For example, they may have been paying less attention to spatial information. However, this explanation for the shift in hemispheric dominance with age needs further experimental study. Thus far, we simply note the occurrence of a shift in hemispheric dominance during development, and draw the parallel to the much more extensively studied shifts in hemispheric dominance that occur in developing chicks (see Chapter 2).

Asymmetries for Control of Sexual Behavior and Sex Hormone Secretion

A population bias for lateralization is not confined to processing at the cortical level. At a lower level of neural organization, in the hypothalamus, the rat brain has a population bias for control of sexual behavior. Implanting estradiol pellets into either the left or right side of the hypothalamus of neonatal female rats causes different effects on sexual behavior in adulthood. Nordeen and Yahr (1982) implanted estradiol into the left or right ventromedial nucleus on day 1 or 2 after birth. Implants in the left side of the hypothalamus were found to suppress lordosis by approximately 35%, whereas implants in the right ventromedial nucleus had no effect on this behavior. Implants of estradiol into the preoptic area on the right side of the hypothalamus elevated mounting by a two-fold factor, whereas implanting the equivalent region on the left side had no effect. That is, the unilateral implants of hormone have asymmetrical effects on the way in which the control centers for sexual behavior develop. It would appear that the two sides of the hypothalamus differ in their sensitivity to steroid hormones (compare with the asymmetry in the levels of the enzyme aromatase, which metabolizes testosterone to estradiol, in the hypothalamus of the chick; discussed in Chapter 2).

The hypothalamus also controls hormone secretion by the pituitary gland, and interestingly there is asymmetry for control of this physiological function. Nordeen and Yahr (1982) found that treatment of the left side of the hypothalamus (ventromedial nucleus or preoptic area) with estradiol suppresses the secretion of the luteinising hormone (LH) and makes the female rats become anovulatory, whereas treatment of the equivalent areas in the right side of the hypothalamus has no effect. Thus, the left side of the hypothalamus controls both female sexual behavior and the secretion of the reproductive hormones.

Pituitary hormone secretion control is mediated by releasing factors (neurohormones) synthesized by neurones in the median eminence and released into the portal blood stream at the level of the hypothalamus. Bakalkin et al. (1984) found that in Wistar rats the luteinizing hormone-releasing factor (LH-RH) occurs at a higher concentration on the right side of the hypothalamus than on the left. This direction of asymmetry appears to be the reverse of that found by Nordeen and Yahr (1982) in the Sprague-Dawley strain; but in another strain, Bakalkin et al. (1984) found the reverse pattern to occur. Also, the direction of this asymmetry is influenced by both unilateral castration and by cold stress.

Another experimental approach has also revealed asymmetry for control of pituitary hormone release. This study, by Fukuda et al. (1984), examined the effects of unilateral lesions of the medial anterior hypothalamus on ovarian compensatory hypertrophy. When one ovary is removed from a female rat, the other ovary enlarges in response to increased output of gonadotrophic hormones to compensate for the loss. These researchers found that a lesion placed in the right side of the hypothalamus suppressed the compensatory hypertrophy irrespective of which ovary had been removed, whereas lesions of the left side had no effect. They used rats of the Wistar strain and therefore their results fit exactly with those of Bakalkin et al. (1984).

Given the lateralized control of pituitary function, it should be considered possible that the effects of the estradiol implants on sexual behavior (Nordeen and Yahr 1982) are indirect ones acting via lateralized changes in hormonal output from the pituitary. Until the plasma concentrations of the sex hormones in these animals are assessed in adulthood, we cannot conclude whether the behavioral effects of the implants are direct or indirect.

Asymmetry of sex hormone effects on the brain may also occur at higher levels of neural processing because there are different numbers of receptors for estrogen in the left and right cerebral cortices of rats (Sandhu, Cook, and Diamond 1986). The receptor number for estrogen is highest in the cortex during the first few days after birth, and then it declines until 25 days of age. During this peak period after birth, the right cerebral cortex of the female has a higher concentration of estrogen receptors than the left. In males, the concentration of these receptors is higher in the left cerebral cortex compared with the right. Presumably this asymmetry in receptor number may determine at least some aspects of functional asymmetry, and possibly influence the asymmetrical development of the cortex in a sexually dimorphic manner (see later). If so, this would be an organizational effect of the sex hormones, since the receptor number asymmetry is present for only the first few days of life.

Structural Asymmetries in the Rodent Brain

Structural asymmetry is a characteristic of the rat and mouse brain. Asymmetries are present in the cortical, hippocampal, and other regions of the left and right hemispheres (Diamond, Dowling, and Johnson 1981; Diamond et. al 1982; Diamond et al. 1983). In male rats of the Long-Evans strain, the cortex and hippocampus are thicker (measured in coronal sections) on the right side than on the left, and similarly in the male mouse, the right neocortex is larger than the left (Rosen et al. 1989). Diamond and co-workers (Diamond et al. 1982, 1983) fixed the brains of rats of various ages, made coronal sections of the cerebral hemispheres, and measured the thickness of the cortex at several sites on a series of selected sections at fixed distances rostro-caudally through the cerebrum (see Figure 3.6). In males up to around 400 days old there was significant asymmetry in almost all of the cortical sites measured, the right side being larger than the left (with differences ranging up to 7%; Diamond 1984). This asymmetry is present even in newborn male rats (Diamond 1985; Zimmerberg and Reuter 1989). It is, however, no longer present in males 900 days of age or older.

Diamond (1985) speculates that at this advanced age the rat may be no longer sexually active, implying that circulating levels of androgens maintain asymmetry in the male. There is evidence from other species that circulating levels of the sex hormones determine the size of various brain structures in adulthood. In songbirds, the volume of the song nuclei increases each spring, although it will be remembered that the same structural changes occur in the nuclei in the left and right hemi-

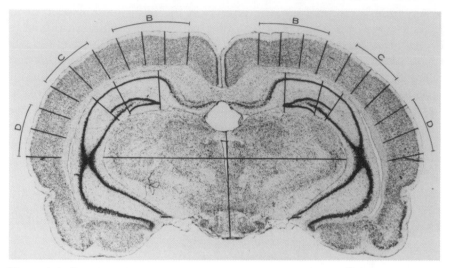

Figure 3.6 Drawing of a coronal section of the rat brain at the level of the posterior commissure showing the sites at which cortical and hippocampal thickness was measured by Diamond and co-workers. Similar sites were also measured in a number of other sections of the cortex. Adapted from Diamond et al. 1983.

spheres even though only the nuclei on the left side control singing (see Chapter 2). In frogs, there is marked asymmetry in the habenular nuclei on the left and right sides of the brain and this asymmetrical structure fluctuates in size with the season of the year (Kemali, Guglielmotti, and Fiorino 1990). Both the left and right nuclei are larger in spring, presumably in response to sex hormone secretion.

Testosterone does play a role in establishing asymmetry in rats by acting during development; males that have been castrated at birth show a reversed asymmetry at 90 days of age, the left cortex being thicker than the right. This pattern of asymmetry is more characteristic of females. Fewer measurements have been made of female brains, but it is clear that their pattern of morphology differs from that of the male. Females, up to around 400 days old, show far fewer significant asymmetries at the chosen sites, but at sites where significant asymmetries do occur, the left cortex is larger than the right (Diamond et al. 1983; Diamond 1985). If the ovaries are removed on day 1 after birth, the asymmetry of the female is reversed so that the right hemisphere becomes larger than the left, although this reversed pattern of asymmetry is not as consistent, or strong, as in the male.

In their paper of 1983, Diamond et al. hypothesized that estrogen may inhibit the growth of the right cortex, and therefore more receptors for estrogen might be found in the right cortex. The same research group later showed that this is in fact the case. As we have mentioned previously, on postnatal days 2 and 3 estrogen receptor levels are high, and in females there are more estrogen receptors in the right cortex (Sandhu et al. 1986).

One of the cortical regions which is clearly asymmetrical in the rat is the primary visual cortex (area 17). Galaburda et al. (1986) have shown that this asymmetry is correlated with differences in the number of neurons. They also showed that brains with greater asymmetry exhibit a lesser total area (left plus right) in sections of the visual cortex. Consistent with this, human brains with symmetry of the planum temporale region have two large plana, each equivalent in area to the larger planum of asymmetrical brains (Galaburda et al. 1987). Therefore, asymmetry appears to result from the loss of neurons, or the production of fewer neurons, on one side. It does not result from the addition of more neurons to one side of an originally symmetrical brain. Thus, if the initial state of the brain is symmetrical, asymmetry must develop through cell death on one side (Rosen, Galaburda, and Sherman 1990). Indeed, by use of radioactive thymidine, which tags dividing cells, Rosen, Sherman, and Galaburda (1991) have shown that neuroblast division does not accompany the establishment of asymmetry in the cat visual cortex.

Diamond and co-workers have measured the thickness of the cortex on histological sections of the brain, whereas other researches have measured weight and volume. Kolb et al. (1982) found that the entire right hemisphere is wider, larger, and heavier than the left, and Sherman and Galaburda (1984) found that the total volume of the right neocortex is significantly larger than the left in male rats. In females, the asymmetry in volume of the neocortex is not so great (Sherman and Galaburda 1985). Collectively, therefore, these studies all indicate the same result of larger right hemisphere structures, particularly in males.

Asymmetry was also found to be present in the hippocampus of the same rats that were used to investigate cortical thickness (Diamond 1985). In males of some, but not all, ages the right hippocampus is significantly thicker than the left. In females, the left is thicker than the right. It should be possible to link the asymmetry in the hippocampus to specific functional asymmetries, as the neurophysiological profile of the hippocampus is better known and, to some extent, its function is better delineated. The hippocampus has been attributed many functions, including roles in memory formation, emotional behavior, spatial performance, and reinforcement, to name a few. As yet, however, we are unable to tie any of the structural asymmetries to these functions.

The association of the hippocampus with spatial ability, which has now been well established in maze tests, is of interest because a larger right hippocampus may imply better spatial ability, in which case it may be manifest in males at particular ages. To speculate further, it may also be associated with demands for territorial behavior at different stages of life. This, of course, would be unlikely to explain the larger right hippocampus in neonates.

It appears to be important to delineate the changes in asymmetry with age. Diamond et al. (1982) found significant hippocampal asymmetry in males aged up to and including 41 days (the difference being in the region of 5% to 10%). At 55, 77, and 90 days there was no significant asymmetry, but then by 108 and 185 days the asymmetry was reexpressed, only to disappear again at 300 and 400 days and return at 650 days. With such a fluctuating pattern it is clear that more ages and more strains of rats need to be tested. The first disappearance of significant asymmetry occurs at the age when sexual maturity is reached, which may indicate an association between hormone levels and hippocampal asymmetry, but the significance (if any) of the other ages when change occurs in unknown. Nevertheless, testosterone levels do influence the development of hippocampal asymmetry, the same as for cortical asymmetry. Castration of males immediately after birth results in the left hippocampus being significantly larger than the right at 90 days (Diamond 1985). Even though the difference is only 2%, it should be noted that normal males show no significant hippocampal asymmetry at 90 days. In females, the left hippocampus is significantly larger than the right by 6% on days 21 and 90. Ovariectomy reduces this difference to 2%, although the left is still significantly larger than the right. Thus, castrated males and females have the same degree of hippocampal asymmetry, and it is in the same direction.

Structural asymmetry has also been found in the amygdaloid nuclei of the S1 strain of rats, selected to be maze-bright, but not in the Long-Evans strain used in the studies showing cortical and hippocampal asymmetries (Melone et al. 1984). In males of the S1 strain aged 55 and 90 days, the right amygdala is larger than the left by some 4-9% depending on housing conditions (females were not tested). Incidentally, the S1 strain also has a thicker cortex on the right side, but in a more limited number of cortical regions than in the Long-Evans strain. In the S1 strain the statistically significant asymmetries are present in the occipital cortex, which may have been strengthened by the selective breeding for superior visuospatial abilities (Diamond 1984).

The amygdala has been associated with a number of functions, including general arousal, agonistic behavior, feeding and sexual behaviors, and reward and punishment. We can, however, only guess what having a larger right amygdaloid nucleus might mean. Melone et al. (1984) drew an association between the presence of amgydaloid asymmetry and superior spatial ability in the S1, maze-bright strain. This makes for interesting speculation which could be explored by investigating more strains of rats.

The corpus striatum region, measured in Long-Evans male rats, also tends to be larger on the right side, although at only one of the 10 ages at which measurements were made (26 days) did this difference reach statistical significance (Diamond 1985). Larger right side structures also include the olfactory bulb. In Sprague-Dawley rats, the total volume of right olfactory bulb is significantly larger than the left (Heine and Galaburda 1986). This is the case for both females and males, although the asymmetry tends to be greater in males. We have discussed earlier the specialization of the left olfactory bulb in Wistar rats for processing of stressful odors (Dantzer et al. 1990). This behavioral asymmetry is not obviously consistent with a larger olfactory bulb on the right side.

Before leaving this section, it is worth mentioning that structural asymmetries also occur in the nervous system outside the brain; for example, in male mice the hypogastric ganglion, which innervates the internal genitalia, is larger on the left side (Suzuki and Arai 1986).

Effects of Enriched and Impoverished Environments on Brain Asymmetry

Perhaps the most important result to emerge from the studies of Diamond and co-workers is that the size and pattern of asymmetry of various cortical regions in adult rats varies according to the environment in which the animal is living.

Rats were housed in three different conditions: an enriched condition, which consisted of a group of 12 animals living in a large cage containing objects that could be manipulated; a standard colony condition, with three rats per small cage and no objects to manipulate; and an impoverished condition, in which each rat was housed singly in a small cage without objects to manipulate (Diamond 1985). After the rats had lived in these environments for varying periods of time (25–55 days, 60–64 days, 60–90 days, and 25–108 days) brain structural asymmetry was studied. In every case, enrichment increased the thickness of the cortex and impoverishment decreased it, relative to the standard, laboratory housing condition. This is a remarkable result, particularly in the case of the second group, which experienced the enriched environment for only 4 days, and the effect appears to be maximal after only 10 days (between 60 to 90 days of age). Overall, these changes occurred on both sides of the cortex, but there were also some changes in the pattern of asymmetry. For example, in males aged 25 to 55 days, asymmetry is present in medial area 10 of the cortex of standard colony animals, but it is not present in the rats reared in either the enriched or the impoverished environment (Diamond 1985). In another area of the cortex (area 18), following enrichment, the left side was larger than the right. This was so for rats of every age sampled. Such a consistent bias was not found in rats of the standard colony or from the impoverished condition.

Females raised in the enriched environment (60–116 days) show the increased thickness of the cortex, but no changes in asymmetry (Diamond 1985). Both hemispheres appear to respond equally to environmental stimulation, but this result should be considered cautiously because only one age group of females has been examined.

Environmental enrichment also increases the size of the amgydaloid nuclei (Diamond 1985). For these nuclei in the S1 strain of rats, environmental conditions do not alter the pattern of asymmetry, but rather the degree of asymmetry. For example, in males that have experienced environmental enrichment for 25 to 55 days of age, the right amygdala is 6% larger than the left; in males housed in the standard colony, the right is 8% larger than the left, and in males housed in the impoverished condition over the same period of time the right is 4% larger than the left. A similar distribution of scores occurs for rats housed in the various conditions from 60 to 90 days of age, which adds weight to the group differences even though they are small. It appears that altering the housing conditions has differential effects on the left and right amgydaloid nuclei, although the direction of asymmetry remains the same for all conditions.

The hippocampus may also be affected by handling, in a way similar to the amygdaloid nuclei. Sherman and Galaburda (1985) measured the weight of the left and right hippocampi in handled and nonhandled rats raised in the enriched or the laboratory environment, and found that the right side weighed significantly more (by 1%) than the left in handled males raised in the enriched environment. No left–right difference was found in nonhandled males raised in standard laboratory cages.

The direct functional significance of these changes in structure has yet to be determined. Quite apart from this, however, the discovery of their dependence on age, sex, and environmental conditions is important. It enables us to recognize the dynamic relationship between brain structure and the environment. Of course, this has been well recognized for the developing brain, which has much greater plasticity (ability to respond to the environment and to repair damage) than has the adult brain. Yet, even in the adult, various regions of the brain retain plasticity and so respond to environmental influences and also to changing hormonal conditions. This remarkable ability of the adult brain to adapt to changing internal and external environments must be kept in mind when we consider asymmetries in other species, including humans.

The Role of the Corpus Callosum in Lateralization of the Rodent Brain, and Factors that Influence Its Development

Many of the earlier concepts of lateralized brain function incorporated the idea that it was present only in the cortex and that a corpus callosum was required to interconnect the two hemispheres so that one hemisphere (the left in most cases) could suppress the other (see Gazzaniga 1974; Denenberg 1981). Gazzaniga and Le Doux (1978) postulated that evolution of the corpus callosum was essential for the appearance of laterality in the brain. They based their argument on evidence of the time that lateralization of language in humans does not develop until the fibers in the corpus callosum are fully myelinated (Gazzaniga 1974). It soon became known, however, that there are morphological asymmetries in language areas in the fetus (Chi, Dooling, and Gillies 1977; Molfese and Molfese 1983; and see Chapter 6). It is not difficult to see that the general hypothesis of Gazzaniga and Le Doux, which claims that a fully functioning corpus callosum plays an essential role in lateralization, was based on the original premise that laterality is unique to humans and their capacity for language (compare with Chapter 2).

Following on from this, Denenberg (1981) developed a model using the corpus callosum to explain his data for laterality in rats; namely, the suppression of the right hemisphere by the left hemisphere via the corpus callosum. In part, Denenberg (1984) based this hypothesized role of the corpus callosum in laterality on the evidence that humans without a corpus callosum have redundant language capabilities in both hemispheres (Chiarello 1980; for review see Jeeves 1990).

Berrebi et al. (1988) have found evidence that handling increases the size of the corpus callosum in male rats aged 110 days (see later), which certainly supports a role for the corpus callosum in functional laterality at the level of the cortex since, as discussed previously, handling unmasks directional laterality in moving off in the open field. Nevertheless, the presence of a corpus callosum cannot be essential for the occurrence of brain lateralization, as some of the acallosal mice tested by Gruber et al. (1991) had laterality of paw preference and, contrary to this, strains which have a corpus callosum can be predominantly ambilateral. Also, laterality in the hypothalamus cannot easily be tied to the corpus callosum unless the laterality is conferred on the hypothalamus by higher centers in the cortex. Furthermore, evidence of laterality in the avian brain conclusively shows that the corpus callosum is not necessary for asymmetry to occur because there is no corpus callosum in the avian brain. We have discussed this fully in Chapter 2. Other left–right connecting pathways do play a role in lateralization in the avian brain, but clearly there is no need for such a large interconnecting structure as the corpus callosum.

Of course, to say that the presence of a corpus callosum is not necessary for lateralization to occur does not imply that the corpus callosum has no role in lateralization. The corpus callosum connects homologous, or corresponding, regions of the left and right hemispheres; clearly it would be able to mediate inhibition of one hemisphere by the other, thereby establishing hemispheric dominance. One function of the corpus callosum is to convey sensory information received by one hemisphere to the other. Another function is to provide each hemisphere with information about the specialized processes being performed by the other hemisphere (Denenberg et al. 1984). While the two hemispheres are carrying out different cognitive functions (the right perhaps being involved with spatial and parallel processing, and the left with sequential and synthetic cognition), the corpus callosum is likely to play a role in bringing these different forms of cognition together, thus unifying the animal's behavior into an integrated and effective whole (Bianki and Shramm 1985).

A body of evidence (reviewed by Denenberg et al. 1984) has shown that the corpus callosum is involved in competitive interactions between the two hemispheres, and also that competitive interaction between the hemispheres affects the development of the corpus callosum. Visual experience, for example, markedly affects the development of the corpus callosum (Innocenti and Frost 1980; Sefton, Dreher, and Lim 1991), as does other sensory stimulation. During development the connections made by the callosal neurons are reorganized and, in general, there is a narrowing down and loss of connections, but some growth also occurs. During the first two weeks of postnatal life in the rat there is a progressive loss of callosal cells from certain regions of the visual cortex, and a gain of callosal cells in other regions of the visual cortex (Olavarria and van Sluyters 1985). Actually, much of the early postnatal reorganization may involve loss of axonal collateral connections, rather than the death of whole cells (O'Leary, Stanfield, and Cowan 1981). By day 12 of age in the rat, the corpus callosum has largely taken on its adult pattern of organization and it should be noted that this occurs before eye opening. Other adjustments of the corpus callosum do, however, occur after eye opening in response to visual stimulation since it is possible to cause anomalous patterns of callosal connections by interfering with visual input in neonatal life (Lund, Chang, and Land 1984; Sefton et al. 1991).

Denenberg et al. (1984) postulated that manipulations that facilitate the development of the corpus callosum should result in a greater degree of brain asymmetry, whereas anything that interferes with its development should result in reduced asymmetry. Of course, this may depend on whether the callosal neurons have primarily inhibitory or excitatory end terminals, and regional variation within the callosum is likely. However, Rosen, Sherman, and Galaburda (1989) have found that brains that are more asymmetrical in terms of cortical volume have fewer callosal connections (a smaller corpus callosum) compared to more symmetrical brains.

These data are not consistent with the previous finding of weaker paw preference in mice with a reduced size of the corpus callosum (see earlier; Ward et al. 1987), although this has not been a consistent finding across strains of mice. In their examination of terminal connections of callosal neurons, Rosen et al. (1989) also noticed that the more symmetrical rat brains have a more "patchy" arrangement of the connections, and this finding is consistent with the report of Cassells et al. (1990) which found that part of the corpus callosum of a weakly lateralized genetic line of mice is made up of a number of discrete bundles.

One might question whether it is development of the corpus callosum which influences the development laterality or vice versa. The developmental process could, of course, involve interaction between both directions of influence. Schmidt and Caparelli-Dáquer (1989) opt for an active role of the corpus callosum in directing the development of asymmetry. They have studied a strain of mice in which the corpus callosum fails to develop in some individuals and is underdeveloped in others. They found an inverse correlation between an asymmetry coefficient (calculated from volume differences between the hemispheres) and corpus callosal area and suggested that callosal neurons from the larger side compete with those from the smaller side and the latter lose the competition and die, resulting in an even smaller hemisphere. This developmental model is consistent with that of Rosen et al. (1990), who proposed that asymmetry may develop from a symmetrical brain by loss of neurons on one side (see earlier). If such a process occurs, one might expect differences in the electrical information passed from the left to right cortex compared to that passed from right to left. In fact, Perez et al. (1990) found lateralization of the interhemispheric responses evoked in anaesthetized rats; repetitive stimulation of the left prefrontal cortex evoked in the right cortex responses of greater amplitude and of less susceptibility to fatigue than those obtained in the left hemisphere from stimulation of the right prefrontal cortex.

Irrespective of the exact processes by which asymmetry is established, the size of the corpus callosum does appear to correlate with some forms of brain asymmetry. Denenberg et al. (1991) have recently devised a method for accurately measuring the cross sectional area of the rodent (and human) corpus callosum. The technique involves computerized calculation of the area of a midsaggital section using a large series of cross-sectional widths. Earlier techniques involved a somewhat less detailed measurement of a series of widths at different sites along a midsaggital section of the corpus callosum (see Figure 3.7). Using the latter method, Denenberg and co-workers have shown that the corpus callosal area of adult female rats is smaller than that of males, and that both handling in early life and sex steroid hormones influence its development (Berrebi et al. 1988; Denenberg, Berrebi, and Fitch, 1989; Fitch et al. 1990; Fitch et al. 1991). Even when an adjustment is made for brain weight, corpus callosum size is greater in male rats than females (Berrebi et al. 1988).

Rats that are handled in early life, using the standard procedure, have an increased size of the corpus callosum, measured at 110 days, but by 215 days this effect is no longer apparent (Berrebi et al. 1988). Handling, it will be remembered, is associated with increased functional lateraliza-

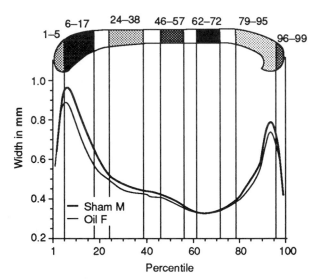

Figure 3.7 Measurement of various widths of a midsaggital section of the rat corpus callosum (at the top of the figure). The data was obtained by expressing, for each individual, the distance along the corpus callosum at which a given width was measured as a percentage of the entire length of that individual's corpus callosum (X-axis). The data given is for males (top curve) and females (bottom curve) at 110 days of age. The curve for the males is significantly higher than that of the females. From Denenberg et al. 1989.

tion in the brain. Therefore, the data is consistent with Denenberg's hypothesis (Denenberg et al. 1981) that an increased corpus callosum size is linked to greater brain laterality. These results were later confirmed by the same group using a more detailed computer method of assessing the area of the corpus callosum (Denenberg et al. 1989). In addition, the latter study found that the anterior 17% of the corpus callosum is larger in males compared to females, irrespective of age or early experience of handling. Handling had the greatest effect on the size of the mid and posterior regions of the corpus callosum. It might therefore be fruitful to further specify the areas of the corpus callosum affected by handling, and to characterize the regions of the cortex connected by neurons which have axons in these regions.

The size of the corpus callosum is also influenced by the levels of testosterone and estrogen circulating in the neonate. Fitch et al. (1990) have measured the size of the corpus callosum in rats aged 100 days. They compared male rats which had been castrated on day 1 of life, females injected with testosterone proprionate on day 4, and females treated with synthetic estrogen (estrogen diethylstilbestrol) on day 4. All of the pups were handled. The testosterone treatment resulted in increased callosal size in the females, but castration had no effect on the callosa of the males. The synthetic estrogen had no effect on callosal size, but the action of an estrogen blocker lead to an increase in the size. These results indicate that both female and male sex hormones affect the development of the corpus callosum and this, in turn, correlates with the influence of these hormones on lateralization. A similar influence of sex hormones on the development of the corpus callosum may occur in humans. A sex difference has been reported for the size (Witelson 1989) and shape (Allen et al. 1991) of regions of the corpus callosum, and the size also relates to handedness.

As castration of the male pups had no affect on callosal size, Denenberg and co-workers (Fitch et al. 1991) decided to investigate the possible effect of testosterone at natural circulating levels (not the pharmacological doses administered in the former studies) in the male fetus prior to birth. To do so, they administered an antiandrogen to the pregnant females during the last 5 days of gestation. The pups were castrated after birth to remove any possible interference from their endogenous sex hormones. The treatment reduced the size of the corpus callosum in males, and so removed the normal male–female size difference. Hence, testosterone does effect the development of the corpus callosum in males prior to birth.

A recent study has shown that testosterone can affect the development of the corpus callosum in females after birth, but only if the females are handled (Denenberg et al. 1991). Nonhandled and handled females were injected once with testosterone on day 4 of life and the size of the corpus callosum was measured at 100 days of age. The handled females treated with testosterone had enlarged callosa, compared to oil-treated controls and nonhandled females treated with testosterone. Thus, circulating testosterone stimulates growth of the corpus callosum only when it interacts with the experience of handling. Although castration on day 1 of life has no effect on the size of the corpus callosum in males, handling increases the size of the corpus callosum in the intact male but has no effect in the chemically castrated male or in the intact female (Fitch et al. 1991). The latter suggests that testosterone primes the brain to the effect of handling after birth. In other words, in males it appears that testosterone action before birth increases corpus callosum size, and after birth testosterone action coupled with handling causes a further increase in corpus callosum size. Handling is known to cause corticosterone release and, presumably, this "stress" hormone interacts with the sex hormones to influence differentiation of the corpus callosum.

Female sex hormones appear to act in the female after birth; ovariectomy in the second week of neonatal life leads to an enlarged corpus callosum (Fitch et al. 1991). That is, the ovarian hormones act in a period when handling also influences the development of the corpus callosum. Ovarian hormones act to reduce callosal size, whereas handling increases it. As previously mentioned, testosterone treatment of females after birth also enlarges callosal size, but this effect occurs only in handled females and not in nonhandled ones (mentioned by Fitch et al. 1991), which adds emphasis to the fact that development of the corpus callosum in females, at least, reflects an interaction between early experience and sex hormone levels.

Presumably, an enlarged callosum contains more axons, indicating that either a greater number of neurons contribute to the corpus callosum or more neurons have branching collateral axons. These extra neurons in the cortex and the extra connections between the hemispheres are stimulated to grow in response to testosterone levels and handling. Estrogen appears to actively promote loss of axons traveling in the corpus callosum, or to prevent the normal growth of these neurones, and in this case the action occurs after birth and in combination with early experience.

Stimulation by handling during neonatal life increases the coupling between the hemispheres (Denenberg 1984). This is evident in terms of behavioral responses as well as in the development of the corpus callosum, for example, the correspondence between behavior controlled by one hemisphere and that controlled by the other hemisphere. In nonhandled rats, those with ablation of the left hemisphere move off in a leftward direction when placed in the open field, whereas those with ablation of the right hemisphere move off toward the right (Figure 3.4). Therefore there is no correspondence between the hemispheres for control of this behavior. By contrast, in handled rats, those with ablation of the left hemisphere tend to move off in the same direction as those with ablation of the right hemisphere; there is significant correspondence between the two hemispheres in this case (Sherman et al. 1980; Denenberg 1984). The nonhandled rats appear to have hemispheres which are operating independently for control of this behavior, and the handled rats appear to have coupled hemispheres. This deduction can be made purely on the basis of the behavioral measures; it does, however, concur with the finding of increased callosal size in handled rats.

In rabbits, too, handling facilitates interhemispheric communication. Nonhandled rabbits have poor interocular transfer of visual information (Van Hof 1979), and, when tested monocularly in the open field, there is no significant correlation between activity scores for the left and right eye conditions (Denenberg et al. 1981). In handled rabbits, there is a strong correlation between the left and right eye conditions, suggesting better functional development of the corpus callosum in the handled animals.

Considered together, these data demonstrate the crucial role played by environmental stimulation on brain development, with a particular focus on the corpus callosum and coupling of the hemispheres. Thus, using rats has led us to the same conclusions reached in experiments using chickens (see Chapter 2). In the chicken, light stimulation during a sensitive phase of embryonic development has a critical influence on the development of structural and functional asymmetry (Chapter 2). Based on this knowledge and the fact that visual experience affects the postnatal development of callosal connections in mammals (Innocenti and Frost 1980), Tees (1984) decided to investigate the role of visual experience on functional lateralization in the rat. However, he used much more drastic conditions of light deprivation than were used in the experiments with chickens. The rats were raised either in a 12:12 hour light–dark cycle or in darkness, until they were operated on at 85 days (ablation of the left or right hemisphere) and then tested behaviorally at around 100 days of age. In the open field, the light-reared rats that received ablation of the right hemisphere were more active than their counterparts with left hemisphere ablation, and also more active than both groups of dark-reared animals. Thus, laterality was present only in the light-reared animals, which presumably means that the light exposure had facilitated development of the corpus callosum. The results for the light-reared groups are the same as those obtained previously by Robinson (1979) and Pearlson and Robinson (1981). Thus, as with chicks, light stimulation influences the development of lateralization in the rat. So far, it is not possible to say if this effect in the rat is a direct effect of light on development of the visual pathways (as is known for the chick) or a change in the general behavior of the animal, possibly the mother–pup interaction. Also, unless light-rearing determines which eye is opened first by the rat pups (as in the case of stress; see later), the effect of light in enhancing laterality is not likely to directly parallel the situation in the chick because the chick embryo is oriented to have only the right eye exposed to light stimulation (Chapter 2). Further study of the role of light stimulation on the development of brain lateralization in rodents would be most interesting.

To summarize thus far, lateralization in the brain develops in response to the interactions of environmental stimulation and hormonal condition (sex hormones, in particular), and at some levels of neural organization this is mediated by changes in the neural pathways that interconnect the left and right sides of the brain. The likely outcome for the whole animal of this dynamic process of development is a brain that has different degrees of left or right hemisphere dominance (possibly for different functions) and different degrees of interhemispheric cooperation. The effects of early hormonal exposure and stimulation are long-lasting to the extent that it appears as though they impose permanent templates on brain laterality and interhemispheric integration.

As yet, there has been no investigation of any potential effects of environmental stimulation on the size of the corpus callosum in adult rats. The assumption has been that the effects of stimulation and hormones are confined to sensitive periods prenatally and neonatally. Certainly, there is a sensitive period for handling confined to early postnatal life (Denenberg 1964), but other forms of stimulation may have an influence on older animals. Indeed, the size of the cortex is influenced by housing adult rats in an enriched environment (see earlier), and therefore it would seem most probable that changes in the corpus callosum might also occur under the same conditions. In fact, it is possible that the cortical changes flow on to changes in the corpus callosum, as some neurones are thrown into prominence and others recede.

Before leaving this discussion of the corpus callosum, we should mention its role in left–right response differentiation in rats. The presence of the corpus callosum appears to confuse discrimination of responses to a stimulus on the left or right side of the animal. Rats with the corpus callosum sectioned learn a left–right response differentiation (turn right if a light is on and left if no light is on) significantly faster than rats with sham surgery (Noonan and Axelrod 1991). The same rats showed no performance deficit in a brightness discrimination. Hence, while the corpus callosum makes an important contribution to brain lateralization, it appears to cause difficulty for the animal in discriminating between left and right in the external world. This confusion may possibly result from the intermixing of lateralized stimuli by the corpus callosum, or from inhibition of one hemisphere by the other, which could conceivably produce a mild form of hemisphere neglect (see Myslobodsky 1983). Without a corpus callosum, rats should perform more poorly in tasks which require the specialized use of one hemisphere, even though their performance in left–right response differentiation is improved. To our knowledge, this has not yet been tested. We have mentioned previously that Zimmerberg et al. (1978) found that rats that have a strong circling preference are better at discriminating left from right. In this case the better left–right discrimination must be based on a greater degree of asymmetry in the dopaminergic systems of the nigrostriatal region; that is, below cortical level and, therefore, possibly independent of callosal connections. If, however, the corpus callosum is involved in this case, better left–right discrimination in stronger circling (and therefore more strongly lateralized) rats would appear to contradict the result of Norman and Axelrod (1991), as a greater degree of lateralization depends on a better developed corpus callosum, which should mean more left–right confusion. This is an issue that could be solved by sectioning the corpus callosum of rats that circle strongly and testing the effects on left–right discrimination

Stress and Environmental and Hormonal Effects on the Development of Lateralization

We have discussed a number of examples in which sex hormones are known to influence the development of structural and functional lateralization. Sex hormones modify the degree and, to some extent, the direction of amphetamine-induced circling behavior, the tail posture adopted by neonatal rats, the population biases for activity levels, and hemispheric asymmetries in cortical size (see earlier). At least some of these effects caused by the sex hormones are brought about by their influence on the developing corpus callosum. We have discussed the interaction between sex hormone level and the effects of handling in neonatal life, which alters the size of the corpus callosum. We have also mentioned that the effects of the sex hormones on brain asymmetry may be a direct action of the hormones on the brain, or, alternatively, an indirect action by altering the behavioral interaction between the mother and the off-spring. Other hormones may have a role to play as a result of the latter interaction. For example, stress in utero or after birth can alter brain asymmetry, and it may do so by altering levels of the stress hormones (corticosterone, in particular). Alonso, Castellano, and Rodriguez (1991) subjected pregnant female rats to stress by restraining them for 2 hours per day over the last week of pregnancy. At 90 days of age, the offspring of these females were tested for side preference in a T-maze. The controls were found to show a population laterality of around 85% right prefer-ence (this is higher than in previous studies using the same measure; compare with Zimmerberg et al. 1974). The male offspring of stressed mothers were less lateralized than the male offspring of control mothers, but the female offspring of stressed mothers were more lateralized than female offspring of control mothers.

Although Alfonso et al. (1991) were unaware of it, their findings corre-late well with those of an earlier report by Fleming et al. (1986) of the effects of prenatally applied stress on sexually dimorphic asymmetries in the cerebral cortex. Pregnant females were stressed in the same man-ner as the other study and the thickness of the cortex was measured in each of the male offspring at more than 100 days of age. In controls, the cortex of the right hemisphere was significantly thicker than the left. In the prenatally stressed animals, the left cortex tended to be thicker than the right, but the differences were not so great and were significant for only one of the three sites measured. Both of these reports (Alonso et al. 1991; Fleming et al. 1986) state that prenatal stress "demasculinizes" the males. It would be better to avoid this term, as it presupposes that there is a standard for masculinity for T-maze behavior and asymmetry of cortex thickness in rats, whereas it is well known that many strain varia-tions and, indeed, substrain group variations occur for a number of measures of laterality in rats (see earlier, and Carlson and Glick 1989).

The effect of stress on brain asymmetry may be a direct action of the glucocorticoid hormones on the developing brain, an interaction between these hormones and the sex hormones which effectively modulates their action on the developing brain, or it may involve long-term changes in mother–infant behavior which persist after the stressor is no longer actually being applied. It may also cause a direct change in levels of the sex hormones themselves. Ward and Weisz (1980) have reported that stressing the mother causes a change in testosterone levels in the male fetus, and it also affects development of the sexually dimorphic nucleus of the preoptic area (Anderson, Rhees, and Fleming 1985). Regardless of the mode of stress action, it presents an example of environmental effects on brain laterality and may well explain some of the variations in results between different strains of rats and, indeed, different subpopulations of these strains.

Handling is generally considered to be a stressful event known to cause secretion of corticosterone, and it may be seen as a postnatal stressor that effects brain laterality. Because of its timing during development, however, it may have different effects than stress experienced during fetal development. For example, Wakshlak and Weinstock (1990) found in rats that stress during fetal development caused a left bias in directional preference but that handling reversed this effect. We have already covered in some detail the way in which neonatal handling interacts with sex hormones to influence the development of the corpus callosum and to change functional lateralization in the brain (e.g., by unmasking control by the right hemisphere). Handling can also impose other lateralities on the developing nervous system. Smart, Tonkiss, and Massey (1986) found that handled pups opened the left eye before the right. Undisturbed pups showed no bias. The same researchers also reported that artificially reared pups, without their mothers and being fed by gastric cannula, were more likely to open the left eye first. Both procedures may be acting as stressors, the researches claim. A lateralized bias in eye opening is a reminder of the situation in the chick embryo, in which the right eye receives light stimulation and the left does not. This establishes both structural and functional lateralities in the brain (see Chapter 2). A similar effect is most likely to occur in rodents. Opening of the left eye first in rats may stimulate development of the right hemisphere and perhaps explain the unmasking of dominance of this hemisphere by handling. Unfortunately the report (Smart et al. 1986) does not state the time difference between opening of the left eye and opening of the right. Nevertheless, the method used was to inspect once daily and note which eye had opened first, from which we can estimate that there is likely to be at least a 24-hour difference between opening of the left and the right eyes. Parallel to the chick, 24 hours is an ample period for developmental asymmetries to be triggered (as little as 2 hours of asymmetrical light exposure is sufficient to establish the direction of functional asymmetry in the chick; Rogers 1982).

In rabbits the right eye opens first, followed several hours later by the left (Narang 1977). The rearing conditions of the rabbits were not stated. The asymmetrical eye-opening leads to the optic nerve fibers for the right eye becoming myelinated in advance of those from the left eye (Narang and Wisniewski 1977). This more rapid myelination has been shown to result from the light stimulation itself (Tauber, Waehneldt, and Neuhoff 1980). Thus, asymmetry of eye opening may lead to lateralization of perceptual inputs and contribute to the establishment of asymmetry in the brain (also see Chapter 4).

In addition to stress having a clear effect when applied in utero or neonatally, it can also influence the laterality of adults. For example, Carlson et al. (1987) stressed rats with inescapable foot shock and found that males which had been right-circling and females which had been left-circling shifted their preferences to the opposite direction. Incidentally, left-circling males and right-circling females showed no change of preference but rather an increased strength of rotation in the original direction. Lateralization changes can therefore occur in response to stress, but the effect is determined by sex and rotational bias. The researchers attribute these changes to lateralized effects caused by stress at the level of the frontal cortex which becomes imposed on the asymmetry in the nigrostriatum (in dopamine; see earlier).

Neurochemical Asymmetries in the Rodent Brain

We have discussed asymmetry in dopaminergic neurotransmission in the nigrostriatal regions on the left and right sides of the brain and the way in which this correlates with the direction of circling, or rotation, and with the direction of tail posture adopted by the neonate. This asymmetry may be modulated by another level of asymmetry in dopaminergic transmission in the medial prefrontal cortex, and this, in turn, may be modified by training the rat to rotate in a particular direction in an operant task (Glick and Carlson 1989). Neural pathways descend from the medial prefrontal cortex to the nigrostriatal region and therefore may mediate this influence. Thus spatial performance may be modified by asymmetrical changes in neurochemistry, since nigrostriatal asymmetry is associated with at least some forms of spatial behavior (see earlier).

There is no population bias in the asymmetry of dopamine neurotransmission located in the nigrostriatal region, but neurochemical asymmetries with a population bias are present in other regions of the brain. This is reflected in the uptake of radioactive 2-deoxy-glucose (2-DG), a nonmetabolizable sugar which is taken up by active neurones and labels sites of neural activity. There is greater uptake of 2-DG in the left medulla pons, left neocortex, right hippocampus, and right diencephalon of neonatal female rats, but not in male rats (Ross, Glick, and Meibach 1981; and see Robinson et al. 1985). A population bias in the dopaminergic system does occur in the striatal region of the cortex: The number of receptors for dopamine is 40% greater on the right side for females, and greater on the left side for males (Drew et al. 1986). As yet, it is not known how this asymmetry in the striata may relate to circling behavior, if at all.

At several brain sites the inhibitory neurotransmitter GABA is asymmetrically distributed with a population bias. It is present at a higher concentration on the right side in the substantia nigra, superior colliculus, and nucleus accumbens, and higher on the left side in the ventral tegmentum, ventromedial thalamus, and caudate nucleus (Starr and Kilpatrick 1981).

Asymmetries in neurotransmission may be present in the hippocampus. For example, the right hippocampus of left-turning rats takes up almost twice as much neurotransmitter or transmitter precursor (serotonin, noradrenalin, and choline) when incubated in a medium containing these substances, than does the left (Valdes, Mactutus, and Cory 1981). Thus, the right side must have more efficient uptake systems or a greater density of nerve terminals which use these neurotransmitters. This directional bias may be considered as consistent with increased metabolic activity of the neurons in the right hippocampus (Ross et al. 1981).

The asymmetry in neurotransmission may also underlie the lateralized state-dependent effects in the rat hippocampus of a procedure known as kindling. Kindling involves regular, low-level electrical stimulation of, in this case, the left or right hippocampus. Stokes and McIntyre (1981) used this procedure and looked at its effects on recall of a step-through avoidance task. They found that rats kindled with electrodes placed in the right hippocampus had recall severely impaired by the convulsion produced by the kindling, but there was no effect on learning. Convulsion of the left hippocampus impaired both learning and recall. The result cannot be interpreted in direct terms of neurotransmission asymmetries, but in some way it is likely to rely on them. Asymmetries in other neurotransmitters, as yet unstudied, may also occur in the hippocampi. Some of these are likely to be present with a population bias, while others may occur at an individual level only. Dopamine concentration is, for example, significantly greater in the side of the hippocampus ipsilateral to side preference determined in a T-maze (Diaz Palarea, Gonzalez, and Rodriguez 1987) and therefore this is an asymmetry not present at the population level.

In the cortex there are also neurochemical asymmetries. Early indications of this came from the investigation of the differential effects of left and right cerebral artery ligations (Robinson 1979) or cortical ablations (Kubos and Robinson 1984). Although equal-sized lesions occurred in the left and right hemispheres, only those in the right led to decreases in noradrenalin and dopamine concentrations in both the ipsilateral and contralateral cortex, substantia nigra, and locus coeruleus. Thus, asymmetry for control of neurotransmitter systems appears to occur at the cortical level, and it imposes neurochemical asymmetry on lower levels of neural organization. The lateralized effects of infarction and ablations have been shown to correlate with lateralized effects on activity levels; hyperactivity results from injury to the right hemisphere only (see earlier). Recent work by Barnéoud, Le Moal, and Neveu (1991) has found that lesioning the left neocortex mainly affects the activity of serotoninergic inputs to the right neocortex, whereas lesions of the right cortex affect the catecholinergic inputs to the left. Other nonlateralized changes in neurotransmitter levels occur in a number of brain regions, but these do not directly concern us here. The brain structures affected by these neocortical lesions are known to be involved in modulating the immune response (Barnéoud et al. 1987).

An asymmetrical distribution of free fatty acids also occurs in the left and right hemispheres (Ginóbili de Martinez, Rodriguez de Turco, and Barrantes 1985). The basal levels of free fatty acids are about 35% lower in the right hemisphere and, relative to this baseline, the right hemisphere produces significantly higher amounts of free fatty acids during electroconvulsive shock (ECS). After completion of the ECS, these free fatty acids are removed more rapidly by the right hemisphere. The free fatty acids involved are thought to be the membrane glycerophospholipids and they could change as a correlate of, or as a result of, the effects of ECS on neurotransmission (Ginbili de Martinez et al. 1985).

Neuropeptide transmitter levels also appear to be asymmetrically distributed between the hemispheres. Brain aminopeptidases are enzymes involved in the degradation of neuropeptides, and the activity of these enzymes is higher in the left hemisphere of the rat (Alba et al. 1986), suggesting that there may be a higher turnover of neuropeptide transmitters in the left hemisphere.

So far, we see a complex pattern of lateralities in neurotransmitters, and neurotransmission, which extends throughout the brain, even to spinal level, where asymmetry has been found for some catecholamine-containing neurones (Singhaniyom, Wreford, and Güldner 1983). The data for neurochemical asymmetries in rats reminds us of similar asymmetries known to be present in the chicken brain; for example, the asymmetry in glutamate receptor numbers that develops as a result of imprinting (see Chapter 2). Indeed, rats have a higher concentration of glutamate in the lateral entorhinal cortex in the left hemisphere compared with the right, and lesions of this region have a greater effect on retention of a visual discrimination task than lesions of the right (Myhrer and Iversen 1990). This result parallels the more detailed findings concerning lateralization of glutamatergic mechanisms in memory formation in the chick.

This study aside, at the present time research into neurochemical asymmetries in the rats is largely demonstrating their existence and documenting sex differences. A next step would be to investigate environmental and hormonal influences on these asymmetries, and to attempt to determine which behavioral functions may be correlated with them. It is certain that, so far, we have only touched the surface of this field, particularly when we consider the wide range of neurotransmitters and putative neurotransmitters in the nervous system. The picture likely to emerge will be an even more complicated one.

Neurochemical asymmetries may correlate with structural asymmetries, but this is not always the case. Brain sites which are structurally symmetrical may have neurochemical asymmetries and therefore use different forms of neurotransmission on the left and right sides.

Asymmetries in the Control of Immune Responses

The central nervous system is known to play an important part in modulating immune responses, and this influence is lateralized. Lesions placed in the left or right neocortex have differential immunomodulatory effects. Barnéoud et al. (1987) placed lesions in the parieto-occipital lobes of the left or right neocortex of mice and tested a number of different aspects of the immune response. Mice with lesions on the right side showed depressed mitogen-induced lymphoproliferation and enhanced antibody production to sheep erythrocytes, compared to mice with bilateral lesions. Mice with lesions on the left side showed no modification of these immune responses. In another study, lesions of the left neocortex were found to depress T-cell functions, whereas right lesions had no effect or enhanced T-cell functions (Neveu 1988; Renoux et al. 1983). Control of these immune responses is, therefore, clearly lateralized.

This laterality may depend on asymmetrical neurotransmitter distribution in the brain, or, as Neveu (1988) suggests, on lateralized control of the secretion of the hormone prolactin, which is asymmetrically regulated by the cortex (LaHoste et al. 1989). Prolactin is involved in modifying lymphocyte reactivity. In rats, ablation of the right hemisphere, but not the left, elevates prolactin levels and depresses T- and B-cell immune responses. Thus, cortical lesions may control immune responses by asymmetrically modulating neuronal circuits at lower levels of brain organization (in the hypothalamus). It will be noted that for T-cell responses, at least, the effect of lesions in mice gives the opposite result to that obtained for rats. In mice, left lesions depress T-cell functions and in rats, right lesions depress them. The authors suggest that this discrepancy may be attributed to species variation or to the sizes of the lesions in each case since various regions of the cortex appear to effect different immune responses. Further experimentation is required to find the answer to this rather important question.

The question of whether individual differences in immune response depend on the degree and direction of behavioral lateralization has been addressed by several recent studies. A correspondence between behavioral lateralization and the immune system was first suggested by the fact that rats which responded to amphetamine by circling to the left exhibited higher lymphocyte stimulation indices than those which turned to the right (Neveu 1988). Immune response in mice has also been associated with their paw preference (Neveu et al. 1988; and see Neveu 1992): Left-pawed mice show higher mitogen-induced T lymphocyte proliferation than right-pawed mice. In addition, after ablation of the right cortex, the difference in T-cell function between left and right pawed mice remains unchanged, whereas ablation of the left hemisphere abolishes the difference (Neveu et al. 1991).

Thus, the left hemisphere is involved with the association between handedness and immune response. These data have direct relevance to the human situation, as there is some evidence (albeit controversial and not supported by all studies) that left-handedness has an association with certain disorders of the immune system (Geschwind and Behan 1982). Even in human subjects without diagnosable autoimmune disease, left-handers are significantly more likely to have one or more autoantibodies circulating in the blood stream (Chengappa et al. 1991). The strongest association for this occurred in males: 41% of left-handed men had autoantibodies, compared to 17% of right-handed men.

Immune function also differs in the high and low genetic lines of mice selected by Collins to be strongly or weakly lateralized for paw preference (see earlier). The strongly lateralized (high) line was found to have lower immune functions than the weakly lateralized (low) line (Fride et al. 1990). Thus, more behavioral lateralization for paw preference means lower immune responsiveness. According to the hypothesis of Geschwind and Behan (1982), testosterone levels may influence both the development of the immune system (via an action on the thymus gland) and brain laterality (via an effect on the left hemisphere). This hypothesis has, however, remained untested, although one might suggest that the lowered immune response in the strongly lateralized mice might be due to such a hormone effect. Alternatively, from the studies of Neveu et al. (1988), showing the effects of cortical lesions on immune responsiveness, we might also hypothesize that stronger laterality (particularly at a population level) influences immune responsiveness directly; that is, not via any hormonal effect but rather by a direct effect of the lesion on neurotransmission (Barnéoud et al. 1991). Indeed, New Zealand black mice, which spontaneously produce pathogenic antibodies directed against red blood cells and DNA, show earlier production of these antibodies if they are left-pawed (Neveu et al. 1989). Also, certain strains of immune defective mice exhibit brain anomalies, such as ectopias, which are abnormal collections of cells in a layer of the cortex (Sherman et al. 1990).

Summary and Conclusions

Rats and mice show two forms of brain organization, differing in terms of their organization in the brain and their likely evolutionary significance. One of these forms of lateralization is present in individuals, but takes a different direction in different individuals so that there is little or no consistent bias within the group or population. Pawedness, direction of circling, and the asymmetry of tail posture adopted soon after birth are examples of this form of lateralization. The other form of lateralization occurs in individuals and in addition, as most individuals have the same direction of lateralization, there is also lateralization of the group or population, and presumably the species as a whole. Examples of the latter are more numerous and include lateralization for activity in the open field, the direction of moving off in the open field and other spatial behaviors, "emotional" or affective behaviors, auditory processing of species-specific calls, sequential versus parallel processing of information, and control of sexual behavior. Briefly, the left hemisphere is specialized for sequential processing and processing of species-specific calls, whereas the right hemisphere is specialized for the control of spatial behavior, affective behavior, and parallel processing.

The first form of lateralization (e.g., pawedness) depends on laterality of neurotransmission, particularly dopamine, at lower levels of neural organization (in the nigrostriatum). Most examples of the second form of lateralization, present at the population level, depend on higher forms of neural processing in the cortex, although this is not so for the lateralized control of sexual behavior, which occurs in the hypothalamus.

While humans show circling behavior much like that of rats and mice (see Chapter 5), asymmetry in humans for cognitive functions, handedness, and language all have more in common with the second form of lateralization in rats and mice as they are present at the population level. Indeed, the division of labor between the hemispheres in rats and mice is notably similar to that of humans; in both, the left hemisphere is specialized for sequential processing and the right for parallel processing. Also, in both rodents and humans, the left hemisphere is specialized for analyzing vocalizations (for humans, language) and the right for controlling emotional behavior (Borod 1989). The rodent brain, therefore, serves as a useful model for studying lateralization. In fact, it may be from studies using rats that we will gain new insights into other functions which may be lateralized in the human brain. Nowhere are these insights more apparent than in the recent research on lateralized control of immune responses in rodents. These discoveries have important implications for brain modulation of the immune system in humans.

The strength of lateralization for pawedness can be altered by genetic selection, but genetic selection cannot alter the direction of the pawedness in the population. For this reason, it is not possible to use genetic selection to generate a group of rats or mice with right-pawedness equivalent to handedness in humans. By raising rats in a biased world this can, however, be achieved through the process of learning. The other similar lateralities of circling and tail posture may respond similarly, as these seem to depend on the same neural circuits as pawedness.

Many of the lateralities that occur in the rodent brain may have structural and neurochemical correlates, since there are abundant examples of these. Nevertheless, it is difficult to make any direct links between the various functional, structural, and neurochemical asymmetries. So far, only generalized parallels can be drawn. However, it is hoped that tighter links can be made between these various forms of lateralization, even though indications are that the jigsaw puzzle which needs to pieced together is very complex indeed. Given the growing list of asymmetries, in future we may find ourselves searching for the "rare" symmetries so that simpler explanations might be pinned on them.

As in the chicken (see Chapter 2), the development of lateralization in the rodent brain is influenced by environmental and hormonal factors. Handling during neonatal life has, for example, been shown to unmask lateralization in rats, apparently by influencing the development of the corpus callosum. The corpus callosum is the structure by which one hemisphere inhibits the other and thereby assumes dominance. An interaction between handling and sex hormone condition during early development influences the size of of the corpus callosum, and thus the amount of inhibition of one hemisphere by the other. We are reminded of a similar interaction between light stimulation and sex hormonal condition in determining the development of the thalamofugal visual projections in the chick (Chapter 2).

Knowledge of these developmental processes adds significantly to our understanding of brain development in a much broader sense than that of lateralization alone. In addition, other forms of stress (other than handling) can influence brain lateralization in rats, and this includes the generation of a population bias for preferentially opening the left eye first, an event which may trigger a train of developmental processes leading to lateralization in the same way as occurs in the chick embryo (i.e., by only the one eye receiving light stimulation during a critical stage of development).

Remarkably, even in the adult, environmental stimulation can alter brain lateralization. Rats which have lived in an enriched environment not only have thicker cortices than those which have been housed in the standard, impoverished environment of the laboratory, but also the direction of the left–right asymmetries of thickness differs in various regions of the cortex. Presumably, this altered structural asymmetry is paralleled by changes in functional lateralization, but this remains untested. The effects of enrichment appear to be, like the effects of handling, dependent on sex hormone levels, as sex and age are variables which influence the result. The adult brain retains some degree of plasticity (ability to adapt to the environment) dependent on the same factors that influence plasticity in the younger brain. These data make us aware that, for other species as well as the rat, plasticity for lateralization may well be present even in adulthood, and therefore we should not view the adult brain as static or with unchanging laterality.

Further Reading

Bianki, V.L. The Right and Left Hemispheres: Cerebral Lateralization of Function. Monographs in Neuroscience, vol. 3. Gordon and Breach, New York, 1988.

Collins, R.L. On the inheritance of direction and degree of asymmetry. In Glick, S.D. (ed.), Cerebral Lateralization in Nonhuman Species. Academic Press, Orlando and London, 1985, pp. 41–71.

Denenberg, V.H. Hemispheric laterality in animals and the effects of early experience. *Behavioral and Brain Sciences*, 1981, 4:1–49.

Denenberg, V.H.; Fitch, R.H.; Schrott, L.M.; Cowell, P.E.; and Waters, N.S. Corpus callosum: Interactive effects of infantile handling and testosterone in the rat. *Behavioral Neuroscience*, 1991, 4:562–566.

Diamond, M.C. Rat forebrain morphology: Right–left: male–female: young–old: enriched–impoverished. In Glick, S.D. (ed.), Cerebral Lateralization in Nonhuman Species. Academic Press, Orlando and London, 1985, pp. 73–87.

Galaburda, A.M.; Aboitiz, F.; Rosen, G.D.; and Sherman, G.F. Histological asymmetry in the primary visual cortex of the rat: Implications for mechanisms of cerebral asymmetry. *Cortex*, 1986, 22:151–160.

Glick, S.D., and Shapiro, R.M. Functional and neurochemical mechanisms of cerebral lateralization in rats. In Glick, S.D. (ed.), Cerebral Lateralization in Nonhuman Species. Academic Press, Orlando and London, 1985, pp. 158–183.

Asymmetries in
Nonprimate Mammals

Introduction

In this chapter we will cover structural and functional asymmetries that have been reported for nonprimate mammals other than rats and mice (see Chapter 3 for discussion of the latter). The aim is to put in place all available pieces of the jigsaw puzzle so that the emerging picture of widespread brain asymmetry can be seen more clearly. Compared to rats and mice, there is far less information about asymmetries in other nonprimate mammalian species. The reason for this is not that rodents have more asymmetries, but rather that they have been more studied.

In all mammalian species, apart from marsupials, the brain has a well-developed corpus callosum which interconnects the left and right hemispheres (see Chapter 3, Figure 3.1 and Figure 4.1D) and which plays an important role in brain asymmetry (see Chapter 3). Additionally, there are the hippocampal and anterior commissures, which might also be involved in brain asymmetry, although this aspect has not been studied. The marsupial brain may present a somewhat special case for the study of asymmetry because it lacks the corpus callosum. The anterior and hippocampal commissures are larger in marsupials, compared to placental mammals, and in one order, the Diprotodonta, an extra interhemispheric connection, the fasciculus aberrans, is present (Figure 4.1). The fasciculus aberrans runs from the dorsal neocortex crossing the midline in the anterior commissure. Thus, the fasciculus aberrans interconnects homologous regions of the neocortex and, together with the anterior commissure, replaces the corpus callosum of placental mammals (Heath and Jones 1971; see also Mark and Marotte 1991). Although the marsupial brain has evolved a different route for the projections which connect the left and right hemispheres, these may well serve the same role in brain asymmetry. Given the different method of reproduction of marsupials and birth of the young at a very early stage of development, these species should be ideal subjects for investigating factors that influence development of interhemispheric connections and their potential role in laterali-

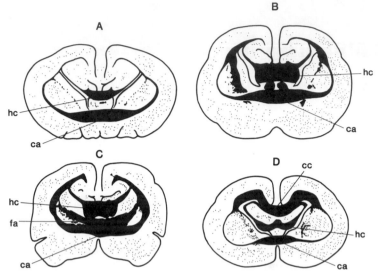

Figure 4.1 This diagram compares the forebrain commissures in various orders of the marsupials (A, B, and C) with those of the placental mammal (D). A. A monotreme (echidna or platypus) showing the hippocampal commissure (hc) and the anterior commissure (ca). B. A Polyprotodont marsupial (e.g., opossum), rather similar to A. C. A Diprotodont marsupial (e.g., brush-tailed possum and sugar-glider) showing the additional commissure, the fasciculus aberrans (fa). D. A Eutherian mammal showing the corpus callosum (cc), which is absent from the marsupials. From Heath and Jones (1971).

zation, and to make comparisons with the studies that have used rats (see Chapter 3). So far this has not been investigated. In fact, to date there has been only very limited study of limb use in some marsupial species. These findings will be discussed along with data for the other mammals.

Structural Asymmetries of the Brain

Kolb et al. (1982) found that similar to rats and mice, the right hemisphere in cats and rabbits is larger and weighs more than the left. The sizes of the hemispheres were measured by perfusing the animals, removing the brains, photographing them, and then measuring various surface dimensions. The brains were then cut down the midline, dehydrated, and weighed. The brains of rats and mice were examined in the same study and larger right hemispheres were also confirmed for these species.

In cats and dogs these size and weight asymmetries are known to be accompanied by anatomical asymmetries of the hemispheric fissures.

Webster and Webster (1975) and Webster (1981) examined the fissure patterns of cats and found asymmetries in 45% of the brains (see Figure 4.2). The patterns of the fissures were classified using a previously compiled plan. In the first study, asymmetries were found more commonly in the visual cortex region of the brain, but this was not confirmed in the later study, using a much larger sample size. In the latter, asymmetries occurred almost as frequently in the anterior region. The anatomical asymmetries were unrelated to the paw preferences of these animals, assessed in a variety of tasks (see later). In fact, the functional significance of asymmetries in fissure patterns is obscure since they may not reflect asymmetries in cortical information processing. These asymmetries may come about simply by chance in the course of brain development. However, the motor and visual cortices of the right hemisphere of rabbits contain more synaptic contact zones than their equivalents in the left hemisphere (Vrensen and De Groot 1974), and this indicates differential information processing correlated with the larger right hemisphere. Goldberg and Costa (1981) have discussed anatomical and physiological evidence for a more diffuse patterning of sensory and motor representation in the human right hemisphere with relatively greater synaptic connectivities in certain structures on that side.

In dogs, the right hemisphere is also larger than the left (Kolb et al. 1982; Tan and Çalişkan 1987b) and it is heavier than the left (Tan and Çalişkan 1987a). Tan and Çalişkan (1987b) examined the surface dimensions of the hemispheres of 25 perfused dog brains (Figure 4.3) and they found the right hemisphere to be significantly larger in terms of length and height, but not in width. The mean lengths of the cruciate sulcus and the Sylvian fissures was not found to be different for the left and right hemispheres, but the Sylvian fissure was lower on the left side. The base of the right planum temporale tended to be larger on the left, but the difference was not significant. The paw preferences of these dogs were previously determined by scoring the paw (or paws) used to remove adhesive plaster from the eyes (see later) and, as in cats, no correlation was found between paw preference and the structural asymmetries assessed. Similarly, the greater weight of the right hemisphere, which is consistent across animals, does not correlate with paw preference (Tan and Çalişkan 1987a).

There is so far very little evidence for structural asymmetry of the brain in other nonrodent and nonprimate mammals, although we feel confident in predicting that many more examples will emerge and these may be present in a variety of structures. In sheep there is asymmetry in the sizes of the lateral ventricles of the hemispheres (Szuba 1965). Casts were made of 100 sheep's brains, and a population bias for the

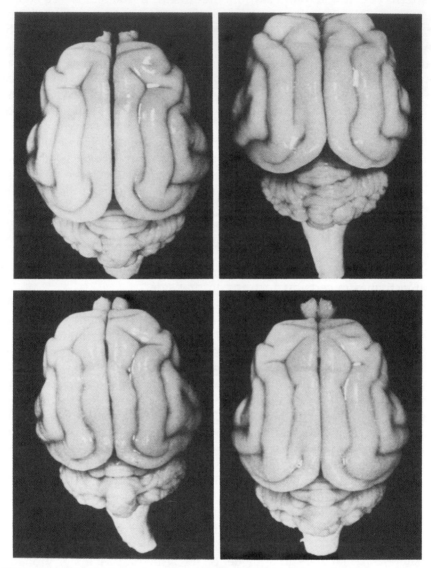

Figure 4.2 Cat brains showing examples of symmetry (photographs at top and bottom left) and asymmetry of fissure patterns (photographs at top and bottom right). Note the discontinuous sulcus on the left hemisphere in top right photograph and the dimple in the anterior of the right hemisphere of the bottom right photograph. From Webster (1981).

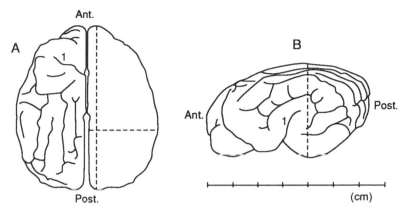

Figure 4.3 Diagrams illustrating the measurements of the hemisphere of dogs made by Tan and Çalişkan (1987). A, Dorsal view of the hemispheres, the dashed lines are the length and width measured. B, Lateral view and here the dashed line indicates the measured height of the hemisphere. The lengths of the cruciate sulcus (no. 1 in A) and the Sylvian fissure (no. 1 in B) were also measured. Measured in centimeters. From Tan and Çalişkan (1987).

right ventricle being larger than the left was reported. Other evidence for brain asymmetry may be inferred from asymmetry in the skull.

Some toothed whales and dolphins have asymmetrical skulls (see Neville 1976; and Figure 4.4). The single nostril (blowhole) is placed to the left side and other aspects of the skull deviate to the right. This asymmetry may be an adaptation for sound communication (compare the skull and brain asymmetry in owls used for this purpose, Chapter 2). Of course, asymmetries of the skull do not necessarily mean that there are corresponding asymmetries of brain structure, although this is frequently the case. In fact, in all of these whales the right side of the brain is better developed than the left (Walker 1980). It also seems that the fissure patterns on the left and right hemispheres differ (Walker 1980). Incidentally, in narwhals usually only one tooth develops into the tusk, and the thread of this tusk is right-handed; this, of course, may have no significance in terms of brain asymmetry.

As far as we know, there has been only one study of asymmetry in the marsupial brain. Haight and Neylon (1978) reported the presence of strong anatomical asymmetry in the thalamus of the brush-tailed possum, *Trichosurus vulpecula*.

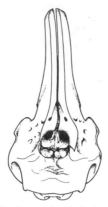

Figure 4.4 The skull of a toothed whale showing asymmetry.

Extra-Brain Structural Asymmetries

As we have seen in Chapter 2, asymmetries in peripheral receptors may give rise to asymmetrical inputs to the brain and so determine lateralized information processing. For example, differences in the refractive indices of the left and right eyes (as in chickens; Chapter 2) lead to differential inputs from each eye. Such asymmetries during development may reflect asymmetrical developmental processes and, according to the stage of development when they occur, they may influence brain development asymmetrically. In Chapter 3 we mentioned that handled rat pups open the left eye before the right, and we discussed how earlier stimulation of the left eye may influence brain development asymmetrically (also clearly shown to occur in the chick; see Chapter 2). In rabbits, the right eye opens first (Narang 1977) and this opening of the eye stimulates myelination of the right optic nerve in advance of the left (Narang and Wisniewski 1977; Tauber, Waehneldt, and Neuhoff 1980). Thus, functioning of the right eye develops in advance of the left. As we discussed for rats and mice in Chapter 3, in rabbits, also with laterally placed eyes, the majority of the optic nerve fibers cross over at the optic chiasma and thus most visual input is relayed to the contralateral hemisphere. Hence, the opening of one eye in advance of the other might well have differential effects on development of the left and right hemispheres. If so, raising rabbits in the dark would be expected to remove, or at least reduce, brain asymmetry just as it does in the chicken (Chapter 2). In fact, this does occur; rabbits raised in darkness have reduced structural asymmetry and, incidentially, increased concentration of the neurotransmitter serotonin in specific brain regions (Alesci et al. 1988).

Even in mammals with frontally placed eyes, in which each eye relays inputs to both the right and left hemispheres, the inputs from the eye to each hemisphere are not identical. The optic nerves from the nasal half of the retina, which cross to the contralateral hemisphere, have larger diameters compared with those which arise from the temporal half of the retina and go to the ipsilateral hemisphere (Bishop, Jeremy, and Lance 1953). The crossed fibers therefore conduct faster and also dominate the uncrossed fibers during binocular stimulation (Walls 1953; Proudfoot 1983). Thus, although each eye of the mammal sends inputs to both hemispheres, the inputs to each are qualitatively different. Consequently, the opening of one eye in advance of the other may also cause differential development of the hemispheres in mammals with frontally placed eyes.

Eye asymmetry and related asymmetry of brain function may be a more widespread aspect of development than presently recognized. In the fetus of the cow, for example, the right eyeball is heavier and larger than the left (Melosik and Szuba 1986). This asymmetry may be inconsequential during fetal development, but we do not yet know whether eyeball asymmetries exist in cows after birth. Alternatively, a large eyeball may possibly reflect earlier opening of that eye, an event which occurs in utero in the cow, and therefore results in differential light input to the brain from the left and right eyes. There is evidence that light can penetrate into the uterus and, if sufficient intensity reaches the developing fetus, asymmetrical developmental events may be triggered, as in the chicken. Evidence for this occurring, however, is not yet available.

Lateralization of Limb Use

Use of the left or right forelimbs has been investigated using dogs and cats tested on a variety of tasks. In both, there is a nonrandom distribution for use of the the left and right paws; that is, pawedness is present.

Cole (1955) trained 60 cats to reach into a glass tube to obtain a piece of meat and scored 100 reaches for each animal. If an individual used the same paw for 75% or more of the reaches, it was considered to show pawedness. Using this criterion, 58% of the cats had a marked preference to use one or the other paw; the remainder were ambidextrous. Of the cats which showed pawedness, 34% were right-pawed and 66% left-pawed (i.e., 20% and 38% of the total population, respectively). According to this study, among the cats that have pawedness, there is a predominance of left-preferent individuals. Another study of limb use in 16 cats found significant preferences in all of the subjects; 6 were right-pawed and 10 were left-pawed (Forward, Warren, and Hara 1962). The cats were tested on three different tasks, each requiring reaching for food from containers or moving a block aside, and paw preference was consistent across the tasks for all subjects.

Contrary to these results, Tan, Yaprak, and Kutlu (1990) found a right bias in paw preference of cats, which they attributed primarily to the females. This contradiction between studies led Tan and Kutlu (1991) to carry out a follow-up investigation using a larger sample size (109 cats, of which 63 were females and 46 were males). The cats had to reach for food through a small hole in the middle of one wall of a cage, a test similar to that used to test for pawedness in mice and rats (see Chapter 3). For the total sample, 50% were found to be right-pawed, 40% left-pawed, and 10% were ambidextrous. Thus, compared to the study by Cole (1955), a much higher proportion of these cats showed pawedness. For the entire group, there was no significant difference between the proportions of left- and right-pawed individuals. In fact, the distribution of pawedness in this study is similar to that obtained for pawedness in mice and rats (see Figure 3.3).

However, differences did emerge when the sexes were examined separately. The distribution of paw preference for males did not differ from that of the entire population (i.e., equal numbers of left- and right-pawed individuals), whereas significantly more females were right-pawed rather than left-pawed (54% of females were right-pawed and 36% left-pawed, the proportion of ambidextrous individuals being the same as for males, 10%). Furthermore, significantly more females than males were strongly right-pawed. Thus, this sex difference in cats differs from that which occurs in rats and mice. For rats and mice, females are more lateralized than males, but for each sex the proportions of left- and right-pawed individuals are equal (Figure 3.3). In other words, cats, but not mice and rats, show a "right shift" in females. A similar pattern of right shift in females has been reported for lower primates and humans (Annette 1985; Ward, Milliken, and Stafford 1992).

A preponderance of right-pawedness was reported to occur in dogs (Tan 1987). A sample of 28 dogs were tested by closing their eyes with sticking plaster and then scoring paw use in attempts to remove the plaster, 100 attempts being scored per individual. The distribution of paw preference was 57% right-pawed, 18% left-pawed, and 25% ambidextrous. By extrapolation from the data for cats, the significant right-paw bias in the total sample may have depended on the females, as there were 19 females and only 9 males in the sample. This task would clearly have been stressful for the dogs and it is possible that this also influences paw use, possibly explaining the higher degree of ambidextrousness in the dogs compared to the cats. Further studies on dogs are needed to address these issues.

The factors which determine pawedness in cats and dogs have been barely studied. However, what is known is that once a paw preference has been established, it is maintained persistently, even when the test imposes difficulties on use of the preferred paw (Warren, Abplanalp, and Warren 1967) or even when extensive lesions are made in the sensori-motor cortex contralateral to the preferred paw (Warren et al. 1972). Extensive practice in manipulation does not appear to significantly influence the development of paw preferences in kittens, and forcing cats to use the nonpreferred paw in certain manipulative tasks does not affect spontaneous paw preference in other situations.

Paw use has been studied in a limited number of species of marsupials. The main study published so far is that of Megirian et al. (1977), who studied the Australian brush-tailed possum, *Trichosurus vulpecula* (a Diprotodont) reaching into a narrow feeding tube to obtain a food reward. They tested 78 wild-caught possums, of which 40 preferred to use the left forepaw, 35 the right, and 3 were classified as ambidextrous. The authors state that the preferences were consistent and predominant, but not always associated with exclusive use of the preferred paw. Unfortunately, the percentage preference per individual was not reported, nor was the sex of the animals tested or the criterion for classifying an individual as showing a preference. The authors did, however, demonstrate the strength of pawedness by paralyzing the preferred limb of some animals. Even with extreme paralysis the possums persisted in trying to use their preferred paw.

A similar study of forepaw reaching has just been completed on a small sample of sugar-gliders, *Petaurus breviceps* (Seccombe and Rogers, in preparation). They were scored for 100 reaches through a hole to obtain food from a dish. Of the eight individuals scored, four were strongly left-handed and four were strongly right-handed. The preference scores were almost 100% in every case. Hand use was also scored in more naturalistic feeding situations, and less strong preferences were found in these cases. In fact, there was little correlation between the limb preference determined in the task requiring reaching through a hole and limb preference for feeding in a more naturalistic setting.

There is only one other brief report of handedness in marsupials, and that is for koalas. Smith (1979) reported a 59% left-hand preference in koalas feeding on leaves. This may not represent a significant bias in the population. The data are also somewhat unreliable because although 140 observations were made, the study included 113 koalas and no mention was made of whether the observations were for repeated data on individuals. Given that many species of marsupials have well-developed hands which they use for fine manipulation of food (see Figure 4 of Ganslosser 1977), some even having opposable thumbs, there is a need to collect much more data on paw use so that clear comparisons can be made with the primate species. Similarly, there would be much interest in collecting data on hand use in placental mammals which use the forelimbs to manipulate food (e.g., bears, racoons, and otters).

Circling Behavior and Side Preferences

In Chapter 3 we discussed the circling or turning behavior in mice and rats. Individuals show a preference for turning either clockwise or counterclockwise, but, as for pawedness in the same species, there is no overall bias of turning direction in the population. Turning has been little studied in other species. Yet, a turning preference is known to occur in whales and dolphins, leftward in confined aquariums and rightward in open sea (see Chapter 5). In either case, the direction of turning is consistently biased within the population and therefore differs from that of rats and mice which show only individual, and not population, turning biases. Some species of baleen whales, although they have no structural asymmetry of the skull, show side preferences at a population level (Warren 1980). From distributions of barnacle encrustations and direct observation, it has been ascertained that some species of whales swim on their sides near the bottom of the ocean with their right side downward (Mathews 1978).

Elephants apparently have a side preference which is manifest in the pattern of wear of the tusks. Individuals show more wear of one tusk than the other, and there have been reports of a right-side bias, although these are not reliable (Warren 1980).

There are two examples of side preferences in ungulates, one for impala and the other for the domestic cow. Jarman (1972) examined the dermal shield of impalas (*Aepyceros melampus*) shot in the Serengeti National Park. One of the features which he reported was the presence of more scars on the right side of the forequarters (Figure 4.5). The asymmetry in distribution of the scars diverged significantly from an even distribution. This must mean that either the attacking impala is more likely to inflict a wound using the left horn, or at least with a left-sided approach, or that the attacked impala tends to turn leftward away from an aggressive encounter so that the right side of its body is more vulnerable to injury. Of course, both mechanisms may operate simulta-

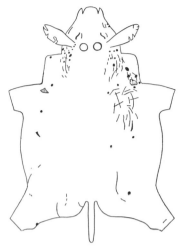

Figure 4.5 A composite diagram constructed from the skins of 48 adult impala showing the sites of scars. Note the greater number on the right side. From Jarman (1972).

neously, as there may be a preference for turning leftward into the attack and a preference for turning leftward away from it, possibly reflecting right hemisphere mediation of emotional responses during agonistic encounters (compare turning biases in humans after unilateral injury; Chapter 9).

The domestic cow shows a left-side preference for lying on a level surface (Uhrbrock 1969; Wagnon and Rollins 1972). This may optimize rumen positioning for the most efficient rumination since this organ is placed asymmetrically in the body cavity. The left-side position is, indeed, adopted only when the animal is chewing and not when lying without ruminating (Grant, Colenbrander, and Albright et al. 1990).

Other side biases are manifest in gait, as in horses, but the field remains open for much more study of a wider range of species.

Lateralization of Brain Function

Webster (1972) demonstrated lateralization of visual discrimination learning in cats that was present at the individual level and highly correlated with paw preference, but not as a consistent bias within the population. Each cat was trained on a series of visual discrimination problems, and then the optic chiasma and corpus callosum were sectioned in the midline. This meant that when the cat was tested monocularly after the operation, the visual input reached the contralateral hemisphere only (see Figure 6.1). On a Go/No-Go task using visual stimuli and requiring withholding responses for 15 seconds, performance was better when the cat used the hemisphere which controlled its nonpreferred paw. The same hemisphere was also better at controlling performance of a simultaneous pattern discrimination task and a form discrimination task.

A significant group bias of lateralized hemispheric function of cats has been demonstrated by Bianki (1981). Regrettably, the English language reports of Bianki's studies of laterality cats, as for his studies of the same in rats (see Chapter 3), lack detailed reporting of the methods used. With this in mind, we take note of his reported evidence for lateralities of the cat brain, present as a significant bias in the population. In a study using 36 cats, he found asymmetry of interhemispheric and intrahemispheric currents of electrical excitation in the motor cortex (Bianki 1981). Electrophysiological recordings were made at many sites in the motor cortex. Higher amplitude positive components of both transcallosal and extracallosal potentials were recorded in the left hemisphere. The area of evoked changes was also larger on the left side.

In a second experiment using 54 cats, Bianki recorded evoked potentials in the visual cortex. Monocular stimulation with a burst of light led to responses of greater magnitude in the right visual cortex in 57% of the animals, in the left visual cortex in 26%, and no asymmetry in 17% of the animals (Bianki 1981). The asymmetry was reported as significant at the 5% level. A third experiment by Bianki investigated noise resistance of the evoked potentials in the visual cortex of the left and right hemispheres of cats (Bianki 1981). The ability of noise to depress light evoked-potentials was greater in the right hemisphere. In other words, the left hemisphere was found to be more resistant to interference from noise.

These asymmetries in evoked potentials may depend on asymmetries in neuronal connectivity and asymmetries in neurotransmitter levels, as is known to occur in rats (Chapter 3) and rabbits (see earlier in this chapter). Consistent with the latter possibility, cats respond to hallucinogenic compounds with changes in electroencephalographic (EEG) activity in the limbic system, and these changes predominate in the left hemisphere (Contreras et al. 1986). Incidentally, the treated cats showed exploratory movements to nonexistent objects, indicating hallucinatory behavior. The hallucinations might well be generated by the left hemisphere. It would be worth investigating whether the cats' exploratory responses are directionally biased. Interestingly, Frumkin and Grimm (1981) review evidence that in humans, barbiturates, via the action of the inhibitory neurotransmitter GABA, may shift hemispheric lateralization in favor of left hemisphere dominance with a resulting increase in loquacity and decrease in emotionality.

These studies by Bianki indicate the presence of a population bias of functional lateralization in cats which occurs independently of paw preference, and which is in contrast to the findings of Webster (1972) in which lateralization correlated with paw preference and thus occurred at the individual level only. Of course, completely different levels of assessment of laterality were employed in each case. In Webster's experiments the cats had to respond to stimuli with motor performance of the forepaws, and this may have determined the form of lateralization revealed. By direct measurement of electrical activity in the hemispheres, Bianki revealed a population-biased asymmetry in both the visual and motor regions of the brain, indicating that population lateralities in the behavior of cats might well be revealed in tasks less dominated by paw use.

Consistent lateralized shifts in EEG also occur in rabbits trained to perform a simple discrimination task for a water reward (Nelson et al. 1977). During reinforcement, the EEG pattern shifted to be lateralized in one direction; in satiation the shift was in the opposite direction. This finding suggests that shifts in hemispheric dominance occur according to the demands of the task. Presumably, these asymmetries reflect hemispheric specializations for processing information.

Lateralization of Sleep

In the conclusion of Chapter 2 we mentioned evidence that birds show lateralized sleep occurring in one hemisphere and not in the other. The same unilateral sleep has been demonstrated in a number of mammalian species. Electrophysiological studies of sleep in the bottlenose dolphin, *Tursiops truncatus*, the porpoise, *Phocoena phocoena*, and also in the northern fur seal, *Callorhinus ursinus*, have shown that the main stage of sleep is a unihemispheric slow wave form (Mukhametov, Supin, and Polyakova 1977; Mukhametov, Lyamin, and Polyakova 1985). The fur seals were recorded when sleeping on land or in water; more unihemispheric sleep occurred when they slept in water. This led the researchers to suggest that unihemispheric sleep is related to an ecological niche and occurs more commonly in marine mammals. Yet this conclusion is not consistent with the presence of unihemispheric sleep in birds. Also, unihemispheric states of arousal appear to be more widespread across animals in a range of ecological niches. In rabbits, for example, there are asymmetries in EEG amplitudes which accompany certain states of arousal. Nelson, Phillips, and Goldstein (1977) found that the side of the brain which has the lower EEG amplitude during normal wakefulness undergoes a greater increase in amplitude when the arousal state shifts towards sleep, either spontaneously or in response to pentobarbitone treatment. Similar lateralized shifts in EEG amplitudes also occur in cats and humans (Goldstein, Stolzfus, and Gardocki 1972). Different lateral asymmetries of cortical activity were found to exist for various states of sleep–wakefulness (i.e., REM sleep, non-REM sleep, and normal wakefulness). There is some suggestion that this asymmetry between the hemispheres in EEG pattern may have a population bias; during REM sleep, in rabbits, cats, and right-handed humans, the right hemisphere increases its relative activity, whereas during non-REM sleep the left is more active (Goldstein, Stolzfus, and Gardocki 1972; Antrobus 1987; Gabel 1988). In cats, the direction of the EEG shift appears to be unrelated to pawedness (Webster 1977), as would be expected if the EEG has a population bias.

Summary and Conclusions

Although much more study needs to be made of the mammalian species before definite conclusions can be reached, the information to date suggests that individual lateralities, particularly of limb preference, are common, and also that population biases exist for hemispheric processing. For cats, dogs, and rabbits, there is evidence that the right hemisphere is larger than the left. In cats and dogs, at least, there is also asymmetry of the fissure patterns on the left and right hemispheres. Other structural asymmetries of the brain or skull have also been reported for sheep, a number of species of toothed whales, and for one species of marsupial. Asymmetries of forelimb use have been recorded for cats, dogs, and three species of marsupials (brush-tailed possums, sugar-gliders, and some limited data for koalas). Strong individual preferences for paw use exist in all of these species. The earlier reports of paw use in cats presented data for a slight left-paw preference, whereas some more recent reports suggest, for both cats and dogs, a right-paw bias within the population for females, but not males. Testing of larger sample sizes on a variety of tasks is now necessary to clarify this issue.

We mentioned that marsupials are a special case because they have interhemispheric commissures which differ from those of other mammals. It would be premature to draw any conclusions about laterality in marsupials, but the small amount of data so far available indicates similarity to other mammals.

Side preferences and circling also reflect laterality. Population biases have been reported for turning in whales, turning in impala during aggressive encounters, and for side preference for lying down in cows. These motor functions may be determined by lateralized information processing by the hemispheres, although a great deal more study is required in this area. Cats, however, have lateralities of brain function that do correlate with individual paw preference, as well as consistent left–right biases of hemispheric specialization for other aspects of information processing. So far, hemispheric specialization has been demonstrated only by the electroencephalographic recording of activity evoked by light stimulation, but not by assessing the behavior of the cats themselves.

Unihemispheric or laterally asymmetric sleep patterns have been found in porpoises and dolphins, fur seals, rabbits, cats, and humans. Apparently this is a rather widespread phenomenon. Whether it relates to lateralization of other functions is as yet unknown.

It is also likely that in the species covered by this chapter, as in rats and chicks, the development of lateralization may be influenced by early experience and by hormonal condition. In rabbits, for example, stimulation in infancy by handling facilitates interhemispheric communication, presumably via the corpus callosum (Denenberg et al. 1981). The corpus callosum is still developing after birth and its further development is therefore susceptible to environmental influences. The known lateraliza tion of eye opening in rabbits may thus asymmetrically constrain the development of the corpus callosum. Its development has in fact been rather well studied in the cat: Specific connections are known to form before birth with further modifications occurring after birth (Innocenti and Clarke 1983; Berbel and Innocenti 1988). So far, however, no attention has been paid to possible asymmetries in these developmental events in cats.

Hormonal condition is also likely to influence the development of lateralization in these species. There has been but one study of this in nonrodent, nonprimate mammals: a lateralized action of androgen on the development of behavior and sex differences has been demonstrated in the Mongolian gerbil, *Meriones unguiculatus* (Holman and Hutchison 1991). In sexually active male gerbils, a sexually dimorphic area of the preoptic region of the hypothalamus controls ultrasonic vocalizations during sexual interactions, and it is the left side which controls the vocalizations. Androgen treatment has an asymmetrical effect in altering the volume of these nuclei and bringing about typical male vocalizations.

To conclude, there is evidence that in addition to rats and mice, other nonprimate mammals exhibit lateralization of brain structure and function. This evidence is, as yet, rather limited and patchily distributed across a handful of species, but the data for rats and mice may well come to be accepted as fairly typical for nonprimate mammals, whatever may be the case with respect to the presence of population biases for limb and turning preference in the other species.

Further Reading

Kolb, B.; Sutherland, R.J.; Nonneman, A.J.; and Whishaw I.Q. Asymmetry in the cerebral hemispheres of the rat, mouse, rabbit, and cat: The right hemisphere is larger. *Experimental Neurology*, 1982, 78:348–359.

Megirian, D.; Weller, L.; Martin, G.F.; and Watson, C.R.R. Aspects of laterality in the marsupial *Trichosurus vulpecula* (brush-tailed possum). *Annals of the New York Academy of Sciences*, 1977, 299:197–212.

Tan, Ü. Paw preferences in dogs. *International Journal of Neuroscience*, 1987, 32:825–329.

Tan, Ü. and Kutlu, N. The distribution of paw preference in right-, left-, and mixed-pawed male and female cats: The role of a female right-shift factor in handedness. *International Journal of Neuroscience*, 1991, 59:219–229.

Walker, S.F. Lateralisation of functions in the vertebrate brain: A review. *British Journal of Psychology*, 1980, 71:329–367.

Webster, W.G. Functional asymmetry between the cerebral hemispheres of the cat. *Neuropsychologia*, 1972, 10:75–87.

Webster, W.G. Morphological asymmetries of the cat brain. *Brain Behavior and Evolution*, 1981, 18:72–79.

Asymmetries in Primates: Manual Asymmetries

Introduction

Limb asymmetries may not be restricted to the forelimbs and a case can be made, especially with nonhuman primates, for studying both performance and preference asymmetries of the lower limbs. As we shall see, apart from a study of apes as to which hindlimb leads in stepping off, this issue has not been addressed. Morphological asymmetries in bone and muscle can provide useful adjunct information to behavioral measures, as can asymmetries in the brain levels of certain neurotransmitters, and possibly related turning biases.

In this chapter we shall review in detail left and right performance and preference asymmetries in primates from the prosimians upward. Findings will be reviewed in the context of an important evolutionary hypothesis based on a change from a left-handed preference for visually guided predation, accompanied by a specialization of the right upper limb for postural support in the prosimians, to a right-handed superiority for fine, sequential manipulation in quadrupedal primates.

The Motor Pathways

Yakovlev and Rakic (1966) and Yakovlev (1972) studied the decussation patterns of the pyramidal tracts in humans. Of 179 fetal and 130 adult specimens, 80% had more fibers going to the right hand from the left motor cortex via the contralateral pathway and from the right motor cortex via the ipsilateral pathway, than there were fibers going to the left side via the respective contralateral and ipsilateral tracts. Subsequently, Nathan, Smith, and Deacon (1990) found that three quarters of cords are asymmetrical and of these, in a further three quarters, the right side is larger (see Figure 5.1). The asymmetry, due primarily to the size of the lateral corticospinal tract (which controls limb musculature), reflects the

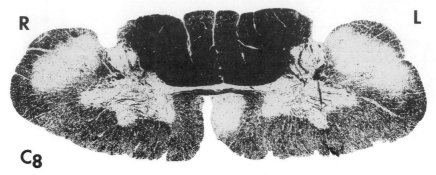

R   L

C₈

Figure 5.1 A cross section of the spinal cord of a male aged 62 years showing marked degeneration of both medullary pyramids due to motor neuron disease. In this example of corticospinal tract asymmetry, the right lateral tract is *smaller* than the left (i.e., the *reverse* of what is normally found). The left anterior tract is large and the right is almost absent. In fact, most of the corticospinal tract has crossed from right to left. Three quarters of spinal cords are asymmetric, and in three quarters of these the right side is the *larger*. From P. Nathan, M. C. Smith, and P. Deacon, The corticospinal tracts in man: Course and location of fibers at different segmental levels. *Brain*, 1990, 113:303–324, with permission.

larger number of corticospinal fibers crossing to the right side. This morphological asymmetry is apparently unrelated to handedness, but correlates with the further observation (Yakovlev and Rakic 1966) that, in three quarters of the corticospinal decussations, the crossing from left to right occurs at a higher, more cranial level than the opposite configuration. This asymmetry also seems to be unrelated to handedness (Kertesz and Geschwind 1971), although it is noteworthy that there are neurochemical asymmetries in the human globus pallidus (Glick and Shapiro 1984, 1985), an extrapyramidal structure closely related to the corpus striatum in the basal ganglia and therefore involved in the initiation of movement; thus dopamine and choline transferase both occur in greater concentrations on the side contralateral to the preferred (usually right) forelimb (and see also Chapter 3 for a discussion of dopamine levels in the nigrostriata of rats).

Turning now to morphological asymmetries of the forelimbs themselves, Schell et al. (1985) found that among 116 dextral white adolescents, biepicondylar breadth of the humerus was significantly larger on the right side. They concluded from further examination of upper arm and calf circumferences that the results were consistent with preferred use of one side of the body and subsequent muscle hypertrophy. Indeed, in a study of tennis players versus nontennis players, Buskirk, Andersen, and Brozek (1956, cited by Falk et al. 1988) found that only the former showed asymmetries in length of ulna and radius favoring the dominant arm. Nevertheless, features may still be larger on the right *regardless* of limb use. Thus Garn, Mayor, and Shaw (1976) found that among 208 dextral and 19 sinistral individuals, the total cross-sectional length and cortical area of the second metacarpal bone was larger on the right side in most individuals, regardless of hand preference. Similarly, Plato, Wood, and Norris (1980) found that the total midshaft length, width, cross-sectional area, and cortical area of the second metacarpal bone was larger in the right hands of both dextrals and sinistrals, even though it only reached significance in dextrals. However, the smaller sample size of sinistrals in both studies make generalizations with respect to the effects of hand preference somewhat dubious.

Falk et al. (1988) studied the forelimbs of 150 rhesus monkeys (*Macaca mulatta*), measuring the right and left humeri, radii, ulnae, and second metacarpals, together with the femora; seven of the ten forelimb dimensions assessed were larger on the right side, significantly so for two measurements of the ulna (its overall length and that of the olecranon process) and two of the humerus (the length of the supracondylar ridge and the diameter of the humerus at the lateral lip of the intertubercular groove). No measurements were larger on the left side. It can be asked whether this hypertrophy reflects a greater use of muscles, for example, for abducting and flexing the arm, or for flexing the forearm and extending the forearm and wrist, judging by the fact that such movements would be mediated by muscles attaching to those structures found to be asymmetrical in size.

Indeed, Helmkamp and Falk (1990) subsequently address this specific issue after reviewing the evidence for directional asymmetries (which they find nearly always favors the right side) in the forelimb bones and muscles of the human fetus, adolescent, and young and mature adult. They also studied 10 forelimb bone measurements in 76 male and 61 female rhesus monkeys, grouped by age and sex. They were thus able to compare the possible contributions of genetically determined adaptations for (possibly asymmetrical) behavior, and progressive developmental responses in the limbs to asymmetrical behaviors. They again found that all significant asymmetries favored the right side irrespective of age and sex. This provided some support for the idea that preferential use of the right limb leads to associated asymmetrical hypertrophies. Indeed, there were numerous instances of right-favoring asymmetries of the muscle attachments. For instance, the (asymmetrical) lateral supracondylar ridge of the humerus is the origin of the forelimb-flexing brachioradialis muscle and the wrist extensor carpi radialis longus. Similar considerations apply to the (almost equally asymmetrical) olecranon process where the common tendon of the forelimb extensor, the triceps brachii, is inserted.

There were also significant sex differences in asymmetries both within and across the age groups of juvenile, subadult, and adult, with numerous age-related changes. Males showed a decreasing asymmetry with increasing age, 40% of the 10 measures were asymmetrical in juveniles, 30% in subadults, and 20% in adults. In females these effects were reversed, with no significant asymmetries for juvenile or subadult females, and 40% of the measures were asymmetrical in adult females. Consequently, the authors conclude that skeletal forelimb asymmetries are not just due to developmental responses to differential limb or hand use, as the patterns of asymmetry fluctuate with age and sex in ways that are inconsistent with such a hypothesis. Thus, as they note, all previous work on forelimb asymmetries should be reassessed (and future work controlled) on the basis of age, sex, and environment. We shall see that whereas most studies until recently denied manual asymmetries in nonhuman primates, especially monkeys, there is now increasing evidence of right-hand preferences and performance superiorities for certain manipulative behaviors, even in monkeys.

Hand Asymmetries: Preliminary Considerations

While the possession of two equally "dextrous" hands might be thought to permit maximum manipulative flexibility, complex manual behaviors may nevertheless be more effective under conditions of manual specialization where one hand is dominant. It is universally acknowledged that skills acquired unimanually do not readily transfer to the opposite hand, even in the presence of intact cortico–cortical commissures. Indeed, the forebrain commissures reduce the likelihood of unnecessary duplication of engrams, as seen both in the split-brain animal studies and with human congenital acallosals, individuals born with incomplete or absent forebrain commissures (Jeeves 1990; Lassonde, Bryden, and Demers 1990). The commissures may also reduce the probability of intermanual competition, as evidenced both by anecdotal accounts from the human split-brain literature (Bradshaw 1989) and the clinical syndrome of synkinesia (synkinesis). See, for example, Cohen et al. (1991).

Individual, as opposed to population-level, hand asymmetries occur with consistent performance both within and between tasks, although when a task has several components there may be a switching of hand preferences between components, or even a change in patterns of asymmetry as a function of practice. The most skilful hand is not necessarily always chosen. This may be a problem in human studies where, for example, questionnaire (preference) measures rather than objective (performance) scores are used. The latter should preferably sample less-practiced skills as well as, or even instead of, highly overlearned movements; the same should apply to animal studies, and any inconsistencies between findings may reflect uncontrolled practice.

Nonhuman primate studies have tended to rely on tests of the sort used with human *infants* (Harris and Carlson, in press), for example, visually guided reaching, even though this particular task has been little used with human *adults*. Available human data (Harris and Carlson, in press) indicate that it may not in fact be an optimal procedure for primate adults. It is generally the case that nonhuman primate asymmetries (hand preferences) are weaker and more variable than our own; they tend to be more task dependent and more subject to chance extraneous factors in the testing environment, for example, location of a manipulandum or incentive, initial posture or behavior stereotypes.

Indeed, differences in testing methods and assessment and lack of reliability between measures may account for further inconsistencies between studies. Nor is there agreement on the nature of statistical treatment and whether individuals should only be regarded as right- (or left-) handed if their performance asymmetries exceed some criterion level. Similarly, many studies have used only three or four animals, especially in recent cognitive tests of language-trained animals, so that generalization is impossible. Only recently have techniques been borrowed from human neurology and neuropsychology; for example, for comparisons of reaction and movement times (Bradshaw et al., 1992), and kinematic analysis where three-dimensional limb speed, acceleration and deceleration, is analyzed when the subject reaches to a target (Meulenbrock and Van Galen 1988).

Most of the major issues relating to naturalistically observed hand preferences in captive (zoo) primates were discussed over 100 years ago by Ogle (1871, cited in Lehman, in press). Ogle could even predict, after long observation, which hand a monkey would use to receive food, and concluded that most are consistently right-handed. He noted that hand preference altered as a function of position, the nearer hand being preferred. Since then, it has generally been argued, mostly from observations of captive animals in zoos or laboratories, that hand asymmetries are weak or absent in nonhuman primates (Walker 1980). Indeed Annett and Annett's (1991) observation of zoos is in this tradition. We shall see that more subtle testing may reveal equally subtle asymmetries.

We can perhaps group theories on manual asymmetries in nonhuman primates into six major categories, several of which are closely interrelated. According to one theory, as concluded by Warren (1980), hand differences are really only present in humans; any effects in infrahuman species are a function of task, practice, and species differences, and any hand asymmetries observed at the level of the individual subject are independent of any hemispheric asymmetry at the morphological (brain) or functional (e.g., ear preference) levels. In a second not dissimilar position still espoused by, for example, Lehman (1989), hand differences may indeed occur, but they are generally consistent within subjects only for certain tasks, and usually do not generalize either across tasks or between subjects to generate a bias at the population level. According to MacNeilage, Studdert-Kennedy, and Lindblom (1987), earlier inconclusive studies are due to the use of subjects that are too young to express stable preferences, and because of tasks that are inadequate to demonstrate underlying hand asymmetries. When these problems are addressed, MacNeilage et al. (1987, 1991) claim in a major hypothesis that in prosimians there was a left-hand preference for visually guided movements, accompanied by (and perhaps stemming from) specialization of the right upper limb for postural support. With the change from vertical clinging to a quadrupedal posture in higher primates, and with decreased importance of unimanual predation, the postural specialization may have evolved into a right-hand superiority for fine, sequential manipulation. Indeed, Beck and Barton (1972) noted that a left-hand preference in reaching for a food reward coincided with a right-hand preference for various manipulatory acts in the same group of 10 stumptail macaques (*Macaca speciosa*). MacNeilage (1990) combined the statistically significant examples in which each animal preferred reaching with its left hand and manipulating with its right, and tested the total against significant examples of the opposite trend (right–reach, left–manipulate), and found the difference in favor of left–reach and right–manipulate to be statistically significant.

Right-hand superiorities, said to be particularly evident in naturalistic settings and with older individuals, may have evolved further with the decreasing demand on the right limb to support a vertical posture, and with the development of the opposable prehensile thumb. In the latter context, however, Fragaszy and Adams-Curtis (1988) found no evidence for differences in population-wide hand biases for reaching and manipulation between species of New World monkeys that do, and do not, display a precision grip. Subsequently, Fragaszy and Mitchell (1990) studied manual preference and performance in seven young adult capuchin monkeys (*Cebus apella*). This is a New World genus of platyrrhine descent, whose members are inveterate manipulators with a precision grip and opposition between the thumb and the other digits. However, the morphology of their hands differs from that of Old World primates in terms of how functional opposition is achieved. This again raises the question: Do they behave like Old World monkeys because of their manipulative and oppositional capacities and therefore show manual lateralization, or do they fail to show it in accordance with their platyrrhine descent and early evolutionary divergence from Old World stock? The animals were observed during routine feeding, searching, and grooming, and in uni- and bimanual versions of a knob-pulling task of various levels of difficulty. Little consistency was observed for directions of manual preference, and only three of the seven animals showed the predicted overall left-hand preference; moreover, the degree and consistency of hand preference in the knob-pulling tasks were not predictable from those occurring during feeding, searching, and grooming. Furthermore, handedness in the bimanual task (where a box had to be held with one hand while its lid was pulled open with the other) failed to correlate with what occurred in the unimanual task (where the box was firmly secured), even though more lateralized animals tended to perform better. The authors appeal to a distinction (and see Fagot and Vauclair 1991) between manual preferences for easy, practiced and familiar tasks, and manual specialization for novel, unpracticed, or difficult tasks. Indeed, Costello and Fragaszy (1988) reported on 419 acts of prehensile grip with one hand by six capuchins; of these, 278 (66%) were right-handed and 141 (34%) were left-handed.

Stronger evidence of right-handed involvement in more complex manipulative acts comes from Masataka (1990), who used a much bigger sample and a more rigorous criterion for lateralization. His subjects were 24 black backed capuchins (*C. apella*), 4 white-throated capuchins (*C. capucinus*), and 3 white-fronted capuchins (*C. albifrons*). Hand preferences were observed for the collection of scattered food remnants, which required sophisticated manipulative behavior. Of the 31 subjects, 25 (81%) showed marked and statistically significant right-hand preferences, 4 (13%) showed equally significant left-hand preferences, and 2 (7%) were ambilateral. Thus there is some evidence in support of MacNeilage et al. (1987) that manipulative behaviors even in New World genera may be associated with right-hand preferences, if the task is sufficiently complex.

Prehensile hands is of course one of the most salient features of primates. An opposable thumb permits precision gripping, and whereas Old World monkey, ape, and human species possess truly opposable thumbs, New World species can at best achieve pseudo-opposition (Costello and Fragaszy 1988). Indeed, the New World genus *Cebus*, a manipulative, extractive forager, may exhibit behavior in that respect almost equivalent to that of the great apes, whereas sympatric species like the squirrel monkey (*Saimiri sciureus*) possesses considerably reduced prehensile powers. The latter species also possesses pseudo-opposability, though the range of finely controlled prehensile responses exhibited by *C. apella* is sufficient even for tool manufacture and use. Even though the trend in Costello and Fragaszy's study (1988) toward a right-hand preference for manipulation was nonsignificant, capuchins favoring the right hand for reaching were nonetheless more likely to employ a precision grip, whereas those subjects who favored the left hand were just as likely to use a power grip.

The issue of task complexity is specifically addressed by Fagot and Vauclair (1991), who distinguish (as in *human* studies, see e.g., Bradshaw 1989, for review) between manual *preferences* and manual *performance* asymmetries (e.g., reaction time and accuracy), and between *relative* and *exclusive* specialization. They emphasize the need for tasks of sufficient complexity, spatially or temporally, to generate asymmetries and note that asymmetries may change or even *reverse* when an initially novel task becomes familiar. Thus simple, familiar, gross, practiced, or repeated tasks, for example, reaching for food, may be insufficient to generate asymmetries (though compare Kuhl, 1988a, and the various studies by Ward's group, all reviewed in this chapter).

The last of the six positions is in fact that of Ward's group. In some respects this is an extension of the position of MacNeilage et al., in that a left-hand superiority for reaching for food is found when considerable postural adjustments (e.g., from an initially awkward stance) are needed to achieve the target, together with a right-hand preference for manipulating food en route to the mouth. The critical feature here is the emphasis on an awkward posture.

An important distinction, as previously mentioned, is that between performance and preference; sometimes these two measures can be combined in an interesting and useful fashion. Thus Hörster and Ettlinger (1985) reviewed their records of 237 rhesus monkeys that were initially trained on a particular tactile discrimination task upon arriving at the laboratory. They fell into three almost equal groups of left-handed, right-handed, and ambilateral individuals. However, the last group attained performance criterion with the fewest trials, followed by the left-handers, and then the right-handers. While conclusions could be drawn that bihemispheric involvement helps, or that one hemisphere is somehow faster or more efficient in such tasks than the other, Hörster and Ettlinger asked what might have happened had the monkeys been *forced* to use their *nonpreferred* hand. Thus it is instead possible that one or other *group of individuals* is simply smarter than the other, rather than that the differences occur at an interhemispheric level. Even then, as Ettlinger

(1988) found, forcing monkeys to use the nonpreferred hand might mean that restriction of the preferred hand would have indirectly impeded performance of the other hand more than restriction of the nonpreferred hand.

Right-Hand Performance Superiorities in Nonhuman Primates

Far more studies have addressed hand preferences than performance differences. Indeed with the exceptions of studies by Fagot, Drea, and Wallen (1991) and Olson, Ellis, and Nadler (1990), which will be separately discussed later, almost no findings of a *left*-hand *performance* superiority seem to have been reported. With respect to a *right*-hand performance superiority, Marchant and Steklis (1986) found that, overall, five chimpanzees *performed* barpresses more quickly with the right than with the left hand for a food reward, although hand *preferences* were quite inconsistent: two individuals preferred the right hand, two the left, and one was ambidextrous.

Preilowski, Reger, and Engele's (1986) studies required *Macaca mulatta* to produce isometric pressures of specific durations between the fingertips under varying spatial and temporal limits. They reported that a significant right-hand superiority was found under the more difficult conditions.

Hopkins, Washburn, and Rumbaugh (1989) studied two rhesus monkeys (*Macaca mulatta*) and three chimpanzees (*Pan troglodytes*). The tasks assessed fine motor control. A computer-generated target stimulus moved on a monitor in a predictable path. A cursor (the computer-generated image controlled by movements of a joystick operated by the subjects) was also displayed on the monitor. As the subject operated the joystick and moved the cursor, the target also commenced moving and had to be caught with the cursor. The right hand proved to be faster in all five subjects, as well as to be the preferred hand, but since subjects were also sometimes forced to use the left hand, right-hand performance superiorities were not thought just to stem from preference or practice effects. Interestingly, there were no significant differences in *reaching* behavior.

Right-Hand Preferences in Nonhuman Primates

A number of other studies report right-hand preferences in addition to the findings just described of Hopkins et al. (1989) and the studies (Costello and Fragaszy 1988; and Masataka 1990) discussed earlier in the context of right-hand preferences for complex manipulative acts in New World genera. Kuhl (1988a) had juvenile monkeys (*Macaca fuscata, M. mulatta,* and *M. nemestrina*—numbers not given) sit in a primate chair and listen to pairs of speech sounds (/ba/,/da/) via earphones and perform a same–different discrimination task. Subjects lifted a centrally located telegraph key to different pairs and held it down if they were identical repeats, without being able to see their hands. They also had to monitor lights governing trial and response initiation. In this complex, continuous, speeded, and manipulative task, all subjects over 10 years of age preferred responding with their right hands either early in training or while still using the other (left) hand, for example, to pick up food. However, when fully trained, they always switched to their right hands.

Fagot (in press) studied from birth four infant Guinea baboons (*Papio papio*) reared by their mothers. In this otherwise rare longitudinal developmental study, he noted that most manual actions were directed towards edible or inedible inanimate objects. Initial manual preferences, present before the end of the second week, were not necessarily consistent with later biases, but were mostly right-sided. However, there were only four subjects, and in this naturalistic task the responses were neither complex nor practiced. Indeed, when a troop of 18 Guinea baboons (*Papio papio*), 10 adults and 8 infants and juveniles, were observed in a field situation (Vauclair and Fagot 1987), only 5 showed a right-hand preference for spontaneous activities, 2 a left-hand preference, and the rest were ambidextrous. However the strength of hand preference was greater for adults, as was the degree of hand specialization in bimanual performance, suggesting that lateralization develops with ontogeny. The authors note that this species has an index of opposability (between thumb and index finger) second only to that of humans, that is, closer to us than that of chimpanzees; one might therefore have predicted on that basis a degree of dextrality comparable to our own.

Butler and Francis (1973), on the other hand, studied eight female baboons of this same species that were sophisticated in tactile discriminations. They were tested for movement preferences (toward or away from the midline) in responding for a reward. Seven out of the eight baboons preferred movement *toward* the midline with the right hand, and five preferred movements *away* from it with the left. (Adductive and abductive movement preferences in humans toward and away from the midline, respectively, are discussed by Bradshaw, Bradshaw, and Nettleton, 1990; abductive responses are typically faster when a simple reaching movement is required, but with serial or continuous movements adduction may be faster.) The authors suggest that the stronger lateralization of the right hand and the weaker asymmetry of the left hand favoring movements in the same direction as, rather than mirror to, the right hand, indicate a right-hand (left-hemisphere) dominance. Such a conclusion, however, is at best inferential.

Turning to gorillas (*Gorilla gorilla*), Fischer, Meunier, and White (1982) found that all four female lowland gorillas displayed a very marked *right*-hand preference for grasping, while usually holding an infant with the *left* hand or arm during tripedal locomotion. Likewise, Lockard (1984) observed a social group of eight gorillas (2 adult males, 2 adult females, 1 subadult female, 1 juvenile male, 1 infant female, and 1 neonatal male) in a naturalistic outdoor setting for a period of 2 years. Five of the seven independently mobile animals showed a right-hand preference for foraging and manipulation, only one subadult female was distinctly left-handed, and the newborn male was again carried for the first three months predominantly in the mother's left arm and hand.

Schaller (1963) reported that in chest beating displays by eight mountain gorillas, the right hand was used first in 59 out of 72 displays, and that all eight animals showed a right-hand preference. Bresard and Bresson (1987) confirmed this right-hand preference in chest beating and, in fact, found that this was true for 24 out of 27 animals. With respect to locomotion, Heestand (1987) observed over 12-hour periods the semi-naturalistic behavior of 70 captive apes in zoos representing four species: *Gorilla gorilla, Pongo pygmaeus, Pan troglodytes,* and *Symphalangus (Hylobates) syndactylus* (siamang). All species showed significant preferences for leading off with the right limb when terrestrially walking or running, despite the absence of observable hand differences during free feeding.

Stafford, Milliken, and Ward (1990) studied the last of these species (siamang) in detail. It is a strongly arboreal and persistent brachiator, and 19 individuals (8 *Hylobates syndactylus,* 7 *H. concolor,* 4 *H. lar:* 5 adult females, 6 adult males, and 4 each juvenile males and females) were assessed for lateral preferences in hand use while consuming the daily ration, and in leading limb in brachiation. With respect to the latter, there were no asymmetries (a finding which is not incompatible with that of Heestand, 1987, since her study addressed *terrestrial* locomotion and not brachiation). However all adult females were strongly right-hand preferent for food reaching (asymmetries were nonsignificant for males and subadult females), and there was a significant positive correlation between age and strength of right-hand preference in females.

In the next section we shall see evidence of a shift from a left- to a right-hand preference for food reaching with increasing age in lemurs (Forsythe and Ward 1988), and also find that right-hand preferences are more characteristic of adult female lemurs than males (Ward et al. 1990), although it will be noted that *left*-hand preferences for food reaching are typically found in such studies. Finally, as so often reported at the level of motor asymmetry, females were overall more lateralized in absolute terms than males. Indeed, Rogers (co-author of this book; see Rogers and Kaplan 1991) notes that while feeding, semiwild orangutans at Sepilok, Sabah, showed some degree of handedness, and that right-handedness was particularly marked in females, especially in older animals. Hand asymmetries were generally more pronounced in older individuals and generally while they were feeding in the propped position. There was no asymmetry in the hand used to touch the body, face, or head *without* finger flexion, whereas there was a highly significant bias to use the left hand to touch the face or head provided that the hand was used *with* finger flexion.

Developmental changes with increasing age (and see also Lehman 1980, and Vauclair and Fagot 1987), and stronger lateralization in females were also reported by Bard, Hopkins, and Fort (1990). They studied seven male and five female nursery-reared chimpanzees from birth through their first three months, assessing them on the presence, strength, and direction of lateralization in a range of behaviors and reflexes. Eight out of the ten animals that exhibited the behavior used their right hand in hand-to-mouth self-calming behavior—a behavior pattern used by both human and chimpanzee infants when fussing or crying to return to a quiet alert state. This asymmetry strengthened during development, as did a nonsignificant bias toward the *left* hand in hand-to-hand grasp (also employed in self-calming). The two behaviors were often interlinked, the left hand grasping the right and placing it in the mouth. (Though this left-hand bias was nonsignificant, unlike the significant right-hand preference for hand-to-mouth behaviors, it is tempting to see it as both logically and chronologically prior when the two behaviors occurred, thereby reflecting a right-hemisphere emotional response.) The stronger female lateralization extended also to a number of infantile reflex behaviors where there was no group-level asymmetry. While we shall continue the discussion of self- (and external) nonmanipulative touching in the next section, we should note here the right-limb preference of captive squirrel monkeys (*Saimiri sciureus*) for face and body touching, and of chimpanzees for environmental touching (Aruguete, Ely, and King, in press).

Left-Hand Preferences in Nonhuman Primates

Turning now to left-hand preferences, we have already noted the observation of Rogers and Kaplan (1991) that semiwild orangutans tended to use the left hand to touch the face or head, provided that the finger was flexed. Dimond and Harries (1984) note that face touching is common in primates—stroking the chin, passing the hand across the brow, and burying the face in the hands—where this cannot be simply attributed to local skin irritation. They studied 14 monkeys for 20 minutes each (7 weeper capuchins, *Cebus grise;* 2 lion-tailed macaques, *M. silenus;* 1 mountain-island macaque, *M. pagensis;* and 4 De Brazza's monkeys, *Cercopithecus neglectus*), 21 apes (8 lowland gorillas, *G. gorilla;* 5 chimpanzees, *P. troglodytes;* and 8 orangutans, *Pongo pygmaeus abelli*), and a sample of humans. They reported that face touching was very rare in monkeys, who displayed a very slight, nonsignificant, left-hand preference. With apes, face touching was very common, especially with chimpanzees, and least with orangutans (although Rogers and Kaplan, 1991, reported frequent face touching by the left hand in the latter species; see Figure 2), and there was a very high left-hand predominance (16 left-hand preferent, 3 right-hand—one from each species—and 2 undecided). The left-hand preference reached significance for the gorilla and failed to do so, probably for lack of power, in the chimpanzees. Humans also showed a very significant left-hand predominance, although their baseline rate of face touching was less than that of chimpanzees.

One would be tempted to invoke the emotional role of the right hemi-

Figure 5.2 Left hand self-touching behavior by a young orang-utan at the Orang-utan Rehabilitation Center, Sepilok, East Malaysia. Photograph by L. J. Rogers and G. Kaplan.

sphere in this context (Bradshaw 1989); however, it should be noted that Suarez and Gallup (1986) dispute the low rate of touching reported by Dimond and Harries, and criticize their use of only a single 20-minute observation session per animal, instead of multiple observations spread over different time periods. Suarez and Gallup employed six 10-minute sessions evenly distributed between morning and afternoon. They studied four pig- tailed macaques (M. nemestrina) and two rhesus monkeys (M. mulatta), and found about ten times the rate of face touching reported by Dimond and Harries; however, whereas three showed a left-hand preference, two were equal, and one showed a right-hand preference. They also rightly criticize Dimond and Harries for numerous computational errors; Suarez and Gallup conclude that there were no differences in the rate of face touching among monkeys, apes, and humans, or in the preference of one hand over the other for touching the face, even in the apes. Indeed, they calculate that instead of 16 preferring the left hand, 3 the right, and 2 without preference, as reported by Dimond and Harries, the figures should really be 8, 1, and 12, respectively.

We must therefore conclude that the issue of hand differences in face touching by primates is as yet unresolved, although see Rogers and Kaplan (1991) who reported a highly significant bias to touch the face or body with the left hand in semiwild orangutans, and Aruguete et al. (in press) who studied hand preferences in two species for simple nonmanipulative touching responses directed either toward the face or body, or toward the inanimate environment. They found that captive squirrel monkeys (Saimiri sciureus) had a right preference for hand and foot responses directed towards their bodies, and no asymmetries for environmentally directed touching; whereas chimpanzees (especially males) preferred the right hand for environmental touching, but displayed no asymmetries for face or body touching.

Box (1977) in an early study observed four male and four female common marmoset monkeys (Callithrix jacchus) for spontaneous hand usage in seven categories of behavior: taking food from a bowl, holding food, climbing up from standstill, reaching up, reaching down, walking from standstill, and pushing or boxing another animal. All scores were taken from an initially unambiguous stance, and thus represented a real choice from an initially symmetrical viewpoint, and were not a consequence of a prior asymmetrical activity or stance. (Indeed, Box is explicit about an important point which may well be overlooked by students of nonhuman performance asymmetries and that is, with caged animals one must ensure that the animals have not developed stereotypes of behavior because of the choice or necessity of occupying one corner of the cage, thus artificially restricting one hand.) Only in the first two behavior categories were there robust (left) hand preferences, which she suggests might reflect ecologically appropriate (to the species) prey-catching behaviors.

Even earlier, Kruper, Boyle, and Patton (1966) reported a significant preference for the left eye by 19 immature rhesus monkeys when sighting food, via a 10-mm diameter tube, which could be taken by either hand. Moreover, the left hand tended to be favored and use of the left hand and left eye to be correlated, although neither effect achieved statistical significance. Recently, Matoba, Masataka, and Tanioka (1991) confirmed Box's (1977) finding, noting that of 46 captive adult marmosets, 43% preferred to pick up food with the left hand, 24% with the right, and 33% were ambidextrous. Furthermore, the percentage of left-hand use by infants correlated with that of the mother, not the father.

The left-hand preference in food-related behaviors as reported by Box (1977) relate closely to similar findings with prosimians later reported by Ward and her colleagues, although it should be noted that Subramoniam (1975, cited by Larson, Dodson, and Ward 1989, and Fagot and Vauclair 1991) made an early observation of left-hand preference for reaching for food in the slender loris (*Loris tardigradus*). Similar reports have been made by Deuel and Schaffer (1987. *Macaca fascicularis* catching moving bait from a sitting posture), Kawai (1967: *M. fuscata* unimanually catching airborne objects), and King, Landau, Scott and Berning (1987: squirrel monkeys catching live goldfish). With the prosimian *Lemur catta*, Masataka (1989) noted that 20 out of 22 individuals showed left-hand preferences across four categories of simple spontaneous acts of reaching by the forelimbs (pickup, push, reach, and grasp). Ward et al. found a left-hand preference for food reaching in a number of prosimian species. With the bushbaby (*Galago senegalensis*), Sanford, Guin, and Ward (1984) found that this effect was particularly strong when a postural adjustment was required, and occurred only under conditions of a bipedal (and therefore intrinsically unstable) posture, rather than with a normal quadrupedal posture. In a subsequent study with the same species, Larson et al. (1989) confirmed these effects, noting that a bipedal posture alone was sufficient to facilitate use of the dominant (left) hand, and that there was also a leftward bias for whole-body turning through 180 degrees from a bipedal stance, although the two effects were apparently uncorrelated. (Dodson, Seltzer, and Ward, in preparation, found a left whole-body turning bias for 14 out of 17 lesser bushbabies, and for 7 out of 8 mouse lemurs, *Microcebus murinus*.) Visibility of prey and angle of reaching seemed not to influence these effects.

With the ring-tailed lemur (*Lemur catta*,) the left-hand preference was most pronounced with males (Milliken, Forsythe, and Ward 1989), adult females tended toward a right-hand bias. To the extent that they used the left hand for reaching food, males also favored a triposture stance (on two back legs and one foreleg) while feeding, rather than an upright or seated posture. This finding is completely compatible with the hypothesis (reviewed previously) of MacNeilage et al. (1987): the right hand preferred for posture and the left for food acquisition. Additionally, left-hand reachers used bimanual manipulation less than right-hand reachers, suggesting that the left hand might be better than the right in acting independently. With the black-and-white ruffed lemur (*Varecia variegata variegata*), Forsythe et al. (1988) again found a left-hand preference for food reaching which was most prominent when postural adjustment of the whole body was necessitated, whereas manipulation of food between the hand and the mouth was quite specific to the lemur's right hand. With the black lemur (*Lemur macaco*), Forsythe and Ward (1988) obtained a left-hand preference even during routine feeding; it was especially prominent in young animals, tending later with increased age to shift to a right-hand preference. There were no differences due to sex or family handedness. Milliken et al. (1989) conclude from such studies that there is a special relationship of the right limb to postural support, since lemurs preferring to reach with their left hands employ the right for postural support, whereas animals preferring to reach with their right hand do not usually employ their left for tripedal postural support.

This conclusion is essentially similar to the hypothesis of MacNeilage et al. (1987) of a right upper-limb specialization for postural support in prosimians, a consequent left-hand preference for visually-guided reaching, which is also found in monkeys, and finally a right-hand preference for manipulation by prosimians (as in the findings of Ward's group), monkeys, apes and humans. Humans tend to hold objects with the left hand while manipulating them with the right; we shall return to these evolutionary questions later.

In a subsequent study, Ward et al. (1990) note three aspects of their earlier lemur work: subjects were comparatively few and often closely related through a single matriline of each species, there was an inverse relationship between frequency of hand use and age with *Lemur macaco* (Forsythe and Ward 1988), and with *Lemur catta* (Milliken et al. 1989) left-handedness was associated with male gender. Consequently, in a mammoth study of 116 males and 78 females without founder effects (the matriline problem) across six species (*L. catta*, *L. coronatus*, *L. macaco*, *L. mongoz*, *L. rubriventer*, and five subspecies of *L. fulvs*), they surveyed feeding behavior (food retrieval, when awkward postural adjustments were not required). As they note, the genus is a more conservative taxonomic unit than the species. Each subject was assessed to determine whether the preferred hand's frequency of use exceeded chance by p<0.001. They found a marked decrease, with age, in use of the mouth with a corresponding increase in use of the preferred hand. Males showed a significant left-hand preference (51% left, 31%, and 18% ambivalent), whereas females showed no overall hand bias. With increasing age, in both sexes the frequency of right-hand use increased, and separately, adults were more strongly lateralized than juveniles. However in 9 of the 10 (sub)species, animals preferring the left hand were more strongly lateralized. As the authors stress, these effects did not require postural adjustments before reaching, despite earlier evidence that reliable expression of hand preference was promoted when prior postural adjustment was necessary.

In the context of males alone showing a left-hand preference, Ward et al. note that according to Geschwind and Galaburda (1985), the right hemisphere of the human fetus leads in development, and the development of the left proceeds over a longer time course; the latter is therefore hypothesized to be more vulnerable to the impact of environmental and hormonal factors such as testosterone. This may account for a consequent increase in sinistrality in human males, and a left-hand preference in male lemurs, as shown here. As the left hemisphere continues to develop, Geschwind and Galaburda predict some degree of takeover by the left hemisphere (right hand) system, again as observed here. Of course, the Geschwind and Galaburda hypothesis in the human context refers to events following a much shorter time course (months), whereas with the lemurs any developmental changes must involve years. Alternatively, as Ward et al. suggest, a shift from left- to right-hand preference in female lemurs may be related to their increasing social status as they grow older.

King and Landau (in press) report on the behavior of squirrel monkeys (*Saimiri sciureus*) from the New World. As arboreal omnivores, they are not naturally fish eating but can be accustomed to a fish diet. Eleven out of fifteen made significantly more left-handed catches of goldfish, and only two were (nonsignificantly) right-handed, while two showed no hand differences. Thus a left-hand preference was shown in a visual task requiring postural adjustment. Hand asymmetries were lost in a static food-reach task, irrespective of posture, and changed to a *right*-hand preference for food reaching (to either static items or moving goldfish) when the animals hung in a vertical cling posture.

Hauser et al. (1991) studied 234 adult and 43 juvenile free-ranging rhesus monkeys (*Macaca mulatta*) lifting the lid of their chow dispenser and extracting food. Overall, the lid was far more often lifted with the left hand and held open with the right, whereupon chow was selected with the left hand. Such results are largely compatible with the MacNeilage et al. hypothesis. However, two major studies with great apes failed to show any asymmetries in food-reaching behavior. Annett and Annett (1991) recorded spontaneous food reaching in 31 captive lowland gorillas (*Gorilla gorilla*). A full range of hand preferences was observed, from strong right to strong left, with most animals intermediate. Despite a high degree of consistency between observations made on different occasions for the same individual, there was no overall population bias. Similarly, Byrne and Byrne (1991) observed the manual dexterity of 44 wild mountain gorillas (*G. gorilla berengei*) of all ages and sexes in their native habitat. There were six naturally acquired feeding tasks of varied logical structure, including multistage performance and bimanual coordination. Very strong hand preferences were evident in five tasks by 3 years of age, but again there were no population trends apart from a slight bias toward right-handed fine manipulation in one group of tasks. Females, however, once again showed stronger hand preferences.

Before we leave the domain of food-reaching behavior, we should note an important and complex study by Olson, Ellis, and Nadler (1990), where hand preferences were assessed in 12 gorillas (*Gorilla gorilla gorilla*), 13 orangutans (*Pongo pygmaeus abelii,*) and 9 gibbons (*Hylobates lar*). Three tasks were used: retrieving raisins from the floor (thrown far enough to require locomotion), collecting raisins attached to mesh high up in the subject's midsagittal plane, and (eight gorillas and eight orangutans only) unfastening a hasp. These three tasks respectively required locomotion, an upright bipedal stance, and fine motor control. The floor retrieval task produced weak and variable effects and will not be considered further; nor will the hasp-unfastening task, since the gorillas' short thick fingers caused them to perform awkwardly and without precision, and the orangutans, despite their long slender fingers also failed to exhibit asymmetrical behavior. However, in the mesh retrieval task, all 6 gibbons showed a left-hand preference, and 10 out of 12 gorillas a right-hand preference, whereas the orangutans exhibited a bimodal distribution. The latter, however, performed the task suspended from three limbs, thus avoiding a bipedal posture and thereby suggesting that the need to adopt such an upright stance facilitates hand asymmetries. (Rogers and Kaplan, 1991, found more hand asymmetries in orangutans when they were in the propped or seated position than when they were hanging.) The authors note that orangutans rarely engage in bipedal locomotion in the wild, unlike the other two species. Thus the degree of natural behavior of a species may correlate with its readiness to exhibit unilateral hand preferences. Indeed, the difference between the (rightward) asymmetry of the gorilla and the (leftward) asymmetry of the gibbon (where one might have expected a right-hand superiority, like that of the gorilla, in a fine motor task) may reflect the gorillas's terrestrial and the gibbon's arboreal habitat; the greater spatial demands of an arboreal environment may predispose a left-hand (right-hemisphere) involvement even in the face of challenge from a task requiring fine motor control. (Note, however, the study of Stafford et al., 1989, where *arboreal* hylobatid apes were *right*-hand preferent in food reaching.) Thus the authors conclude that hand preference may at least partly depend on the degree to which a species locomotes bipedally in its natural habitat, the degree to which it is naturally arboreal, and the postural instability induced during task performance. Natural bipedal locomotion, they say, may encourage hand asymmetries, an unstable posture during testing may facilitate their detection, and an arboreal lifestyle may favor right-hemisphere (left-hand) processing. Nevertheless, we note that in the Rogers and Kaplan (1991) study many of the orangutans were right-handed, despite being a highly arboreal species.

Turning now to behaviors that are more complex than reaching for food, Westergard (1991) observed the tool using behavior of five New World capuchins (*Cebus apella*) and four macaques Old World (*Macaca silenus*). In both groups, all but one showed a significant hand preference, and of these all but one again preferred the left. The task was an artificial one designed to elicit behavior functionally analogous to the manufacture and use of tools (twigs and stems) by chimpanzees in the wild while foraging for termites. The task involved inserting a probe into a dipping dish, removing it, and feeding therefrom. The animals modified their implements with the coordinated actions of their hands and teeth.

In a more complex task, Fagot, Drea, and Wallen (1991) found a significant left-hand population bias in 21 out of 29 rhesus monkeys when solving a haptic discrimination task. The bias was strongest when the animals hung from a vertical three-point posture, or while sitting; it was weaker in a visual version of the task and was absent if the animals employed their usual tripedal posture. Unlike the findings of Ward's group, there was therefore no evidence that a tripedal posture facilitated asymmetric performance, although a vertical three-point posture certainly did so. Nor were there any of the sex differences often noted by Ward's group, although adults were more left-handed. As we saw, Ward and colleagues often noted a greater degree of lateralization in older individuals.

Similarly, Fagot, Drea, and Wallen (1991) and Fagot and Vauclair (1991) observe that lateralization may be stronger with higher-level or more complex tasks, although we have reviewed many successful demonstrations with highly practiced or simple behaviors. With both gorillas (Fagot and Vauclair 1988a) and baboons (Fagot and Vauclair 1988b) simple reaching was not associated with hand asymmetries, whereas in a visuo-spatial adjustment task where a sliding window had to be carefully aligned with an aperture to obtain food, strong left-hand superiorities were observed in both species. Subsequently, Vauclair and Fagot (in press) extended these findings and observations with some extra subjects (baboons), reiterating an important conclusion that hand asymmetries in nonhuman primates (and, we believe, in humans as well, see Bradshaw 1989) depend on postural constraints, the nature of the task and its demands, and levels of novelty, familiarity, and practice. Their claim that lateralization may be stronger with higher-level or more complex tasks is reminiscent of similar observations with humans, vis-á-vis deeper levels of processing (Moscovitch, Scullion, and Christie, 1976).

We noted earlier that Fischer, Meunier, and White (1982, and see Lockard 1984) found that all four female lowland gorillas (G. gorilla) usually held an infant with the left hand or arm during tripedal locomotion. Manning and Chamberlain (1990) extended these observations, noting that 14 out of 18 apes (10 chimpanzee, 4 gorilla, and 4 orangutan mother–infant pairs) exhibited an appreciable excess of left-sided cradling. Humans, too, cradle infants and books with the left hand (Saling and Kaplan-Solms 1989). If this is not just a strategy to leave the right forelimb free, we might have to conclude that a left-side cradling preference perhaps originated at least as early as the common ancestor of African apes and humans, 6–8 million years ago.

Manual Behavior in Humans

Do humans ever show a left-hand preference in reaching behaviors? There are repeated claims in the literature of a period in early infancy when the left hand is preferred for visually guided reaching (Bresard and Bresson 1987; McDonnell 1978, and see Humphrey and Humphrey 1987, and Liederman 1983, for review), giving way later, around 9 months, to a right-hand preference. Many earlier studies, for example, that of Seth (1973), have been criticized on statistical grounds (Young 1977) or have failed to include the necessary baseline control of the absence of a target; thus the left hand may simply move more overall.

Alternatively, an excess of left-hand responses in the neonatal period could stem from the asymmetric tonic neck reflex (ATNR); here, limbs tend to extend on the side toward which the child faces, with leftward turns being more effective in eliciting the ATNR than rightward turns, according to Liederman (1983). When Liederman attempted to control for these factors in infants between the ages of 3 and 10 weeks, he nevertheless found a *dextral* bias in arm extension, especially when the head was turned to the right rather than the left; moreover, he obtained greater right-arm activity even in the absence of stimulation. On the other hand, Carlson and Harris (cited by Harris 1983) found that female infants increased their use of the right hand between the ages of 4 and 9 months, whereas males showed a consistent left-hand preference. Indeed, Humphrey and Humphrey (1987) found that females between the ages of 5 and 8 months tended to use their right hands when unilaterally reaching for objects on the end of a rod, whereas males of the same age had no consistent hand preferences; by 9 to 12 months, both sexes preferred the right hand, suggesting that the left hemisphere of females may develop earlier than that of males. In fact, according to Harris and Carlson (1988) and Michel and Harkins (1985), most infants by 9 to 12 months prefer to reach with the right hand. Harris and Carlson go on to suggest that obligatory reaches across the midline might particularly reveal preemption by the dominant hand. When children 24 to 39 weeks of age were tested under such conditions, not only was there an overall preference for the right hand, but the right hand reached across the midline to the left side on 50% of the trials, while the left hand reached across the midline to the right side only 11% of the time.

It must be concluded that the weight of evidence supports a normal preference, certainly in adults, for reaching with the preferred right hand, although early in infancy there may be a transient left-hand preference. Indeed, even in *adults* there are reports of a *left*-hand *performance* superiority under certain conditions with single aiming movements; for example, when the movements are well practiced (Flowers 1975; Oldfield 1969; Provins 1967), or made in the absence of visual feedback (Guiard, Diaz, and Beaubaton 1983), or even in the presence of reaction time (RT) rather than movement time (MT) measures (Bradshaw, Bradshaw, and Nettleton 1990, Experiment 1; Haaland and Harrington 1989). Other similar findings have been reported by Carson et al. (1990): a left-hand RT superiority in aiming for targets when extinguishing the house lights; by Brown et al. (1989): the left hand was faster by RT, not MT, in aiming tasks; and by Goodale (1990b): whereas left hemisphere damage affects MT in such tasks, right hemisphere damage affects the RT component.

The left hemisphere has long been considered to control praxis and movement performance (Kimura 1982), whereas the right may mediate arousal and attention processes important in movement initiation (Heilman, Watson, and Valenstein 1985). On the other hand, robust *right*-hand superiorities typically appear in such serial paradigms as Fitts' reciprocal tapping task (1954) and Annett's pegboard task (1973), especially with more difficult movements or late in the critical, final decelera-tion phase (Annett et al. 1979; Goodale 1990b), or when errors must be minimized (Todor and Cisneros 1985), with MT measures (Bradshaw et al. 1990, Experiment 1; Goodale 1990b), and when advance information and memory load is maximal (Bradshaw et al. 1990, Experiment 2).

If one can summarize the human data, the right-hand superiority is more apparent where there are high accuracy demands involving fine corrective movements late in a sequence, often, but not necessarily, under visual control. The right hand may be more consistent in timing the application of varying forces. These considerations receive more detailed treatment by Elliott (1991).

Turning Biases: Human and Animal Studies

In a free-field situation, rats tend to turn (rotate) spontaneously (at night) or after drug treatment (in the daytime) in a preferred direction, either clockwise or counterclockwise (Glick, in press; Glick and Shapiro 1984, 1985; and see Chapter 3). These turning biases are closely related to forelimb preferences and their neurological bases may reflect chemical and pharmacological asymmetries of the dopaminergic nigrostriatal sys-tem. Unilateral damage to this structure typically leads to ipsiversive turning, and in normal animals there are asymmetries in dopamine concentrations, metabolites, release, uptake, and receptors. Indeed, d-amphetamine (a dopamine agonist) leads to increased rotation in a direc-tion contralateral to the side with the higher concentrations of dopamine. The left side of the globus pallidus (a related structure) is usually larger and correlates positively with a rightward turning tendency (the predom-inant direction), and negatively with a leftward tendency.

Bracha et al. (1987) further reviewed these endogenous asymmetries, for example, in dopaminergic function within the basal ganglia in rats and cats and their relationship to spontaneous and drug-induced circling. They note that there are similar left–right dopaminergic (and choliner-gic) asymmetries in the human globus pallidus. Kooistra and Heilman (1988) volumetrically measured the human globus pallidus on the two sides and found that the left side was larger in 16 out of 18 brains. This structure may be intimately concerned with determining the direction, amplitude, and velocity of movement (i.e., its parameterization). In light of the chemical and anatomical asymmetries of similar structures in the rat and their association with rotational direction and paw preference, it seems possible that a larger left globus pallidus in humans may determine the direction of motor dominance. The fact that the morphological asymmetries reported by Kooistra and Heilman were present before the age of 18 months suggests that they are not the consequence, but more likely the cause, of hand preference, although other factors, such as asymmetries in pyramidal decussation, are almost certainly also involved. Moreover, very early environmental factors may also play a role; thus in Chapter 2 we reviewed the effects of light in determining morphological asymmetry in the chick brain.

Corbin, Williams, and White (1987) reported a 21-year-old dextral male schizophrenic with tardive dystonia who tended to turn in a clockwise (rightward) direction while walking, and speculated on the role of asymmetries in dopamine concentrations which are thought to occur in schizophrenia. Indeed, Crawley et al. (1986) noted a pronounced lateralization in the density of dopaminergic D2 receptors in some schizophrenics, but not in controls; the striatal levels on the right side were much higher. Such findings, however, are disputed by Reynolds et al. (1987), who found no evidence of asymmetry in dopamine (or any other neurotransmitter markers) in the striatal region of schizophrenics, although they *did* find that dopamine levels were higher in the left amygdala of schizophrenics.

Reynolds et al. found in normal subjects at postmortem that there was a significantly raised dopamine receptor density in the right putamen. (In schizophrenics there was an even greater right preponderance possibly as a consequence of neuroleptic treatment.) These findings contrast with the higher levels found on the *left* in normal rats (Schneider, Murphy, and Coons 1982).

Do normal (i.e., nonschizophrenic) humans show turning biases analogous to those evidenced in rats and schizophrenics? Bracha et al. (1987) monitored, in natural circumstances, human rotational movements during the working day without their subjects' awareness. As has been previously observed, females had the higher average rates of rotation; males who were right dominant for hand, foot, and eye rotated more to the right than to the left, whereas similarly right dominant-females showed the reverse effect, as did mixed-dominant males. The few mixed-dominant females had no clear bias. Glick (in press) found that both male and female children between the years of 7 and 9 exhibited predominantly left turning biases. (Glick also presented findings of hemi-Parkinsonian patients rotating toward the hemisphere that has less dopaminergic activity; that is, a left-sided rotation in patients with right-sided symptoms, and a left-sided rotational preference in all 10 schizophrenics studied.)

On the other hand, Bradshaw and Bradshaw (1988), instead of using free-field monitoring, got left- and right-handed males and females, under conditions of reduced sensory input, to deliberately rotate exactly two complete 360-degree turns clockwise or counterclockwise, or to walk in a straight line. In the first task, dextrals, especially females (again!), showed a significant rightward bias, while sinistrals showed a nonsignificant leftward tendency. While trying to walk, blindfolded, in a straight line, all four groups deviated to the right, although again more so for females and dextrals. It is not clear why the different studies (and sexes of subject) should produce such conflicting findings, although paradigmatic differences (deliberate versus unconscious turning) might be partly responsible.

While a rightward turning bias may be apparent in rats and possibly in humans, why is it that according to informal observation we tend to walk around enclosed areas (halls, exhibitions, and ships' decks) in a counterclockwise (leftward) direction? Ridgeway (1986) notes that dolphins, too, have a marked preference in tanks and aquariums for counterclockwise swimming, and to favor the right eye (which would result in a leftward deviation on approaching an object). Their echolocation beam is also slightly offset to the left, which could result in leftward swimming, though in all the toothed whales there are marked cranial asymmetries, with the nasal opening shifted to the left of the midline and with several skull structures larger on the right side. A *behavioral* explanation for human and dolphin biases for leftward turning within enclosed spaces, while having a *rightward* turning tendency in the free field, would speculate that on entering such an enclosed area with a rightward turn, the subject encounters a barrier to the right and thus has to follow it around in a counterclockwise direction.

Just as in a previous section we discussed asymmetries in reaching, we note here that a right bias has been found for the leading limb in initiating the human neonatal stepping reflex (Melekian 1981; Peters and Petrie 1979). We cannot make clear statements about the corresponding situation in adults (although compare Seltzer, Forsythe, and Ward, 1990) since experience in marching (where stepping off with the left leg is often obligatory) may confound matters.

A Critique of the Evolutionary Hypothesis of MacNeilage et al. (1987)

So far, we have considered primate asymmetries at the motor level, that of hand performance and preference asymmetries. The natural context is that of Ward and her group: a left-hemisphere specialization for whole-body postural organization in prosimians, and a right-hemisphere specialization for left-hand preference in reaching for food. As such, this underlies the postural-origins theory of MacNeilage et al. (1987) previously reviewed. It should be noted, however, that subsequently these authors (MacNeilage et al. 1988) shifted ground somewhat in a revised approach. Left-hand reaching preferences in prosimians, and the likelihood that the earliest prosimians were vertical clingers and leapers, led to two complementary specializations: a left-hand (right-hemisphere) *perceptuo*motor specialization for unimanual predation with *visual, spatial,* and probably *auditory* perceptual components (italics added), and a left-hemisphere specialization for whole-body postural organization, particularly involving the right side of the body. Later, the right side of the body "became the operative side for forceful interactions with the environment associated with invasive foraging," with the advent of quadrupedalism, which thereby freed the upper limb from the postural demands of vertical clinging. (In many ways this is a curious inversion of the argument, to be considered later, that the advent of bipedalism in hominids might have freed up the hands for tool using behavior.) Thus the *left* side undertook support, while the right hand became favored for manipulative and practiced acts. Finally, according to this revised scenario which now attempts to incorporate cognitive asymmetries (to be reviewed next) into the general scheme, the left hemisphere specialization for whole-body postural organization becomes associated with a left-hemisphere specialization for vocal communication and language.

There are however still some major problems with such a scenario. As Peters (1988) observes, not enough emphasis is placed upon the mouth, which in nonhuman primates plays a role equivalent to that of the preferred hand, while the two hands support the mouth in a manner comparable to how the *nonpreferred* hand in humans supports the preferred hand. (As we saw earlier, the *foot* in parrots plays the same role as the hand in nonhuman primates: the foot holds and positions the food so that the beak can manipulate it. It is the beak, not the foot, which is the parrot's primary agent of manipulation, such that the parrot's foot equates with our left hand.) Thus in nonhuman primates, the hands are subordinate to the mouth. Peters speculates on the possibility, therefore, of a lateral specialization for jaw control in nonhuman primates, linking with the left hemisphere, so that both the right hand and articulatory oral control eventually come under the left hemisphere in humans (compare Kimura 1982, and our earlier discussion of left hemisphere mediation of praxis).

Indeed, as MacNeilage et al. (1988) observed, if primate hands are subordinate to the left-hemisphere driven mouth, one would predict a right-hand preference in hand–mouth interactions. They note that this was reported in yellow baboons by Post, Hausfater, and McCuskey

(1980); it may also be recalled that we reviewed similar right hand involvement in transporting food to the mouth by prosimians in several of
Ward's studies, and in some but certainly not all recent reports of observations with great apes. However, as we shall see in the next section, the
main objection to the theory of MacNeilage et al. (that hand differences of
whatever sort in early prosimians drove the later evolution of hominid
and human lateral asymmetries) is the clear evidence of widespread
perceptual and cognitive asymmetries in apes, monkeys, birds, and rodents; hemispheric specialization must precede handedness, and indeed
hand preferences in humans, apes, monkeys, and prosimians may be
epiphenomenal to a more fundamental underlying asymmetry.

There seems little doubt now, that the manual behaviors of nonhuman
primates are lateralized, despite the fact that many papers with negative
findings may never get published and many of the published papers
reporting significant hand differences also include in the data unaccept-
ably high numbers of ambilateral individuals. Indeed, in a meta-analysis
of all available published sources for *Pan, Gorilla, Pongo,* and *Hylobates,*
Marchant and McGrew (1991) conclude that most studies are seriously
deficient and are intrinsically unlikely to demonstrate the expected effects. Most studies have focused on hand rather than foot asymmetries
(despite the importance in apes of the lower limb for holding and climbing), on captive rather than wild individuals, on immature individuals
(when older animals tend to be more lateralized), on small sample sizes,
on a small sample of trials and tasks, on simple rather than the probably
more lateralized complex tasks, on spontaneous and unskilled rather
than induced or skilled responses, and on contrived and artificial situations. They note that the typical study of functional laterality in apes is
only of hand preference in captive chimpanzees based on a few immature
individuals performing a few trials on a single task which is simple and
consists of one-handed, nonsequential gross movements, with little use
of sex as an independent variable. There is typically hardly any distinction between performance and preference asymmetries—hardly an opti-
mistic picture

Another argument against the evolutionary primacy of hand asymmetries in primate evolution, is the fact that more human individuals are
left-hemisphere dominant for language than are right-handed (Bradshaw
1989), indicating that handedness is probably secondary rather than
primary to language. Indeed, there are yet other cognitive asymmetries
in *humans* (quite apart from apes) which are not dependent on language
(Bradshaw and Nettleton 1981). In any case, the dissociation of hand
preference (and brain lateralization for praxis) from language lateralization is a further indication that handedness is an epiphenomenon of

brain asymmetry (Witelson 1988), and not as MacNeilage et al. would have it, the precipitating cause. We must look instead to asymmetries of a perceptual or cognitive nature for a prior cause.

Summary and Conclusions

Humans typically show a right-hand superiority in tasks which entail high accuracy demands with fine corrective movements, frequently, but perhaps not necessarily, under visual control. The right hand may possibly be more consistent in timing the application of varying forces. It is, of course, usually stronger, although this may simply reflect greater levels of use. The same explanation can partly, but not completely, explain morphological asymmetries that favor the bones of the human (and even rhesus monkey) forelimb. Other relevant mechanisms include corticospinal asymmetries and asymmetries in dopamine concentrations in the basal ganglia. Thus not only are there more corticospinal fibers leading to the right side of the body, but the crossing from left to right occurs in humans at a higher level than from right to left; the situation with nonhuman primates is unknown. Similarly, there are greater concentrations of the neurotransmitter dopamine, on the side contralateral to the preferred hand, in the basal ganglia, a structure known to mediate movement control and turning. Indeed, both humans and rats may exhibit free-field rightward turning biases, even though in an enclosed space informal observations of humans, and studies of dolphins, suggest counterclockwise turning. Nevertheless, even this phenomenon might stem from an initial rightward turn on entering the arena (or aquarium), thereafter constraining leftward turns upon encountering successive walls.

Primate studies are hampered by poor control of age, sex, practice levels, task difficulty, artificial versus natural behaviors, wild versus captive animals, uni- versus bimanual tasks, uni- versus multicomponent tasks, criterion levels for lateralization of individual subjects, number of subjects, number of trials, number of tasks, performance versus preference measures, and so on. That said, the hotly debated evolutionary hypothesis of MacNeilage et al. (1987, 1991) offers the best explanatory and predictive synthesis to date, although Ward and her colleagues have adapted various aspects of it. According to this account, in ancestral prosimians a left-hand preference for visually guided reaching is accompanied by (and may stem from) a specialization of the right upper limb for postural support. With higher primates, the change from vertical clinging to a more quadrupedal posture and a decreased reliance on unimanual predation has resulted in previous postural specializations evolving into a right-hand superiority for fine, sequential manipulative movements.

It is certainly true that a huge number of studies, many but not all by Ward's group, find left-hand preferences in prey catching or food retrieval, especially from an unstable posture, in prosimians and certain species of New World monkeys. The left hand may also be preferred for haptic judgements or visuospatial discriminations in rhesus monkeys, baboons, and gorillas. Like ourselves, the great apes cradle infants with the left arm, thereby leaving the right hand free.

The right hand is superior (in a few studies) and preferred (in many) in more complex, manipulative responses by macaques and chimpanzees. The great apes may also prefer the right hand in chest beating, grasping, and food reaching, and may lead off with the right lower limb in terrestrial walking. However, sex and age should not be ignored in drawing these conclusions. Indeed Milliken et al. (1991), in a recent and major study of the small-eared bushbaby, *Otolemur garnettii*, confirmed tendencies observed earlier with other prosimians and found that adult females showed a strong right-hand preference in food-related behaviors, and in adult males a left bias, with less consistent lateralization in juveniles. Any evolutionary model will ultimately have to accommodate such observations, together with evidence of a greater incidence of left-handedness in human males (Bradshaw 1989). It is difficult not to invoke a role for a lateralized effect of (male) sex hormones (Geschwind and Galaburda 1987), although other effects of environmental stimulation must also be considered in light of the ample evidence for such effects in determining lateralization in chicks and rats (see Chapters 2 and 3).

In the next chapter we address the question of cognitive asymmetries.

Further Reading

Bradshaw, J.L. Animal asymmetry and human heredity: Dextrality, tool use and language in evolution—10 years after Walker (1980). *British Journal of Psychology*, 1991, 82:39—59.

Fagot, J., and Vauclair, J. Manual laterality in nonhuman primates: A distribution between handedness and manual specialization. *Psychological Bulletin*, 1991, 109:76—89.

MacNeilage, P.F.; Studdert-Kennedy, M.G.; and Lindblom, B. Primate handedness reconsidered. *Behavioral and Brain Sciences*, 1987, 10:247—303.

Marchant, L.F. and McGrew, W.C. Laterality of function in apes: A meta-analysis of methods. *Journal of Human Evolution*, 1991, 21:425—438.

Ward, J.P. (Ed.) Current Behavioral Evidence of Primate Asymmetry. New York: Springer, 1992.

Asymmetries in Primates: Cognitive Asymmetries and Asymmetries in Brain Morphology

Introduction

Studies of behavioral asymmetries in humans have relied on three major lines of evidence: observation of clinical patients with brain injuries or disease that is lateralized to one or other hemisphere, studies of the behavior of patients whose commissures have been divided for the relief of otherwise intractable epilepsy, and finally experiments with normal human subjects. Animal work, in comparison, has both advantages and disadvantages compared to the human situation. It is obviously far easier to place a discrete lesion exactly where the experiment demands it in animal subjects; with humans, except under rare or chance circumstances, we have to await the event of suitable unfortunate cases. On the other hand, considerable effort and ingenuity may be needed to devise experimental techniques and behaviors in animals that can even approximate those that can be performed with, and by, human subjects. In this chapter we shall encounter some instances of just such ingenuity, although unfortunately the resultant tradeoff is the very small number of available subjects so that findings may not generalize to a wider population.

There are numerous examples of animal behavioral asymmetries which seem to parallel the human situation; we must be wary of ascribing common underlying mechanisms because there is always a 50% chance of the same effects occurring. Moreover, analogous rather than homologous mechanisms might be responsible. Finally, it is tempting, though perhaps dangerous, to explain away *discrepancies,* especially effects in the *reverse* direction, as being due to differences in strategy, experience, and so on. However, the absence of language in nonhuman primates is a factor which cannot be ignored. Indeed, we shall see that the Hopkins and Morris group have taken advantage of the recent availability of "language-trained" chimpanzees to ask whether this variable may influence asymmetries in a variety of tasks.

In the second half of this chapter we shall review the evidence for morphological asymmetries (of extent and sulcal), comparable or otherwise to our own, in nonhuman primates (New and Old World monkeys, and apes) and fossil hominids. Again, we would caution against inferring a necessary causal link between structure and function; it will be seen, however, that while precursors (preadaptations) to our own periSylvian (i.e., speech related) asymmetries may be apparent with modern measuring techniques, even in New World genera, which have long been isolated from the main line of hominid evolution, it is only with fossil *Homo* that we find close precursors to our own situation. A connection with evolving communicatory behaviors seems almost inescapable.

Tactual Studies in Nonhuman Primates

We have already noted in the last chapter, under left hand superiorities, the study of Hörster and Ettlinger (1985): 78 rhesus monkeys (*Macaca mulatta*) spontaneously using the left hand outperformed 77 which chose to use the right hand in acquiring a tactual discrimination task. This task involved discriminating a cylinder from a sphere, and the measure was the number of trials to criterion. Thus the right hemisphere, as with humans (Bradshaw 1989) seemed to be the more proficient. Alternatively, as the authors observed, tactual information primarily reaching the right hemisphere may more readily gain access to the opposite hemisphere than can inflow primarily reaching the left hemisphere. We also noted in the last chapter a significant left-hand population bias in rhesus monkeys solving a haptic discrimination task (Fagot, Drea, and Wallen 1991).

Visual Studies in Nonhuman Primates

We also described in the last chapter left hand superiorities by gorillas and baboons in a visuospatial adjustment task, where a sliding window had to be carefully aligned with an aperture to obtain food (Fagot and Vauclair 1988a,b). However for some reason, most of the visual studies on possible lateralization in primates have focused on face processing in one form or another. Face processing is, of course, for us second only to language processing in terms of complexity, and perhaps for that reason it has been (at least partly) assigned to temporo-parietal regions of the minor hemisphere opposite to and equivalent to speech areas in the left hemisphere. Face processing in nonhuman primates, given their reduced communicative ability compared to our own, may well surpass in complexity that which is required by their call systems. In any case, we not only identify individuals primarily by the face—and the same may well be true of apes and monkeys—but we communicate emotions even more powerfully via the face than via the body or limbs.

There is little doubt that the output side of human speech is predominantly mediated by the left hemisphere, as evidenced by the aphasiological literature (see e.g., Benson and Geschwind 1985; Kertesz 1985; Ojemann 1983). However, this is not necessarily true of all aspects of language; there may be major contributions from the right hemisphere with respect to the connotative aspects of comprehension; for example, humor (Bihrle, Brownell, and Gardner 1988), metaphor (Gardner et al. 1983), imagery and concreteness (Chiarello, Senehi, and Nuding 1987; Lambert 1982a,b; McMullen and Bryden 1987), certain aspects of reading (Coltheart 1983, 1985), emotion and affect (Ross, 1985), and so on (Code, 1987; Searleman 1983; Zaidel and Schweiger 1984, although see Gazzaniga and Smylie 1984; Patterson and Besner 1984a,b).

The idea that our right hemispheres are specialized for the processing of faces in a complementary fashion to left-hemisphere language processing has had a similarly checkered history. Prosopagnosia, clinically a most interesting syndrome where there is an acquired inability (through brain trauma) to recognize faces, was originally associated with posterior right hemisphere damage. After a period during which the need for bilateral injury was stressed, with each hemisphere perhaps differentially specialized for different aspects of face processing, the pendulum of opinion seems now to be swinging back to a major right hemisphere involvement in prosopagnosia proper (although see Damasio and Damasio 1986).

We have an amazing capacity to recognize perhaps thousands of faces, often after only a single exposure, despite the "paraphernalia" of hats, spectacles, and beards, the changes wrought by aging, and the aging of our own memories. Face recognition depends on both whole-face "gestalt" recognition, where the right hemisphere may play a major role, and piecemeal feature-by-feature analysis, which probably appeals to left hemisphere mechanisms. A prosopagnosic cannot recognize a familiar face, even his or her own, although it is seen *as* a face. Any subsequently successful recognition may depend on other sensory modalities (e.g., sound of voice), or distinctive features (presence of a mole in a characteristic location). The syndrome is quite separate from any inability to discriminate between *unfamiliar* faces, and may not in fact even be restricted to faces. It may include other potentially ambiguous generic stimuli which require context-related individuation; there are therefore problems in recognizing individual members which closely resemble each other (like faces) in a particular category, and where successful recognition depends on association and context of an "episodic" (i.e., arbitrary) nature. Thus farmers, car enthusiasts, tailors, and birdwatchers have recently been reported as displaying prosopagnosia-like disabilities in recognizing individual cows, cars, clothes, or birds, respectively (Damasio, Damasio, and Tranel 1986). Other generic stimuli of a potentially ambiguous nature which prosopagnosics may experience difficulty with include food, animals, abstract symbols, coins, handwriting, furniture (e.g., chair types), playing cards, and personal possessions.

While this seems to indicate a problem of evoking the specific historical context of a potentially ambiguous stimulus, it is not just a matter of episodic memory failure per se; there is no confusion between one's own and anothers' possessions, and cars are misrecognized by make rather than by owner. Nor do pure prosopagnosics usually have any accompanying deficit in other visuospatial capacities; e.g., discriminating unfamiliar faces, figure–ground segregation, stereopsis, and recognition of objects from noncanonical (i.e., nonstandard) directions or when unevenly illuminated or incompletely presented. They usually do not experience problems of spatial localization or unilateral neglect, although they do typically also suffer from a color deficit.

Debate continues about the lateralization of prosopagnosia. The early studies finding a right hemisphere association tended to employ discrimination of unfamiliar faces in an attempt to equalize task difficulty between patients, and therefore were not strictly testing for prosopagnosia (the failure to recognize *familiar* faces). While the need for a bilateral lesion is frequently emphasized (Damasio and Damasio 1986), the right hemisphere component nevertheless seems to be consistently occipitotemporal, whereas the left hemisphere component is usually quite variable. One possibility is that bilateral lesions are required because, with unilateral right hemisphere lesions, left hemisphere feature-analytic strategies may suffice. Indeed, prosopagnosics *do* tend to rely on certain distinctive features in a manner reminiscent of left-hemisphere processing.

Thus there may be an asymmetric spatial distribution and temporal occurrence of bilateral involvement. In fact, prosopagnosia may be more common when right hemisphere lesions follow rather than precede left hemisphere injury, and a nonpaired right hemisphere injury (i.e., one without any accompanying damage to the left hemisphere) may more commonly lead to prosopagnosia, even if transient, than a nonpaired left-hemisphere lesion. We may therefore rely more on right-hemisphere mechanisms, especially perhaps for the important configurational aspects of faces, so that a previous left-hemisphere lesion would be relatively silent until the occurrence of a second right-hemisphere lesion. We can now address the issue of the lateralization of face processing in nonhuman primates.

Overman and Doty (1982) presented L ⅃ and Я R face composites to monkeys (*Macaca nemestrina*, N = 6) and humans for matching to a normal original. Humans, as we saw, usually select L ⅃ photographic composites (where L refers to the side of the photographic image, not the owner's face) as apparently being more similar to the original than Я R composites, and this effect was confirmed with humans by Overman and Doty. No asymmetries in performance were obtained with monkey face composites, and monkeys showed no asymmetries with either human or monkey faces. The task, however, was hardly a naturalistic one, and other approaches are more likely to succeed.

Hamilton and Vermeire (1983) severed the optic chiasm and forebrain commissures of 18 rhesus monkeys (*Macaca mulatta*). When visual information is presented to either eye of such split-brain animals, it is transmitted only to the ipsilateral (same side) hemisphere, and so either hemisphere can be separately tested by restricting input to one eye, for example, occluding one and testing via the other (see Figure 6.1). In Hamilton and Vermeire's study, hand preferences and sex were approximately matched and the animals were tested with each hemisphere in speed of learning to discriminate photographs of unfamiliar faces of other monkeys with varying expressions. Overall, neither hemisphere showed an advantage, nor was there any differential effect from the hemisphere ipsilateral or contralateral to the preferred hand. (In a previous, similarly unsuccessful, laterality study involving visual discriminations which were based on sequential presentations of stimuli that varied in color and form, Hamilton and Vermeire (1982) did find a significant correlation between each subject's preferred hand and the hemisphere which most rapidly learned the task.) A fault in the experimental design was the inclusion in the stimulus set, of faces drawn from several species of macaque. In their later studies, Hamilton and Vermeire *did* obtain asymmetric performance when monkeys were presented with face stimuli, and, as with humans, it was usually the *right* hemisphere which predominated.

Thus Ifune, Vermeire, and Hamilton (1984), in a carefully conducted study, showed colored videotape recordings of monkeys, people, animals, and scenery to eight split-brain rhesus monkeys (*M. mulatta*),

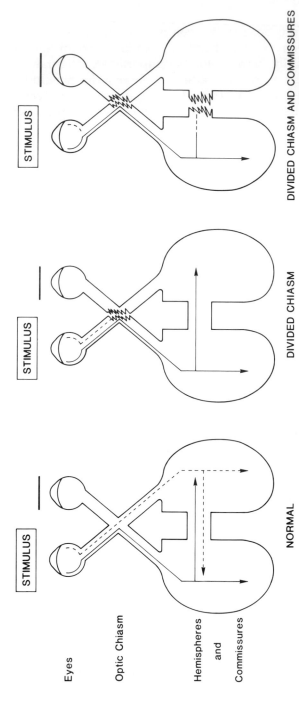

Figure 6.1 Flow of visual information, with monocular stimulation, in the normal intact animal, in one with a divided chiasm, and in one where both chiasm and commissures are divided. In the first two preparations, both hemispheres can receive appropriate information from a single eye, either directly (Normal) or via the corpus callosum (Divided Chiasm). With division of both chiasm and commissures, only the ipsilateral hemisphere can receive input from a single eye. From J. L. Bradshaw, Hemispheric Specialization and Psychological Function. Chichester: Wiley, 1989, with permission.

balanced for sex, preoperative hand preferences when reaching for food, and operative procedure. The number of facial expressions (lip smacking, which is a form of greeting, appeasement, or submission, and fear grimacing, threatening, and yawning, which occur in tense or stressful situations) elicited from the right hemisphere was significantly greater than when the left hemisphere was stimulated. The monkeys also tended to view longer via the right eye–right hemisphere system.

In the first of three papers which demonstrated a double dissociation, that is, complementary and opposite specializations by the two hemispheres for different kinds of stimuli—always the strongest evidence of cerebral lateralization (Bradshaw 1989; Chapter 5)—Hamilton and Vermeire (1985) again tested split-brain monkeys. They obtained a significant *left* hemisphere advantage when the subjects discriminated between pairs of lines differing in slope by 15°; for example, 15° and 30°, 60° and 75°, 105° and 120°, 150°, and 165°. Only one line appeared at a time so that a memory component was involved. Indeed, Mehta and Newcombe (1991) found that in humans damage to the left hemisphere was more disruptive of this capacity than right hemisphere damage. When Hamilton and Vermeire instead presented photographs of monkey faces for discrimination, a powerful right hemisphere superiority emerged, which was especially strong under conditions of retention. In 12 animals that could be tested on *both* tasks, 8 showed the predicted opposite specializations for line slope (left hemisphere) and faces (right hemisphere), and 11 showed *relative* effects in the expected directions; that is, a greater right-hemisphere superiority for faces rather than for lines, or a greater left-hemisphere superiority for lines over faces.

In a later paper, Hamilton and Vermeire (1988a) provide more details and indicate that no asymmetries emerged when the animals discriminated between geometric patterns instead of faces. The right-hemisphere superiority for faces persisted on retesting six months later with both old and new face stimuli, indicating that processing involved identities and expressions, rather than the presence or absence of irrelevant details.

Finally, Hamilton and Vermeire (1988b) reported on hand asymmetries in these same monkeys, which had otherwise shown complementary hemispheric specialization at cognitive levels. They were tested on reaching for food, reaching into a bottle for food, and opening a matchbox by pulling a tab and retrieving food. This last task involved bimanual manipulation and reaching. There was no evidence of a left-hand preference for any of the reaching behaviors, simple or complex, or of a right-hand preference for manipulation, contrary to the predictions of MacNeilage et al. (1987) or of Ward (see, e.g., Milliken et al. 1989), even though at an individual level, subjects tended to use opposite hands to reach and to retrieve. Hamilton and Vermeire speculate that the monkeys may have been too young (they were juveniles, wild caught, and experimentally naive on testing), or that the tasks were insufficiently complex or did not require enough postural readjustment. Another possibility (see Chapter 3), is that lack of expected hand preferences stemmed from the fact that the animals were split-brain.

Hamilton and Vermeire (1991) review their earlier findings and make these additional observations. While monkeys readily learned to discriminate inverted faces, this was achieved, as with humans, without evidence of hemispheric asymmetry, as if the stimuli were now treated simply as geometric patterns. Secondly, even though 16 out of 25 monkeys showed complementary specialization (the left hemisphere showing superior ability for processing line orientation, the right for faces), the complementarity, as judged by the negligable correlations between the two indices of dominance, appeared to be statistical (i.e., independent) rather than causal (see e.g., Bryden 1982, for similar findings with humans). Thus the lateralization of certain functions in one hemisphere appears not to force complementary functions to be lateralized in the other. Finally, and again as with humans (Moscovitch 1979), cognitive mechanisms deeper in the processing hierarchy of each hemisphere appeared to be more clearly lateralized than mechanisms based on more immediate, shallow, peripheral, or perceptual processes.

If there is a right-hemisphere mediation of face processing, which structures may be involved? Yamane, Kaji and Kawano (1988) recorded data from single cortical neurons in the inferotemporal gyrus (ITG) of monkeys trained to discriminate three selected human faces from a large number of different (foil, distracter) faces. They identified neurons which did not respond to nonface stimuli, but which did respond to faces. These face neurons responded to the configurational information contributed by the combination of distances between facial elements or features, for example, eyes, mouth, eyebrows, and hair.

Other evidence of cells that are responsive to faces and to parts of faces (eyes, mouth, hair, gaze direction, distance between eyes or between eyes and mouth) comes from Harries and Perrett (1991) and Simone (1991) who used real or plastic faces, photographs, and line drawings. In comparison, Baylis, Rolls, and Leonard (1985) identified neurons with responses selective to different faces which were located in the middle and anterior part of the *superior* temporal sulcus (STS); information about a particular face appeared to be present across an ensemble of neurons, rather than just one. One possibility is that there are two poulations of cells responsive to faces, one group responding more to expression (superior temporal) and the other to personal identity (inferior temporal).

Perrett et al. (1988) also noted that such face-sensitive neurons in the STS were commoner in the upper bank and fundus of the structure in the *left* hemisphere; fewer cells responded on the right side. These cells processed upright faces faster than inverted ones—humans, too, can make discriminations between upright faces faster than when they are inverted and the configurational aspects are thereby obscured (Diamond and Carey 1986). The authors identified cells which were differentially selective for the sight of different faces. While the species of monkey is not reported, four out of five showed bigger effects with cells in the left than in the right hemisphere. However, it is noteworthy that the authors claim that their tasks encouraged piecemeal (feature analytic) processing strategies rather than the configurational approach required by, for example, Yamane et al. (1988). There is abundant evidence in humans that a piecemeal strategy of processing faces can bring about right-field (left-hemisphere) superiorities, whereas the imposition of a configurational approach, with the same stimuli and subjects, will induce a left-field (right-hemisphere) superiority (Bradshaw and Sherlock 1982; Patterson and Bradshaw 1975).

Other evidence of a left-hemisphere superiority in a visuospatial task, which initially might be expected to appeal to right hemisphere processors, comes from Jason, Cowey, and Weiskrantz (1984). Nine male rhesus monkeys (*M. mulatta*) were trained in discriminating the relative vertical position of a dot on a square background. Four then received an occipital lobectomy of the left hemisphere together with a splenial transection to isolate that hemisphere from geniculo-striate visual input from the other side; the other five animals instead received a right occipital lobectomy together with splenial transection. Performance thresholds were determined both before and after surgery, and percent change in thresholds was calculated. Left-lesioned animals were all worse affected than those lesioned on the right. This task is typically more impaired in humans after right rather than left posterior damage (Taylor and Warrington 1973), suggesting that asymmetries for processing dot position in monkeys may be analogous rather than homologous to our own lateralization in this regard.

In a study using only two chimpanzees (*Pan troglodytes*), Hopkins and Morris (1989) investigated lateralization for visuospatial processing, with presentations to one or other visual field in intact, language-trained animals. They were first taught to manipulate a joystick controlling the movement of a cursor on a computer monitor to a central fixation point; this procedure trained them on holding central fixation. In the experimental task, they were then taught a visual discrimination task based on the location of a short line contained within a geometrical form. If a short line lay above a longer one, the joystick was to be moved upwards with the index finger and downwards if the relative positions of the two lines were reversed. Each hand was separately tested. Both subjects showed a left-field bias, by reaction time, with a range of exposure durations between 15 and 226 msec.

The Hopkins and Morris group has, in fact, recently pioneered a number of studies with two and sometimes three language-trained chimpanzees. They found that left hemisphere activation occurred in both animals when there was priming with meaningful symbols used for interspecies communication, whereas *nonmeaningful* symbols produced bilateral activation (Morris et al., in preparation). Morris and Hopkins (unpublished) generated chimeric-composite human faces as described in Chapter 1 and using human subjects; faces which were happy on one side and sad on the other. Two of the chimpanzees significantly (and one nonsignificantly) chose as the happier face the face where the smiling half fell in the left visual field, exactly like humans. Overall, this asymmetrical pattern occurred on 80% of the test sessions, the effects across the three animals were in the same order of relative magnitude as that found for right-hand preferences in various handedness tasks. All these and other studies are reviewed by Morris et al. (1992). They note that despite the small number of animals, the manual and cognitive asymmetries closely resemble our own; it is unclear whether language training is in any way responsible. The effects could of course simply stem from the enriched social environment, or could be present in (untested) animals that have not been exposed to language-like behaviors.

Hopkins, Washburn, and Rumbaugh (1990) presented complex visual forms unilaterally to ten humans, three chimpanzees, and two rhesus monkeys in a matching-to-sample task where a joystick had to be moved, and both accuracy and reaction time measures were obtained. The three chimpanzees initially showed a significant left-field (right-hemisphere) superiority in accuracy, which declined over time and gradually reversed, whereas the humans and monkeys failed to show such asymmetries. The reaction time data indicated a right-field (left-hemisphere) superiority for the two monkeys; with the chimpanzees, this measure again revealed an early left-field (right-hemisphere) superiority which later reversed. The humans showed a left-field (right-hemisphere) superiority only in the third of five blocks of trials, and only when the right hand was used. Inspection of their data reveals at best a confused pattern, their subject numbers were small, and the report is of interest more for the sophisticated techniques used and the attempt to test human and nonhuman subjects in a comparable task. Otherwise it is premature to draw too many conclusions from this study.

Similar comments apply to Hopkins, Morris, and Savage-Rumbaugh's (1991) study: Two language-trained chimpanzees were laterally presented with prime or warning stimuli (lexigrams, i.e., geometrical forms representing words in the chimpanzee's acquired lexicon) which were either meaningful, familiar, or meaningless. After an interstimulus interval, the command signal appeared, at which point a manual reaction-time response was required. Priming with *meaningful* warning stimuli resulted in a right-field (left-hemisphere) superiority, whether the various primes were presented in blocks of trials or randomly intermingled. No asymmetries were found for familiar or meaningless warning stimuli. While the parallels with the lateralization of language in humans and the effects of priming and activation are intriguing, again the findings are limited to two subjects, despite the ingenuity of the procedures used.

Auditory Lateralization in Nonhuman Primates

Studies of laterality in humans have concentrated more on the visual and auditory modalities than the tactual. However, methodologies in the auditory and tactual modalities have more in common with each other than either has with vision, and have been approached from similar procedural and theoretical standpoints, for example, use of competing simultaneous stimulation of the two ears (dichotic) or hands (dichaptic). In animal studies, monaural stimulation has been preferred, probably for practical reasons. With humans, dichotic presentation is still the norm, due largely to Kimura's (1967) hypothesis. According to her, one hemisphere, usually the left, is specialized for speech and language and auditory inputs are more strongly represented in the contralateral hemisphere. Dichotic stimulation initially lateralizes the input, as ipsilateral information is partially or completely suppressed by the contralateral input. Information from the nondominant ear or hemisphere is transferred across the forebrain commissures, particularly the corpus callosum, to the speech-dominant hemisphere for processing. There it meets, and perhaps competes with, direct, contralateral right ear inputs. Ear differences, therefore, may, according to this model, reflect information loss during interhemispheric transfer to a solely and uniquely specialized hemisphere. However, the findings of often quite powerful ear asymmetries, even with monaural presentations under certain circumstances (Bradshaw 1989; Bradshaw and Nettleton 1988), have led to a rejection of the theory as a complete explanation.

Dewson (1977) forced seven cynomolgus macaques (*Macaca fascicularis*) to form conditioned associations between acoustic and visual stimuli in an operant nonspatial auditory discrimination task, pressing a red panel to a pure tone and a green one to noise; there was a short-term memory delay between the termination of the auditory signal and the illumination of the manipulanda panels. Four animals with lesions in the left superior temporal gyrus (corresponding to Wernicke's receptive speech area in humans) performed very poorly, while two with lesions in the right superior temporal cortex, and one with damage to the primary auditory area of the left hemisphere, showed no defects. These differential laterality effects only occurred when delays between the auditory signal and panel illumination were of variable rather than of constant duration, suggesting that something even beyond short-term-memory storage was required. The small number of subjects should be noted in this early and pioneering study.

Again, only four subjects (baboons, *Papio cynocephalus*) were used by Pohl (1984) in an auditory gap-detection task requiring the resolution of brief silent intervals within bursts of noise. Such temporal resolution was hoped to be analogous to the processing required when humans discriminate consonant–vowel syllables which differ in their temporal features, when a right ear advantage is typically obtained. In fact, the same four animals were also used earlier (Pohl 1983) in discriminating pure tones, three-tone musical chords, vowels and consonant–vowel syllables. Looking across the two experiments, the same three animals showed a clear left-ear advantage (the other gave a clear right-ear advantage) to the chords and to the auditory gap-detection task. We shall return shortly to the results for vowels and consonant–vowel syllables.

In a comparatively early study, in the context of a monkey's ability to determine species-specific vocalizations, Hupfer, Jürgens, and Ploog (1977) studied the effect of lesions in the auditory cortex of New World squirrel monkeys (*Saimiri sciureus*). Eleven were trained to discriminate species-specific calls from other complex sounds; seven then received bilateral lesions to the auditory cortex. Small lesions within the superior temporal gyrus failed to affect performance, whereas lesions destroying three quarters of the auditory cortex had a detrimental effect. Lesions destroying the entire auditory cortex completely abolished performance.

Turning to laterality studies, Petersen et al. (1978; and see also Beecher et al. 1979, for more details) found that all five Japanese macaques (*M. fuscata*) showed a significant right-ear advantage in monaurally discriminating between two classes of species-specific communicative "coo" sounds. These were the smooth early high (SE) and smooth late high (SL) coo sounds which differ in the relative temporal positions of the peak frequency attained. Four comparison monkeys, which were not

Japanese macaques, were also tested, but only one showed any evidence of an ear asymmetry. Because it was not clear that all animals (i.e., including the controls) were in fact attending to the peak position cue, in a later study, Petersen et al. (1984) repeated the experiment with two Japanese macaques and two comparison animals of different species (*M. nemestrina* and *M. radiata*); again only the two Japanese macaques showed the right-ear advantage. This time, however, a generalization test was conducted with a novel set of SE and SL vocalizations, which demonstrated that all subjects were attending to the same features of the calls. The authors concluded, though again on the basis of a very small sample, that whereas all animals had the requisite sensory capacities to perform the discrimination, and were attending to the same acoustic dimensions, lateralized mechanisms only came into play when the stimuli were of species-specific significance. This analysis, if valid on the basis of the numbers studied, would suggest that, *mutatis mutandis*, in *humans*, left hemisphere lateralization may reflect processing of a communicatory significant nature, rather than merely an underlying specialization for sequential, analytic, time-dependent processing (although see Chapter 3, and Bradshaw and Nettleton (1981) for contrary views).

Heffner and Heffner (1984, 1986) again trained Japanese macaques to discriminate between two subtypes of species-specific "coo" vocalizations in a conditioned shock-avoidance task. There were seven SEs and eight SLs, as used by Petersen's group. Five animals were then subjected to a unilateral left temporal lobe ablation, including all of the primary and secondary auditory cortex; five others received similar treatment on the right side. All were then retested, and then two from each group were lesioned on the other, intact, hemisphere. Unilateral left lesions consistently led to an initial impairment in performance (although the fact that coos could still be discriminated from noise suggests that performance deficits reflected either biological relevance or acoustic complexity), with some subsequent recovery. Right-sided ablations had no such detrimental effects, whereas bilateral damage led to a permanent deficit. Neither unilateral nor bilateral ablations dorsal to and sparing the auditory cortex had any detrimental affect on this task.

Thus this analogue of Wernicke's area in humans may predominate in the monkeys' left hemisphere in the perception of species-specific vocalizations. (Wernicke's area is generally thought to lie in the secondary auditory association cortex just dorsal to the primary auditory cortex, see e.g., Benson and Geschwind 1969, although compare Bogen and Bogen 1976.) Damage to this region in humans has a major effect. The fact that effects were comparatively minor and transient in monkeys may reflect the simple, artificial nature of a task where coos only constitute a small part of the animals' repertoire. Consequently, Heffner and Heffner (1986) rightly caution us against concluding that lateralization is necessarily weaker in monkeys. It is however interesting to note that the subsequent recovery after the initial left hemisphere lesions, with later total loss after bilateral damage, indicates that the right hemisphere can take over this function.

Human speech sounds have also been studied in the context of primate lateralization. In an early study, Dewson, Pribram, and Lynch (1969) found that monkeys were unable to discriminate vowel sounds after bilateral lesions of the auditory cortex; the effects of unilateral lesions were not assessed. As we previously saw, Pohl (1983) employed, among other stimuli, monaural presentation of vowels and consonant—vowel syllables (/ba/ versus /pa/) differing in their temporal characteristics. In both tasks, three of the four baboons (*Papio cynocephalus*) gave a clear left-ear advantage, and one a clear-right ear advantage. There is, in fact, evidence in humans that the processing of vowels in the absence of other acoustic information (for review, see Bradshaw, Burden, and Nettleton 1986), and the discrimination of voicing contrasts (e.g., /ba-pa/, /da-ta/, and /ga-ka/ in terms of voice-onset time, see, e.g., Morse et al. 1987) may be mediated by the right hemisphere. On the other hand, there is typically a right-ear (left-hemisphere) advantage for the categorical perception of stop consonants varying in place of articulation, for example, /ba-ga/, /ta-ka/. (As Liberman et al. 1967, showed, humans can readily discriminate between two categories of speech sounds, e.g., /ba-pa/, but may perform at chance levels when trying to discriminate between two speech tokens, *within* a category, that differ acoustically to the same extent as the two contrasts *between* categories; this phenomenon of categorical perception of human speech sounds extends to human infants and even other species.)

Morse et al. (1987) presented 15 1-year-old rhesus monkeys (*M. mulatta*) with voicing contrast stimuli; the animals exhibited categorical discrimination for such contrasts as /ba-pa/. Electrophysiological indices (averaged evoked responses, AERs) varied systematically as a function of voice-onset time. There were four levels of voice-onset time, giving a perceptual range from /ba/ to /pa/. A late component of the AER recorded over the temporal region of the right hemisphere of all 15 monkeys discriminated between stimuli in a categorical manner, just as with humans. Thus monkeys, like us, seem to process voicing contrasts categorically in the right hemisphere. We shall return later to the issue of categorical discrimination by humans and other species.

Morphological Asymmetries in the Human Brain

During evolution, with the development of an increasingly complex behavioral repertoire, much of it mediated by asymmetrically represented cortical mechanisms, the cortex not only increased in size, but increasingly differentiated (see e.g., Kaas 1987). Mammals emerged from the reptiles 250 million years ago and, after the great extinction at the end of the Cretaceous, underwent an enormous radiation. Turtles may have changed little from the putative basic reptilian stock, and insectivores like hedgehogs may still closely resemble the original mammalian prototype. However cats, dogs, and monkeys evolved to have a greatly expanded neocortex, especially in the sensory, perceptual, cognitive, and motor regions, with cortical fields divided into functionally distinct areas. As we shall see, many of these areas, at least in the primates, exhibit lateral asymmetries of structure, perhaps reflecting the functional asymmetries just described. Of course, we cannot necessarily *assume* that the larger of a pair of anatomically asymmetrical regions is the one specialized for a given function, but it is a good working hypothesis, especially when taken in conjunction with converging behavioral evidence.

Geschwind and Levitsky (1968) noted that the gross anatomical asymmetry of the human brain is visible even to the naked eye. The left lateral ventricle is three times as often wider and longer than vice versa, especially in males (LeMay 1984). In particular, the anterior portion of the right hemisphere and the posterior portion of the left hemisphere protrude further, and are wider, than their counterparts (Galaburda et al. 1978; LeMay 1976), especially in dextrals (see Figure 6.2). These effects may be visible as petalias or impressions on the inner table of the skull produced by the bulging of the adjacent cerebral lobe. We shall see shortly that similar morphological asymmetries in the brains of fossil hominids can be inferred from endocasts (endocranial casts), whereby a natural infilling or cast of the inside of the skull preserves the petalias. Whereas the left-occipital petalia is more pronounced than its right-frontal counterpart, the brain may be described as exhibiting an overall counterclockwise torque. Indeed, it should be noted that in terms of overall *volume*, the right frontal and left occipital lobes are larger than their counterparts (Weinberger et al. 1982).

Bear et al. (1986) confirmed these morphological asymmetries favoring the right frontal and left occipital poles, using CT scans in 66 adult outpatients without diagnosed abnormalities. They noted that males showed greater degrees of asymmetry than females, who in fact frequently reversed with respect to the general trend. (Cognitive, but not motor,

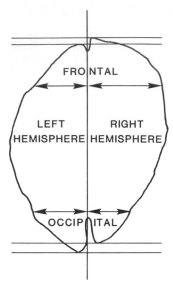

Figure 6.2   Asymmetries of frontal and occipital regions. The wider right frontal and left occipital regions, and the more extensive left occipital and right frontal poles, give the human brain an overall counterclockwise torque. From J. L. Bradshaw, Hemispheric Specialization and Psychological Function. Chichester: Wiley, 1989, with permission.

asymmetries are also generally stronger in females, see e.g., McGlone 1986.) They also noted that a reduction or reversal of the usual pattern was associated with nondextrality (and see also LeMay 1977; although also see Chui and Damasio 1980, and Koff et al. 1986).

Asymmetries in the periSylvian (speech related) cortex are prominent in humans, and have been observed for a century (Galaburda 1984; LeMay 1985). Thus the Sylvian fissure, that prominent sulcal landmark separating the temporal from the frontoparietal regions, is longer in the left hemisphere and continues further in a horizontal direction on that side before tending upwards (Cunningham 1892, and Ebersteller 1884, cited in Galaburda 1984; Rubens, Mahowald, and Hutton 1976; Yeni-Komshian and Benson, 1976), see Figure 6.3. The distribution of this asymmetry again varies with handedness, with weaker effects typically occurring in nondextrals (Hochberg and LeMay 1975; LeMay and Culebras 1972). The greater length of the Sylvian fissure on the left suggests that the temporal and parietal opercula, respectively constituting the floor and roof of the posterior portion of the Sylvian fossa, are also larger on the left side (Geschwind and Galaburda 1985). Over 100 years ago (Heschl 1878 and Pfeiffer 1936, cited in Witelson and Kigar 1988) it was

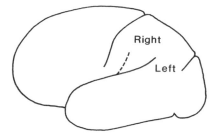

Figure 6.3 Left (solid line) and right (broken line) Sylvian fissures, the right shown superimposed upon the appropriate part of the left hemisphere. The Sylvian fissure is generally longer, but terminates lower, on the left. From J. L. Bradshaw, Hemispheric Specialization and Psychological Function. Chichester: Wiley, 1989, with permission.

demonstrated that the part of the temporal operculum known as the *planum temporale*, a triangular region caudal to the first transverse auditory gyrus of Heschl on the superior surface of the temporal lobe, and within the Sylvian fossa, was larger on the left side. Geschwind and Levitsky (1968) subsequently confirmed a greater extent of the lateral edge (outer border) of the left *planum temporale*, see Figure 6.4, with 11%

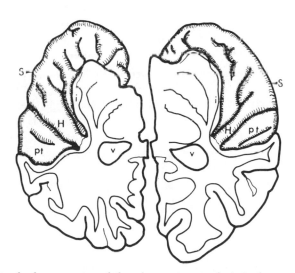

Figure 6.4 Standard asymmetry of the planum temporale (pt), the roughly triangular region lying posterior (inferior in the diagram) to the transverse auditory gyrus of Heschl (H) and bounded laterally by the Sylvian fissure (S). This pattern of asymmetry is seen in about two-thirds of normal human brains. Insula (i); Lateral ventricle (v). From N. Geschwind and A. M. Galaburda, Cerebral Lateralization: Biological Mechanisms, Associations and Pathology. Cambridge, Mass.: Bradford/MIT Press, 1987, with permission.

of the brains reversing, and 24% showing no measurable differences (and see also Rubens et al. 1976; Geschwind and Galaburda 1985; Witelson and Pallie 1973). As Galaburda (1984) observed, the region contains higher-order auditory association cortex and is a typical site of injury resulting in Wernicke's aphasia; it is therefore an important region in the mediation of language comprehension. However, asymmetries here precede language development and are present even by the thirty-first week of gestation (Chi, Dooling, and Gillies 1977). Thus the posterior asymmetry of the Sylvian fissure reflects the asymmetry of the *planum temporale*; however, it also reflects left-favoring asymmetries in the inferior parietal lobule, especially in area PG, which lies mainly on the angular gyrus (see e.g., Galaburda 1984). Area PG is a multimodal sensory region connecting to frontal speech regions and the language-relevant posterior thalamus. When it undergoes damage, anomic aphasia and reading and writing disturbances typically result.

Turning now to the frontal lobe, the left frontal operculum is said to be more folded than the right, especially in regions affected during Broca's aphasia—the *pars opercularis* of the inferior frontal gyrus (Galaburda 1984). The architectonic areas of the *pars opercularis* and *pars triangularis* are usually larger on the left (Galaburda 1984; Geschwind and Galaburda 1985; although see Wada, Clarke, and Hamm 1975). Indeed, agreement on the extent and even direction of frontal asymmetries is much less strong than that for temporoparietal differences (see, e.g., Koff et al. 1986, and Witelson and Kigar 1988); reliance should perhaps be placed more on cytoarchitectonic differences, the characteristic arrangement of neurons, cell densities, dendritic branching, and arborization in different structures which render them distinguishable from their neighbors, than on gross surface landmarks such as gyri and sulci. Thus an area known as Tpt, a major portion of the *planum temporale* and adjacent posterior superior temporal gyrus and commonly involved in the presence of Wernicke's aphasia, usually reflects (or determines) any asymmetries also present in the *planum temporale*. Area Tpt also connects to inferior prefrontal regions in the frontal operculum. Indeed, leftward asymmetries in all three regions, area PG on the angular gyrus of the inferior parietal lobule, area Tpt on the *planum temporale,* and the *pars opercularis* of the frontal lobe, may be linked within a single subject (Eidelberg and Galaburda 1984) and participate in a single speech system.

Recently, Kertesz et al. (1990) employed magnetic resonance imaging (MRI) techniques to determine linear and area measurements of hemispheric, frontal, temporal, and parietal regions of 52 dextral and 52 sinistral normal, healthy males and females; the sample was therefore fully balanced for sex and handedness. MRI provides excellent distinction between grey and white matter, producing images that approximate those obtained at autopsy. They found that the overall area of the right hemisphere was significantly larger across all groups, although in sinistrals the left frontal width and area was larger. As in the general findings reviewed previously, dextrals had larger right anterior frontal, and left parietal and occipital widths. The phenomenon of "counterclockwise torque" (see above) was in this case revised to a third dimension: the upper and rostral portions of the hemispheres were larger on the right, and the lower and caudal parts were equal or slightly larger on the left. Complex sex and handedness effects emerged, particularly with respect to parietal regions.

Finally, just as we saw in Chapter 3 that pharmacological asymmetries have been noted in subcortical structures in rat brains, for example in the basal ganglia (extrapyramidal structures controlling movement) and probably underlying turning biases, so too in humans have higher levels of dopamine been noted on the left in various basal ganglia structures. This suggests a pharmacological substrate for manual and even speech dominance (Glick, Ross, and Hough 1982).

Morphological Asymmetries in the Brains of Nonhuman Primates and Fossil Hominids

Many of the anatomical asymmetries previously reviewed also appear in Old World monkeys, suggesting that an expansion of prefrontal and parietal integration cortices in the left hemisphere occurred in an early primate ancestor common to monkeys and ourselves. Thus during primate evolution, the Sylvian fissure appears to have migrated from a nearly vertical to a horizontal orientation. This may be due to the increased development of the inferior parietal lobule, especially in the left hemisphere, which in humans has come to subserve speech after a lengthy process of preadaptation.

It is clear that the human brain is the most asymmetric, functionally and anatomically, and that primate evolution is characterized by increasing asymmetries, either as a direct consequence of increasing brain size and demands upon limited processing space, or as an independent phylogenetic trend. De LaCoste, Horvath, and Woodward (1988) studied the Lemuridae, a family of strepsirhines perhaps the furthest removed from humans of the four major radiations of primates; the other three are the ceboids, the cercopithecoids, and the hominoids. They studied *Microcebus murinus* (n = 6), *Cheirogaleus medius* (n = 2), *Lemur albifrons fulvus* (n = 2), *Lemur mongoz* (n = 1), and *Lepilemur ruficaudatus* (n = 3), filming an anterior to posterior series of coronal sections of the cerebra. Despite considerable interspecific variability, in two thirds of the species, the retrocalcarine cortex was unexpectedly the most asymmetric part of the brain. (These findings would have predicted the frontal and temporo-parietal association cortex to be the most asymmetrical.) Generally, there was a left-side advantage, especially in the three nocturnal species (*M. murinus, Ch. medius,* and *Ch. major*), suggesting that some aspect of visual processing was responsible.

Turning to gross asymmetries in the skulls and brains of gorillas and baboons, Groves and Humphrey (1973) noted that whereas there were no such asymmetries in the skulls of 138 coast gorillas (*Gorilla gorilla gorilla*), in the case of the two subspecies of mountain gorilla (*G. gorilla graveri*, n = 17, and *G. gorilla beringei*, n = 38), the left side was significantly longer with, in the latter subspecies, an obvious lopsidedness of the sagittal crest and heavier tooth wear on one side. The authors speculated that all such skull asymmetries may be secondary to asymmetry in mastication, with a tendency to chew more on the left side, though they do not consider possible underlying neurological reasons for this. They noted that the coast gorillas, who have a less-developed masticatory structure, did not show this asymmetry. With respect to baboons, Cain and Wada (1979) studied the brains (fixed in formalin) of six adult *Papio papio* and one *Papio anubis*. Six of the seven showed an asymmetry of the frontal poles (see Figure 6.5), with the right the longer by an average of 1.4 mm; the seventh had poles of equal length.

Turning to a systematic investigation of petalial asymmetries, LeMay, Billig, and Geschwind (1982) measured endocranial casts of 25 apes, 21 Old World monkeys (6 species), and 43 New World monkeys (14 species). They found asymmetries largely similar to those of humans; that is, prominence of right frontal and left occipital poles in the chimpanzee and gorilla, but not in the orangutan; less striking asymmetries, but in the same direction, in Old World monkeys; and no significant effects in New World species, except that left occipital petalia tended to be more common than right. Holloway and de LaCoste-Lareymondie (1982) brought together Holloway's earlier reports on petalial asymmetries in 190 hominoid endocasts, living species and extinct progenitors and collaterals of the human race. All taxa showed asymmetries to varying degrees. Of the pongids, *Gorilla* showed the greatest asymmetry in left occipital petalia, *Pan paniscus* less so, and *Pan troglodytes* and *Pongo pygmaeus* not at all. But even *Gorilla* usually failed to show both wider and longer occipital measurements, whereas *Homo sapiens, H. erectus* (whose status as a grade they recognize, without reference to controversies on its actual

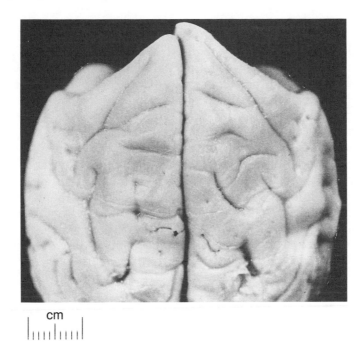

Figure 6.5   Brain of a baboon illustrating asymmetry of the frontal poles as found also in humans. From D. P. Cain and J. A. Wada, An anatomical asymmetry in the baboon brain *Brain, Behavior and Evolution*, 1979, 16:222–226, with permission.

genetic relationship), *H. sapiens neanderthalensis, Australopithecus africanus,* and *Australopithecus robustus* all showed the fully modern combination of left occipital and right frontal petalias and lobe width, which was absent in the pongids. Thus, generally, all the hominid taxa, *Australopithecus* and *Homo,* showed the same patterns; these differed from those of the other hominoids (i.e., pongids), even including gorillas, which here resembled hominids most closely. The hominid pattern seems therefore to have persisted for around 3 million years, and if this petalial pattern correlates with dextrality, as seems to be the case with modern humans, then dextrality must date back 3 million years. We shall later review additional archaeological evidence for such a proposition.

Falk et al. (1990) digitized and measured the lengths of cortical sulci using three-dimensional computer technology on 335 endocranial casts from rhesus monkeys, *M. mulatta.* As Witelson and Kigar (1988) observe, values obtained directly from the brain and those obtained from photographs may not be exactly comparable due to the loss of convexity information in two-dimensional photographs. Thus some earlier monkey studies using photographs failed to show Sylvian fissure asymmetries, whereas later three-dimensional technology did. The frontal lobes were found to be directionally asymmetrical, with a significant elongation of the right orbital and dorsolateral frontal lobe. In particular, the rectus (principal) sulcus and the lateral orbital sulcus were both longer in the right hemisphere, whereas the central sulcus was longer in the left. The lateral end of the central sulcus was caudal more often on the right, and the medial end was caudal more often on the left. There was a substantial predominance of right frontal petalias (49% versus 26% left frontal petalias, with 24% symmetrical), but the occipital lobe displayed nearly equal frequencies of right (38%) and left (37%) petalias (24% were symmetrical). These frontal findings mirror the human situation, although with humans, occipital petalias, in the reverse direction to the frontal, are generally stronger.

However, Cheverud et al. (1990), in a subsequent study of 403 rhesus macaque endocasts, found that right frontal petalias were more commonly associated with left occipital petalias than would be expected by chance, just as with human brains (see Figure 6.6). They also found a significant but moderate heritability for frontal petalias, whereas the heritability estimate for occipital petalias was relatively low and not statistically significant. While they observed that a major function of the prefrontal cortex of macaques involves visual memory, one comment is

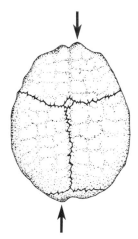

Figure 6.6   Endocast of a rhesus macaque skull exhibiting a right frontal and a left occipital petalia (both arrowed). From J. M. Cheverud, D. Falk, C. Hildebolt, A. J. Moore, R. C. Helmkamp, and M. Vannier, Heritability and association of cortical petalias in rhesus macaques (*Macaca mulatta*). *Brain, Behavior and Evolution*, 1990, 35:368–372, with permission.

that these regions are also concerned with motor intentionality and that motor asymmetries continue to be more prominent than sensory or cognitive asymmetries, both in humans (Bradshaw 1989) and in other primates.

The problem of inadequate interpretation of two-dimensional representations of a three-dimensional convex surface is particularly salient with respect to left—right asymmetries of relative length and height of the Sylvian fissure. Moreover, early studies often merely measured straight line distances between the two end points of a sulcus, even though sulci are sinuous, not straight, and lie on a convex surface. Falk et al. (1990) also note that the right frontal petalia causes more of the Sylvian fissure's length to lie on the orbital (under) surface on the right side than on the left, thus making the right hemisphere sulcus *seem* shorter than that of the left, even when they really happen to be of equal length.

Yeni-Komshian and Benson (1976) found that chimpanzees, like us, have longer Sylvian fissures on the left hemisphere, although the differences were less extreme. Rhesus monkeys showed no such asymmetries. The authors cite Fischer (1921), who stated that whereas 50% of his chimpanzee brains possessed longer Sylvian fissures on the left, only 17% had longer fissures on the right. Cunningham (1892, cited in LeMay 1985) reported such effects in both the chimpanzee and macaque, and also noted that as in humans, the Sylvian point (the end of the Sylvian fissure) was higher on the right in chimpanzees, orangutans, mangabeys, and baboons, but not in macaques. LeMay and Geschwind (1975) also failed to find this effect in New and Old World monkeys, but did so in the chimpanzee, gorilla, and orangutan.

Heilbroner and Holloway (1988) studied relatively large samples, approximately balanced as to sex (a precaution not always taken previously), of formalin-fixed brains from five New and Old World monkey species, *Macaca mulatta*, *M. fascicularis*, *Sanguinus oedipus*, *Callithrix jacchus*, and *Saimiri sciureus*, the first two from Old World stock, and the last three from the New World. Only in *S. sciureus* did the overall length of the left Sylvian fissure fail to exceed that of the right, and even in that species the effect held if, instead, just the anterior portion of the fissure was measured. As they observed, measurements taken from photographs failed to show these three-dimensional effects, thus accounting perhaps for some of the earlier confusion.

Falk et al. (1986), using appropriate three-dimensional technology, succeeded in demonstrating asymmetries in Sylvian fissure lengths similar to humans (left longer) in 10 specimens of the rhesus monkey, *M. mulatta*. Interestingly, Falk et al. (1990) failed to replicate their own earlier finding. Heilbroner and Holloway (1988) also made two other observations: *M. mulatta* and *M. fascicularis* exhibited no asymmetries in Sylvian fissure length *posterior* to the central sulcus, whereas in humans this segment typically shows the largest effects due to asymmetries in the language-mediating temporal planum, a structure absent in other primates. Also, Heilbroner and Holloway failed to find any asymmetry in height of the Sylvian point in their two *Macaca* samples (they do not mention the other three New World species in this context), unlike the findings with *Pan*, *Gorilla*, and *Pongo*. So perhaps the left temporoparietal cortex started changing prior to the divergence of the hominids from the pongids, and after the appearance of the first hominoids.

In the context of fossil hominids, LeMay and Culebras (1972) studied the endocast of *Homo sapiens neanderthalensis* ("Neanderthal Man") from LaChapelle-aux-Saints of 30 to 50 thousand years ago. The endocast showed the imprint of a right Sylvian fissure (length and height of the point) exactly similar to our own. We shall return later to the question whether this taxon possessed full speech. Even more interesting, the same observation applied to a *H. erectus* specimen, Peking Man, of 300,000 years ago. We might finally note that Heilbroner and Holloway (1989) assessed possible hemispheric asymmetries in the frontal, parietal, and temporal cortex in 60 specimens of *M. mulatta* and *M. fascicularis*. They found no significant left–right asymmetries in the mean lengths of four sulci (the superior temporal, the interparietal, the horizontal arcuate, and the principal sulcus) in visual processing areas, nor in the sulcus adjacent to the region cytoarchitectonically homologous to human motor–speech areas. Nor was there asymmetry in the area of the dorsal cingulate gyrus; however, in both species a small, unnamed parietal-lobe sulcus showed greater left hemisphere development.

Summary and Conclusions

If we combine across hand, ear, and visual field asymmetries, and across studies with lateralized presentations to otherwise normal animals, and studies involving the effects of lateralized brain injury or ablations, we can draw the following conclusions in terms of apparent hemispheric mediation. Monkeys and, in some cases, apes often evidence an apparent right-hemisphere mediation of haptic and certain visuospatial discriminations, including face processing. In this respect there is a close correspondence with humans. However, a left-hemisphere mediation has been reported for the discrimination of line slopes (as indeed may also be the case with humans) and of the relative vertical position of a dot on a square background. The latter at least seems contrary to what we find with humans, although it may be tempting—if dangerous—to seek to invoke possible differences in processing strategy.

A number of auditory studies (frequently monaural, for practical reasons, and thereby out of line with mainstream human work) have found apparent left hemisphere involvement in a variety of auditory discriminations at least analogous to what is typically found with humans. In particular, the discrimination of *species-specific* vocalizations seems to have a strong left-hemisphere involvement in macaques, as shown by a number of converging studies. Conversely, though again homologously (or analogously?) to the human situation, the right hemisphere seems to be involved in macaques (and humans) in the categorical discrimination of voicing contrasts.

The Hopkins and Morris group have reported a series (unfortunately, though necessarily, with *extremely* small sample sizes) of studies with language-trained chimpanzees. Effects comparable to those of humans, and possibly as a consequence of the animals' exposure to such language training, have been reported in a series of tasks. The tasks are noteworthy for their extreme ingenuity with respect to stimuli, procedure, and responses, often closely mimicking human tachistoscopic-reaction time or computer tasks. Only time and much further research will tell whether these findings and conclusions can be generalized.

With respect to morphological asymmetries in the primate brain, increasingly sophisticated measurement techniques are pushing back, in evolutionary terms, the likely appearance of petalial and periSylvian asymmetries. Thus a right frontal extension, if not a corresponding extension of the left occipital lobe, seems secure in apes and Old World, though maybe not New World, monkeys. Asymmetries in the Sylvian fissure may, in certain dimensions, even extend to New World monkeys if recent techniques can be trusted. However, the absence of other peri-Sylvian asymmetries and human-style lobe *width* asymmetries in all except fossil hominid (and human) brains, suggests that increased development of language-related structures in the inferior parietal lobule of the left hemisphere did not occur until the advent of the hominids.

Further Reading

Cheverud, J. M.; Falk, D.; Hildebolt, C.; Moore, A. J.; Helmkamp, R. C.; and Vannier, J. Heritability and association of cortical petalias in rhesus macaques (*Macaca mulatta*). *Brain, Behavior and Evolution*, 1990, 35:368–372.

Falk, D.; Hildebolt, C.; Cheverud, J.; Vannier, M.; Helmkamp, R. C.; and Konigsberg, L. Cortical asymmetries in frontal lobes of rhesus monkeys (*Macaca mulatta*). *Brain Research*, 1990, 512:40–45.

Galaburda, A. M. Anatomical asymmetries. In N. Geschwind and A.M. Galaburda (eds.), Cerebral Dominance: The Biological Foundations. Cambridge, Mass.: Harvard University Press, 1984, pp. 11–25.

Hamilton, C. R., and Vermeire, B. A. Complementary hemispheric specialization in monkeys. *Science*, 1988, 242:1691–1694.

Harries, M. H., and Perrett, D. I. Visual processing of faces in temporal cortex: Physiological evidence for a modular organization and possible anatomical correlates. *Journal of Cognitive Neuroscience*, 1991, 3:9–24.

Heffner, H. E., and Heffner, R. S. Effect of unilateral and bilateral auditory cortex lesions on the discrimination of vocalizations by Japanese macaques. *Journal of Neurophysiology*, 1986, 56:683–701.

Heilbroner, P. L., and Holloway, R. L. Anatomical brain asymmetries in New World and Old World monkeys: Stages of temporal lobe development in primate evolution. *American Journal of Physical Anthropology*, 1988, 76:39–48.

Hopkins, W. D.; Morris, R. D.; and Savage-Rumbaugh, E. S. Evidence for asymmetrical hemispheric priming using known and unknown warning stimuli in two language-trained chimpanzees (*Pan troglodytes*). *Journal of Experimental Psychology: General*, 1991, 120:46–56.

LeMay, M. Radiological, developmental, and fossil asymmetries. In N. Geschwind and A. M. Galaburda (eds.), Cerebral Dominance: The Biological Foundations. Cambridge, Mass.: Harvard University Press, 1984, pp. 235–342.

LeMay, M. Asymmetries of the brains and skulls of nonhuman primates. In S. D. Glick (Ed.), Cerebral Lateralization in Nonhuman Species. New York: Academic Press, 1985, pp. 235–245.

Petersen, M. R.; Beecher, M. D.; Zoloth, S. R.; Green, S.; Marler, P. R.; Moody, D. B.; and Stebbins, W. C. Neural lateralization of vocalizations by Japanese macaques: Communicative significance is more important than acoustic structure. *Behavioral Neuroscience*, 1984, 98:779–790.

Witelson S. F., and Kigar, D. L. Asymmetry in brain function follows asymmetry in anatomical form: Gross, microscopic, postmortem, and imaging studies. In F. Boller and J. Grafman (eds.), Handbook of Neuropsychology, Vol. 1. New York: Elsevier, 1988, pp. 111–142.

Hominoids and Hominids: The Evolution of *Homo*

Children, behold the chimpanzee!
He sits on the ancestral tree
From which we sprang in ages gone.
I'm glad we sprang; had we held on
We might, for aught that I can say
Be horrid chimpanzees today.

(Oliver Herford, cited by Raijmakers, 1990)

Introduction

Contrary to conventional wisdom until very recently, we now see a continuum of many nonhuman species, and ourselves, in manual and cognitive asymmetries, and in lateralization at the neuroanatomical level. Lateralization seems to have occurred very early in evolution and the basis for its manifestation in primates was already present in rats. The very recent discoveries of lateral asymmetries in the primates are paralleled by equally recent developments in the field of human evolution, not just from new archaeological finds, but in part occasioned by new dating techniques, for example, thermoluminescence, and in part by progress in molecular biology. Thus the nature and even the dates of the divergences of ancient lineages can be estimated from similarities and differences in the genomes of extant individuals. Indeed, progress in linguistics has shown remarkable parallels with molecular biology; not only can extinct and ancestral languages be recreated and dated from commonalities in extant tongues, but the two techniques, linguistic and molecular, can be correlated in terms of geography and prehistory. We are now able, dimly, to perceive the entire human trajectory from the dawn of life several billion years ago, through the evolution of the vertebrates, the mammals and the primates, to the australopithecines and various grades—or species, the jury is still undecided on this issue—of

225

humans, to anatomically modern people of the last hundred or so millennia.

In this chapter we shall briefly review our phylogenetic relationship to the great apes and describe the rise of the australopithecines, our immediate ancestors, and of the various grades or species of *Homo*. We shall discuss evidence for the advent of bipedalism, of manual dexterity, tool use, and ultimately language, and the origin and spread of anatomically modern human beings. This will set the stage for the discussion, in the following chapter, of the growth of intellect as manifested in the minutiae of tool using and linguistic behaviors, fire, burial, and art, and their physiological and neurological correlates.

The Rise of the Primates

The universe may have formed more than 15 bya (billion years ago); the Earth is around 4.5 billion years old; and the oldest sedimentary rocks of about 3.75 bya record the cooling and stabilization of the Earth's crust (see Figure 7.1). Within these rocks the ratio of isotopes of carbon, ^{12}C to ^{13}C, is characteristic of organic activity, as if life appeared as soon as the earth was cool enough. Stromatolites, mats of sediment trapped and bound by bacteria and blue-green algae, and actual cells have been found in the earth's oldest unmetamorphosed sediments dating to 3.5 to 3.6 bya in Africa and Australia (Gould 1991a). In the Burgess Shale of the Canadian Rockies, modern multicellular animals make their first uncontested appearance in the fossil record, 530 mya (million years ago), shortly after the Lower Cambrian explosion of around 570 mya, and with the sudden simultaneous appearance of virtually all major groups of modern animals, including chordates. *Pikaia gracilens* is the first recorded chordate member of our own immediate ancestry. With a segmented notochord possessing the characteristic zigzag bands of muscles (myotomes), it in many ways resembles *Amphioxus,* the archetypical model of prevertebrate chordates. Fossils of true vertebrates, initially represented by jawless (agnathan) fishes, first indisputably appear in the Middle Ordovician, though fragmentary indications appear as early as the Upper Cambrian.

During the Triassic period of the Mesozoic era, around 200 mya, the class Mammalia evolved from mammal-like therapsid reptiles. Cretaceous insectivores around 100 mya gave rise to placentals and possibly even primates, a period marked by the evolution of angiosperms (flowering plants), which with nectar, fruit, berries, nuts, and associated insects provided a new ecological niche for rapid evolutionary development. The end of the Cretaceous, around 65 mya, was marked by a

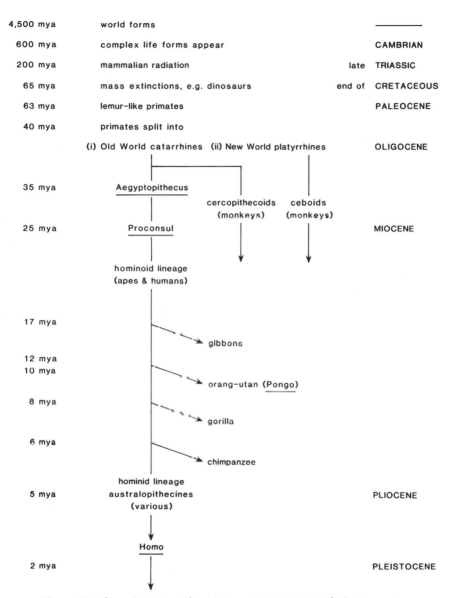

4,500 mya	world forms		
600 mya	complex life forms appear		CAMBRIAN
200 mya	mammalian radiation	late	TRIASSIC
65 mya	mass extinctions, e.g. dinosaurs	end of	CRETACEOUS
63 mya	lemur–like primates		PALEOCENE
40 mya	primates split into		
	(i) Old World catarrhines (ii) New World platyrrhines		OLIGOCENE
35 mya	Aegyptopithecus		
	cercopithecoids ceboids (monkeys) (monkeys)		
25 mya	Proconsul		MIOCENE
	hominoid lineage (apes & humans)		
17 mya	gibbons		
12 mya 10 mya	orang-utan (Pongo)		
8 mya	gorilla		
6 mya	chimpanzee		
5 mya	hominid lineage australopithecines (various)		PLIOCENE
	Homo		
2 mya			PLEISTOCENE

Figure 7.1 Some important dates in our more remote evolutionary past.

catastrophic event involving mass extinctions, the nature of which (an extraterrestrial impact, mass vulcanism, or climatic change) is the subject of much debate (Kerr 1991); in any case, the elimination of the dinosaurs opened the way for further evolutionary development. Unfortunately, as Klein (1989) notes, there are major temporal and geographical gaps thereafter in the fossil record, and much associated uncertainty concerning primate evolution. However, the beginning of the new Cainozoic era (the Paleocene and Eocene periods) was warm and forested (Harrison et al. 1988), with thereafter an increase in seasonality and cooling, and a shift towards savannah grasslands in Africa, where most of the primate evolution occurred. In fact, primates are not known for certain as fossils until the Cainozoic; at its commencement they became relatively abundant, with a lineage developing which was broadly similar to the living lemuroids of Madagascar. In the Oligocene, platyrrhine (whose living members include marmosets, tamarins, and various other New World species) and catarrhine lineages with larger brains appeared. Indeed, Old World anthropoid catarrhines, today comprising cercopithecoid monkeys (e.g., macaques and baboons) and hominoids (apes and humans) evolved as forest primates in Africa, spreading later to Asia (where species still survive) and Europe.

Thus the extant species of apes consist of the southeast Asian hylobatids (gibbons and siamangs—the "lesser apes"; Eisenberg 1981) and the great apes of Africa (*Gorilla* and *Pan*) and Asia (*Pongo*). Most living catarrhines are arboreal. By the middle of the Pliocene (3–4 mya), the faunas of East Africa largely resembled modern ones. From 2.4 mya, there was increasing fluctuation in habitat due to the gathering momentum of glacial–interglacial cycles, leading to a mosaic pattern of habitats favorable to the evolution of primate species that could range over considerable distances in search of a varied diet (see Figure 7.2). In the last million years, glaciation further intensified, aridity increased, the habitat opened up yet more, migratory herds grew, and the human species rapidly evolved. However, as Harrison et al. (1988) observe, it may be oversimplistic simply to conclude that early hominids exploited newly expanded savannahs as the climate cooled, while *Homo* responded to the challenge of the ice ages with an expanded brain. Of course, there were earlier periods of glaciation in the Earth's history, the most recent cycle commencing with the formation of continental ice sheets around 3.5 mya, that have been waxing and waning ever since. The last major interglacial peaked 120 kya (thousand years ago), and the last major melt started 11 kya, with seas now lower than on the previous occasion. Various factors combined to initiate the current series of glaciations, including variations in the earth's axial tilt, variations in the eccentricity of the earth's orbit around the sun, the precession of the earth's axis, and continental plate movements and the growth of mountains which altered the circulation of oceanic and atmospheric currents, thereby permitting freezing at the poles (Gould 1991b).

The origin of the anthropoids (monkeys and apes) may have been

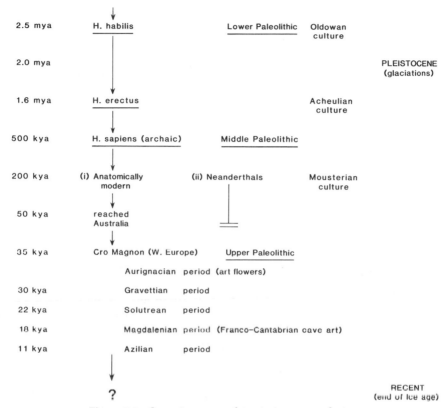

Figure 7.2 Some important dates in human evolution.

causally associated with a move from nocturnal to diurnal activity rhythms (Pickford 1986). Most prosimians, among the most "primitive" of extant primates, are nocturnal with poor color vision and less well-defined stereoscopic vision than the anthropoids. Thus the visual fields of the two eyes overlap less in prosimians (except perhaps in the case of the bush baby) than in more "advanced" primates, where improved stereopsis is reflected in increased encephalization. Stereopsis requires close alignment of the eyes even during chewing, when, in the absence of a postorbital septum, bulging of the jaw muscles can mechanically interfere with ocular alignment. Anthropoids were the first primates to develop such a septum to isolate the eyes from the temporal muscles (Pickford 1986); it is first apparent in the fossil record in *Aegyptopithecus* of the early Oligocene Fayum deposits of Egypt, dating to at least 35 mya and predating the split into cercopithecoid and hominoid lineages

(Harrison et al. 1988; Klein 1989). According to Simons (1985), *Aegyptopithecus* was a good structural ancestor to the east African Miocene genus *Proconsul* (23–17 mya), itself the probable ancestor of the great apes and humans. It was a generalized arboreal frugivorous quadruped with marked sexual dimorphism in size, which may suggest male rivalry for females. According to Lewin (1989a), *Proconsul* possessed, mosaic fashion, the characteristics of modern apes in shoulder, elbow, cranial, and dental characteristics, as well as the foot and lower leg bones, and of modern monkeys in the long trunk, and arm and hand bones. Klein (1989) concludes that such features indicate palm- rather than knuckle-walking, together also with underbranch suspension; that is, perhaps a variety of locomotor repertoires above and upon the ground.

After *Proconsul*, there is another large hiatus in the fossil record until the emergence of *Sivapithecus* (now subsuming *Ramapithecus*) in west and south Asia between 18 and 12 mya. This large hominoid, now demoted from its earlier presumed status as a hominid ancestor of the human lineage, was largely arboreal, frugivorous, and sexually dimorphic with a dentition superficially resembling that of hominids, and a general appearance similar to that of modern apes. Indeed, it may well have been the ancestor of the living Asian orangutan, *Pongo pygmaeus*. DNA hybridization distance techniques have been used in attempting to date subsequent evolutionary divergences. The techniques gained acceptance in the 1980s as a powerful adjunct to the more traditional approaches of physical taxonomy, and exploit the fact that species diverging from a common ancestor steadily accumulate mutations in their DNA. The double-stranded DNA of the two species to be compared is heated to cause separation into the two component single strands. The single-stranded DNA is then mixed and allowed to cool, forming a hybrid molecule—a helix—which consists of the complementary strands of the two species. Wherever mutations have occurred, the DNA fails to match up, weakening the bonds that normally hold the helix together. The hybrid molecule is then reheated; since its bonding is weakened by mutational mismatches, it separates at a lower temperature proportional to the evolutionary time that separates the two species.

Such DNA hybridization techniques (and see reviews by Harrison et al. 1988, and Lewin 1989a) indicate that the ancestor of the gibbons split off first, at least 17 mya; then a split between the Asian *Pongo* and African *Pan* lineages occurred around 12 mya, followed by the gorilla divergence between 7 and 9 mya. We shall discuss in detail the likely chimpanzee–human split of between 5 and 8 mya, noting at this point that these findings and likely dates have now been confirmed by a quite different technique involving mitochondrial (mt) DNA, by Ruvolo et al. (1991).

Lewin (1989a) reviews the likely appearance of the immediate common ancestor of chimpanzees and humans. He sees it as a sexually dimorphic frugivore (with females weighing around 30 kg, and males maybe twice as heavy) operating in a polygynous social structure, with long arms and legs, a generally hominoid body, and arboreal habits like a chimpanzee. It perhaps included knuckle-walking in its locomotor repertoire. This would have permitted transformation into both classical knuckle-walking in the case of chimpanzees, and bipedalism in the case of the hominids. Harrison et al. (1988) note that the fact that both chimpanzee and *Gorilla* knuckle-walk does not necessarily mean that their and our common ancestor did; none of the known Miocene hominoids or Pliocene hominids show any features suggesting knuckle-walking, and homoplasy (i.e., parallel or convergent evolution) is common in primate evolution. On the other hand, Lewin (1989a) doubts that chimpanzees and gorillas could have independently evolved such a complex set of anatomical adaptations from a non-knuckle-walking ancestor; however he also notes the intrinsic improbability of the only other alternative, that the common ancestor of humans and African apes was a knuckle-walker (an adaptation lost in the hominids), given the absence of even vestigial signs of knuckle-walking in the anatomy of living or fossil hominids.

The phylogeny of extant hominoids has been extensively studied by various molecular techniques; for example, microcomplement fixation, amino acid sequences, DNA hybridization, restriction enzymes, and recently portions of mitochondrial and nuclear DNA (see Saitou 1991). However, as yet full consensus has not been achieved on the branching order of human, chimpanzee, and gorilla, or on the divergence time among these three species, because different tree-making methods can, with the same set of data, produce rather different trees. Nevertheless chimpanzees (splitting off between 5—8 mya) may be even be more closely related to us than to gorillas (Falk 1987; Gibbons 1990), a conclusion at variance with at least some interpretations (of a much more ancient divergence) of the paleontological evidence, leading to the following comment (Lewin 1988a) attributed to Sarich: "I know my molecules had ancestors, but the paleontologists can only hope that their fossils had descendents."

Lovejoy (1981) even speculates, given the evidence that we share with chimpanzees 98% or more of nonrepeated (i.e., "useful") DNA, that a viable hybrid between chimpanzees and humans might be possible (and see also Miyamoto, Slightom, and Goodman 1987). Saitou (1991) estimated the evolutionary distances between ourselves, the apes, and several species of monkeys, from a variety of mitochondrial and nuclear DNA sequence data, and applied the neighborhood-joining method to these distances to reconstruct the molecular phylogenesis. He concluded that there is a progressive decrease from us in evolutionary distance from rhesus monkey, orangutan, and gorilla to chimpanzee, and that we are closer to chimpanzees than any of the others are to each other; however, he emphasizes the need to examine more genes to firmly and conclusively establish the branching pattern of the human, chimpanzee, and gorilla triad. It is, of course, counterintuitive that we might be closer to chimpanzees than they are to the gorilla; the latter two species look more alike, both are hairy, walk on all fours or knuckle-walk, have short legs and long arms, brachiate, are less dextrous manually than ourselves, and have smaller brains and larger canine teeth with thin enamel. However, the apparent chimpanzee–gorilla similarities may not necessarily reflect closeness of kinship, or even perhaps convergent evolution, so much as perhaps retention of ancestral (plesiomorphic) traits which we have lost, although the problem of knuckle-walking remains unresolved.

Such considerations have led to suggestions (Lewin 1988a) that we abandon the traditional view whereby humans are the sole occupants of the family Hominidae, with the great apes (chimpanzees, gorillas, and orangutans) constituting a second family, Pongidae. Instead, humans and apes should perhaps be treated as separate genera within one family, Hominidae (see Figure 7.3). If this approach were to be adopted, we would have to drop the term Hominid, as used both currently and in the rest of this book, for humans and their direct ancestors, down to and including *Australopithecus*.

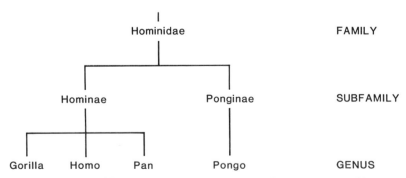

Figure 7.3 A possible new taxonomy to encompass the great apes and humans.

The Australopithecines

While *Aegytopithecus* and *Proconsul*, which lies on the lineage that led both to the apes and hominids, were as we saw, forest-living arboreal quadrupeds, the first true hominids, the australopithecines, were small-brained bipeds (for review, see Harrison et al. 1988; Klein 1989; Lewin 1989a; Simons 1989). We shall discuss possible reasons for this locomotor shift which, although it long antedated an increase in brain size, was nevertheless at the very least an essential preadaptation for the fundamental changes in behavior, for example, tool use, which were to follow. Four or more species may have evolved between 5 and 1 mya in south and east Africa as a consequence of climatic changes fragmenting habitats and isolating different sections of a once continuous population. (Such evolutionary pressures may not only have increased the rate of speciation, but also, with the consequent need to move considerable distances over the ground carrying offspring and perhaps food, they may have directly contributed to the advent of bipedalism.) All the species are associated with woodland and savannah communities; two major subdivisions are important, the gracile (*Australopithecus africanus*) and the robust (*A. robustus* followed by *A. boisei*). However, it should be noted that the systematics and biogeography of hominid species are complicated by problems of dating, fragmentary remains, probable sexual dimorphism, and different taxonomic and personal philosophies. We may have evolved from the gracile lineage, or alternatively our own lineage may have sprung, together with both groups of australopithecines, from a common ancestor currently designated *A. afarensis* (although see the following discussion on whether the latter itself really subsumes two separate species, themselves gracile or robust).

The australopithecines seem to have possessed an apelike schedule of ontogenetic development, with short maturation times and an apelike pattern of dental development. All species have an increased brain size, although it remained small compared to our own. They all had large, long-fingered hands (indicating retained arboreal potential) and arms shorter in relation to torso length than is found nowadays in comparably sized apes, but with lower limbs only fractionally longer than would be found in such apes. Simons (1989) concludes that their overall anatomy forms a connecting link between apes and humans; while foot and hip seem to have been well adapted to bipedal walking (perhaps even better than our own in some respects), and with no evidence of knuckle-walking, their stride would have been short and running speed slow; they may well have continued like their predecessors to sleep in trees or on cliff faces. Like ourselves, they have been variously reconstructed, depending on the author's personal predilections, as gentle vegetarians or bloodthirsty carnivores (Harrison et al. 1988).

Harrison et al. (1988) provide a useful summary comparison of the four main australopithecines, whose status is generally accepted, together with *Homo habilis*, our almost certain direct ancestor who overlaps chronologically with three of the australopithecines.

A. afarensis occurred from 4 (or even 5) mya to between 2.6 and 2.9 mya. It possessed a brain size of around 400 cm^3, with a body weight of around 35 kg (although other authorities suggest generally larger values for all the australopithecines, and see the following discussion on sexual dimorphism and the possibility of more than one species having been included within this taxon). Despite its relatively long arms, short legs, bipedalism, and climbing adaptations at the level of finger and hand bones, and wrist, elbow, and shoulder joints, it may well have been the ancestor of all subsequent hominids. While we shall discuss evidence for its bipedalism, it should be noted that its hand anatomy was roughly comparable to our own in overall size, whereas the relatively more curved fingers suggest a powerful grasp, with a grip that was less precise than our own but probably superior to that of the apes. Again, *A. afarensis* falls nicely between the two taxa, modern human and African ape. The almost complete fossil "Lucy" is perhaps its best known specimen and it almost certainly made the dramatic footprint tracks found at Laetoli in Tanzania (Boaz 1988).

A. africanus, occurring between 3 and 2 mya had a slightly larger brain (450 cm^3), though with a body weight (35 kg) and postcranial skeleton largely unchanged. There were differences between the two species in facial structure, perhaps reflecting diet.

A. robustus, between 2 and 1.5 mya, possessed a still larger brain (around 500 cm^3), though this may well have been needed just to drive the larger body (around 45 kg).

A. boisei, between 2.5 and 1.3 mya, possessed a somewhat larger brain (500—530 cm^3), but was even more robust than *A. robustus*. (Its larger body, weighing around 50 kg, would have required yet more neural tissue to drive it; see discussion on allometry.) Its unique dentition sets it apart from both apes and other hominids, although some claim that *A. boisei* was simply a geographical variant of *A. robustus* and assign both species to the genus *Paranthropus*, to emphasize their distinction from other australopithecines on the human line. Its unique dentition and changes around the face and top of the cranium suggest a harsh vegetarian diet.

H. habilis, included here for sake of comparison with the australopithecines, occurred between 2.3 and 1.5 mya, with a body weight around 50 kg and a robustness probably exceeding our own. It possessed a considerably enlarged brain (around 700 cm³), as befits a species which probably lay on our direct lineage, and which produced the first undisputed stone tools of the Oldowan industry. (Susman, 1988, notes that *A. robustus* was more humanlike in both hands and feet than *A. afarensis;* its hand bones show evidence of manipulative abilities compatible with tool use, and indeed tools were found nearby. However as Klein (1989) observes, the fact that only one tool tradition, the Acheulian, occurs despite the long contemporaneity of the robust australopithecines and *H. erectus,* argues against the former using tools. Its foot bones indicate that it was bipedal and less arboreal than *A. afarensis.*) Competition between *H. habilis* and the australopithecines could have led to the latter's demise.

A possible "consensus" view reflecting current evidence and thinking about hominid evolution is as follows: *A. aethiopicus,* not previously mentioned, is included as a possible member of the lineage leading to the robust taxon (Walker et al. 1986). Dates (appearance) are approximate, and not everyone agrees on the independence of *A. robustus* and *A. boisei,* on the insertion of *A. africanus* on the *H. habilis* line, or on the lineage from *H. habilis* onward (see Figure 7.4).

McHenry (1991a) reassesses the question of body-weight sexual dimorphism in *A. afarensis* and concludes that it is in fact *below* that seen in modern gorillas and orangutans, although greater than that found in extant *Pan* and well above that of *H. sapiens.* He further concludes that this degree of sexual dimorphism fits in well with the idea that *A. afarensis* lived in large kin-related multimale groups with females that were not kin related (Foley and Lee 1989), and removes one objection (variation in size) to lumping together all the Hadar hominids into the single species (*A. afarensis*).

Boaz (1988) in his major review of the status of *A. afarensis* also rejects the widely held belief that more than one species is represented at Hadar and Laetoli. He notes that while it is the most primitive hominid species presently known, it seems closely allied morphologically to *A. africanus.* Like the other australopithecines, it had large cheek–teeth dentition relative to its body size (postcranial megadontia), with larger anterior dentition (incisors, canines, and premolars) relative to posterior dentition than in any other known hominid. Such dentition indicates a diet of fruit and leaves, whereas other robust features of teeth and jaw suggest much chewing. Its cladistic relationship with the *A. robustus/boisei* clade is considered by Boaz to be unclear, as all shared characteristics are primitive. However, after considering various phylogenetic hypotheses, he

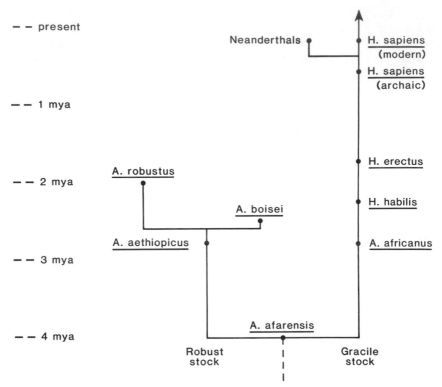

Figure 7.4 A possible "consensus" view of hominid evolution.

favors it as a species ancestral to *A. africanus,* with the latter itself ancestral to *H. habilis,* although he does not reject the alternative and widely held possibility that both of the latter sprang independently from *A. afarensis.*

McHenry (1991b) studied the femoral lengths of 31 Plio-Pleistocene hominids from between 3.1 and 0.7 mya, to estimate stature via femur–stature ratios, which have been found very accurately to predict height even in fossils. He obtained the following values (cm), female followed by male: *A. afarensis,* 105, 151; *A. africanus,* 115, 138; *H. habilis,* 118, 157; *A. robustus,* 110, 132; *A. boisei,* 124, 137, and (African) *H. erectus,* 160, 180. He notes the strong sexual dimorphism in the heights of *A. afarensis* and *H. habilis,* with much smaller values for *H. erectus,* and that by 1.7 mya, individuals at least as tall as modern humans were not uncommon. However, he urges caution in the use of some of the these figures because of uncertain attribution of femora to species and sex categories.

Conroy, Vannier, and Tobias (1990) note that two parameters for inferring evolutionary relations among early hominids are brain size and cranial circulation patterns. They conclude that venous outflow patterns in *A. africanus* resemble those of early *Homo* in accommodating the demands of bipedalism and an upright body posture, and differ from *A. robustus* and *A. afarensis*, two upright bipedal species that may *not* be on the direct human lineage. They observed that the gracile and robust australopithecines therefore evolved different cranial venous outflow patterns in response to the realigned hydrostatic pressures of bipedalism, enlarging the occipital marginal sinus system to increase blood flow to the vertebral venous plexus. By contrast, *A. africanus* (and its descendent *Homo*) preferentially delivered blood to the vertebral venous plexus via extensive anastomotic channels, including emissary veins, around the foramen magnum. We shall see later, in connection with Falk's (1990) "radiator" theory, that this may have helped cool the enlarging brain in an *A. africanus–Homo* lineage. (Both Falk 1990, and Conroy et al. 1990, note that the other older alternative circulatory pattern may still occur.)

The Evolution of an Upright Bipedal Posture

A major question is the nature and extent of the bipedalism apparently exhibited by *A. afarensis*. First, however, an even more fundamental issue needs to be addressed: What were the initial advantages of such a mode of progression? As Lovejoy (1988) observes, all other primates are basically quadrupedal; bipedalism deprives us of speed and agility, and greatly limits our ability to climb trees, thereby depriving us of many important primate foods, like nuts and fruits. We are not, of course, totally unique with respect to bipedalism, any more than we are unique in our ability to make and use tools, or to communicate by means of language. Chimpanzees, orangutans, and gibbons (and even dogs and bears) can occasionally and briefly walk upright on their two hind limbs, but only we do so habitually, with a confident striding gait of heel strike followed by a stance phase, then a push-off with the toes, a swing phase, and finally once again a heel strike in a repeated cycle (Lewin 1989a). This led to a freeing-up of the hands, permitting objects to be carried and tools to be manufactured and used, but this does not mean that bipedalism evolved "to permit" such activities; at most an upright bipedal posture was a preadaptation which *allowed* such behaviors to develop.

Chimpanzees cannot extend the knee joint to produce a straight leg and lock the knee to stand upright without having to continue to expend considerable muscular effort. They have a waddling gait because the femur does not slope inwards to the knee, as in our own case, with the result that their feet have to be placed apart, rather than in line as with us. Their anatomy is therefore a compromise between adaptations for tree climbing and knuckle-walking—an energy-inefficient mode of progression. As we shall see, bipedalism, so far from being inefficient, in fact is a much more efficient solution than the knuckle-walking bipedalism of the chimpanzees, and that fact alone may have led to its evolution.

Lovejoy (1988) compares the bipedal and quadrupedal anatomy of walking. In the latter case, the center of mass lies well forward of the propulsive hind limbs, with the result that extending the hind limbs against a substrate to propel the body forward requires a large horizontal component. On the other hand, the bipedal posture places the center of mass almost directly over the foot, and to propel the upright trunk we must reposition the center of mass ahead of one leg; however, because of the upright posture and the extension of the limbs to maintain it, there is much less latitude for further extensions for forward propulsion. Consequently, with a change from a quadrupedal (chimpanzee-like) to bipedal (human) mode of locomotion, much of the lower spine, pelvis, femur, knee joint, foot, musculature, and muscle attachments require redesigning. The gluteus maximus is greatly enlarged to prevent the trunk from pitching forward, the pelvis is redesigned (with obstetric consequences; Lovejoy 1975), and there are changes to the femoral neck to alter leverage, and so on. Thus the demands of stabilizing the human pelvis and controlling the limbs occupy several muscle groups, which serve for propulsion in the chimpanzee.

Marzke, Longhill, and Rasmussen (1988) note that the unique attachment of the human gluteus maximus to the dorsal ilium, apparent also in the australopithecines, may have permitted trunk movement which was exploited by early hominid foraging. Experiments with humans when digging, lifting heavy objects, throwing, clubbing, and so on, show extensive electromyographic activity during these behaviors, controlling flexion and rotation of the trunk on the femur during heavy use of the forelimb. The gluteus maximus positions the trunk and maintains it at angles optimal for retrieving and carrying loads. It also contributes to controlled rotation of the trunk, initiating movement and then braking movement of the trunk when the latter is used for leverage, and for accelerating the movement of hand-held tools, for example, digging sticks. Marzke et al. conclude that attachments for the gluteus maximus on the australopithecine ilium indicate that this taxon could effectively have used the trunk for leverage in foraging, lifting, digging, pounding, clubbing, and throwing; such foraging behaviors that exploit the leverage of the trunk, forelimb and hand-held tools may, as we shall see, have been a factor in the origin of hominid bipedalism.

An immediate question, of course, is from what did human (and australopithecine) bipedalism evolve? While the ancestral form might possibly have knuckle-walked like a chimpanzee, it might instead have been much more *arboreally* adapted; if so, as Lewin (1989a) observes, the transformation from quadrupedalism to bipedalism would have been less drastic. One possible scenario commences with an erect *body*, as found in vertical clinging and leaping prosimians, advancing through quadrupedalism in monkeys and apes, to brachiation in apes. Brachiation provides an easy step to bipedalism, with all its other evolutionary consequences. It should however be reiterated that the other major milestones in human evolution—tool use and an elaboration of material culture, an enlarged brain, speech, symbolism, art, music, and so on, all seem to have appeared independently at widely different times. Bipedalism greatly antedated these other developments, by at least 1 million years, and while it might have provided an initial preadaptive link in a long evolutionary chain with many links, it can in no way be said to have *directly* led to tool use (or manual signaling) by, for example, "freeing up the hands" (Frost 1980).

A considerable number of evolutionary scenarios have been offered with respect to "why" bipedalism evolved—leaving aside questions about invoking teleology in evolution—and some have already been touched on. Nor should we perhaps seek a single-factor explanation; increasingly in medicine and in ecology, multifactorial interactive causation is being invoked, and similar processes are likely to operate in evolution in general, and in the question of the origins of bipedalism in particular. We have already seen that so far from it being an inefficient and costly mode of progress, under certain circumstances bipedalism might provide decided locomotor advantages.

As Lewin (1989a) observes, in the 1960s a popular scenario was that of "Man the Hunter" (Ardrey 1976), a biped which, though less efficient than a quadruped at maximum speed, nevertheless possessed considerable stamina for following game at relatively slow speeds. (According to Calvin, 1983, the upright posture would also have permitted throwing missiles at prey.) This kind of approach gave way to views of "Man the Scavenger," following in the wake of migratory herds and opportunistically taking weak, sick, or dying animals (Sinclair, Leakey, and Norton-Griffiths 1986), thereby occupying an unfilled niche for a mammalian scavenger that would need to use stone tools to butcher through tough hides. However, Leutenegger (1987) rejects these propositions, observing that *A. afarensis* was fully bipedal by 3.5 mya or more, whereas the dental, gnathic, and cranial evidence of all australopithecines, then and later, indicates vegetarianism; the earliest clear evidence of butchering and meat eating is associated with early *Homo* around 2 mya. Thus meat eating postdates bipedalism by around 1 million years. Similarly, Verhaegen (1987) observes that vultures possess sharp beaks, and canid, felid, and hyaenid carnivores have long canines and do not need to carry stone tools.

A related idea, which however accommodates the evidence for vegetarianism, is that of Rodman and McHenry (1980): Bipedalism arose as an efficient foraging strategy for traveling between trees and other similar food sources which had become increasingly widely separated on the savannah as a consequence of late Miocene aridity. On the other hand, Jolly and Plog (1987) note that an upright bipedal posture would permit the reaching and gathering of otherwise inaccessible seeds, pods, and fruit from, for example, thorn scrub, by a terrestrial animal larger than a baboon. Presumably, any retained arboreal capacity, of the sort invoked for the australopithecines, would not have helped with the sort of food trees envisaged (see Wrangham 1980). Similarly, such an upright posture (as adopted regularly by meerkats) would also aid in predator avoidance, by enabling the creature to see over vegetation.

While Lewin (1989a) and the others reject the idea of "Man the Hunter," Zihlman and Tanner (1978; and see also Zihlman 1981) turned to the idea of "Woman the Gatherer," the female setting off, perhaps carrying an infant and digging stick, to forage for the more dependable (as is evident still in modern hunter–gatherer societies) vegetable food sources to be found on or beneath the ground. Whether (earlier or later in evolution) the male traveled afar to bring back protein or high energy food, or the female foraged for staple vegetable food, or a mixed economy developed with food sharing between the sexes, Lovejoy (1988) sees bipedalism as accompanying a set of behavioral adaptations by the time of the australopithecines. These included the family, lasting monogamy, and care of the infant by both parents. This, he says, would have expanded the mother's ability to nurture each infant and to reproduce more often. However, as we shall see, sexual dimorphism and other evidence in *A. afarensis* (and *Homo habilis*) argue instead for a polygamous society. The recurring suggestions of a division of labor between the sexes (spatially oriented hunting and foraging by the male parent, and developing the offsprings' social and communicatory skills by the female) are reminiscent of two claims; on the one side, that males may be slightly superior to females with respect to spatial abilities and correspondingly inferior verbally (Halpern 1986; McGlone 1986), and on the other, that in females there is more focal and anterior representation of speech and manual praxis (Kimura and Harshman 1984). Kimura (1983) speculates that this configuration might improve the precision and speed of fine motor skills and speech control, important abilities, she says, for a home-based parent. On the other hand, the male would have benefited from a more diffusely represented posterior (i.e., perceptual) mediation of spatial skills for foraging and hunting where fine manual dexterity and fluent speech is less important. We emphasize the speculative nature of this theory and observe that males also require dexterity in the manufacture of hunting implements, and in language for planning and coordinating the hunt itself. Moreover, any sex differences could themselves stem from environmental factors and different social roles.

One of the more bizarre hypotheses (Morgan 1984) invokes major periods of drought associated with the Messinian event: Hominoids retreated to the seashores and made rapid adaptations for marine life; for example, loss of body hair, changes in salt metabolism, weeping (salt excretion), the diving reflex (apnea and bradycardia), the presence and distribution of fetal hair, the deposition of subcutaneous fat in females, face-to-face copulation (as in most aquatic mammals), and upright bipedalism where similar muscles are used for swimming. It might also explain our universal fascination with the sea. However, the paleoclimatic evidence is at best extremely weak, other explanations can be appealed to for most of the above adaptations, and the almost uniquely human capacity to sweat would not help heat control under water (and see Wheeler 1984).

Perhaps the newest and most controversial theory relating in part to the origins of bipedalism comes from Falk (1990)—her "radiator" theory of heat control. (It should be noted that Wheeler, 1984, observed that bipedalism is also a thermoregulatory device, removing much of the body from close proximity to the heated ground, and exposing less skin to the sun's vertical rays at noon, the hottest time of the day. This would remove a major barrier to further brain growth, a large brain itself being a considerable source of metabolic heat, and would permit exploitation of a new ecological niche, the hot noontime savannah.) According to Falk, in the case of apes, the blood returning from the brain passes directly to the jugular vein, as indeed it does with reclining humans. An erect posture, from the time of the first hominids, would have imposed a large gravitational or hydrostatic load on the returned jugular blood flow. She notes two different evolutionary solutions for the problem of venous drainage (see Figure 7.5). In the case of the robust australopithecines, which were not ancestral to *Homo* (e.g., *A. afarensis*, perhaps *A. robustus* and *A. boisei*), there were unusually large occipital and mar-

A   B

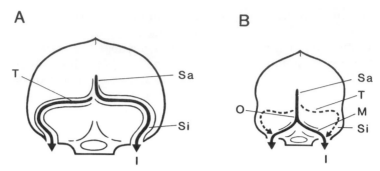

Figure 7.5 Occipital views of typical cranial venous systems in modern humans (A) and some early hominids (B). A, Blood drains from the rear of the skull via the superior sagittal sinus (Sa) and the transverse (T)—sigmoid (Si) system, and exits via the internal jugular veins. B, The transverse-sigmoid system is reduced or absent, where much of the drainage is via the occipital (O)—marginal (M) sinus system to the vertebral plexus of veins adjacent to the foramen magnum. The O-M sinus can also drain to the internal jugular vein depending on postural and respiratory constraints. Redrawn with permission from D. Falk and G. Conroy, The cranial venous system in *Australopithecus afarensis. Nature*, 1983, 306:779–781. Original version's copyright © MacMillan Magazines Ltd., NATURE, 1983.

ginal sinuses at the back of the head. When the creature was erect, blood would have flowed preferentially in the sinuses and out of the skull to veins in the vertebral plexus surrounding the spinal column instead of directly to the interior jugular vein. With the gracile australopithecines, ancestral to ourselves, and *Homo*, she notes that the occipital and marginal sinuses are much smaller. When the creature was (or is) erect, blood instead flows mostly via the parietal and mastoid emissary network to the vertebral plexus. This system, which became more elaborate and frequent over time, is said to be more efficient, especially in hot conditions, acting as a radiator for heat loss from the brain, thereby preadaptively releasing a constraint on brain size. (Brain size stayed small in the robust australopithecines with the first system.) Falk also speculates that these australopithecines lived out in the hot savannah, whereas the robust australopithecines inhabited shady forest regions where a less efficient cooling system would suffice.

It should be emphasized that this hypothesis, if confirmed (and see the associated commentaries) does not provide a prime mover for the evolution of a larger brain; indeed, that question will be addressed separately in a later section. Instead, in addition to permitting occupation of a hot niche, it removes a physiological constraint on an increase in brain size. Indeed, the venous system probably evolved in response to the demands of an upright posture (although Whiten, 1990, observes that apes are truncally erect when moving and feeding but lack this circulatory adaptation), rather than to provide a cooling radiator; this can be achieved more efficiently in other ways, for example, the elephant's large and vascular ear flaps and the counter-current system found in the nasal cavity of camels and donkeys which serve to cool the brain.

We should now return to the question posed earlier, how "modern" was bipedalism as evinced by *A. afarensis*? Klein (1989) notes the forward-placed and downward-directed foramen magnum, the humanlike foot possessing an arch and a nonopposable big toe, the leg structure which centers the body over one leg while the other moves forward, and the shape of the pelvis with a short, broad, backward-extended iliac blade. Lewin (1989a) concurs, noting also the valgus angle of the knee, and adds that such adaptations would allow a full striding gait, essentially, if not perhaps in every respect, similar to our own. While *A. afarensis* seems therefore to have been a fully committed terrestrial biped, Tuttle (1990) claims that it nevertheless possessed down-curved toes like chimpanzees and gorillas. Similarly, Jungers (1988) notes that modern humans possess exceptionally large hindlimb and lumbo-sacral joints for their body size compared to the apes, reflecting our ability to travel long distances without joint damage. In *A. afarensis*, a modest degree of hind limb joint enlargement had already occurred but not to the modern level, leading Jungers to conclude (and see also Simons 1989) that the adaptation of *A. afarensis* to terrestrial bipedalism was incomplete. Indeed, he reviews evidence that a small femoral-head diameter is found in *H. habilis* and may have persisted until *H. erectus*. On the other hand, according to Lovejoy (1988), the *afarensis* pelvis in terms of extremely flared ilia is even more adapted for bipedalism than our own, and the femoral neck is even longer than ours, with an abductor advantage even greater than our own. He concludes that adaptations for such an efficient level of upright bipedalism would *not* have permitted tree climbing as well; the femoral neck under such circumstances would have been subjected to unacceptable bending stress, and the nonopposable great toe, while an excellent propulsive lever for terrestrial walking, was quite unsuitable for arboreal behavior. Similarly he sees the hands and arms as being less suited for climbing. In his later publications (Latimer and Lovejoy 1989, 1990a,b), Lovejoy comes to essentially similar conclusions, from an examination of the heel bone (calcaneus), the metatarsophalangeal and tarsometatarsal joints in *A. afarensis*, African apes, and modern humans: *A. afarensis* was fully ground-adapted, retained no arboreal habits, and in some respects was even more adapted for bipedalism than we are.

Of course, the small diameter of the *A. afarensis* head at birth, not yet bigger than that of a chimpanzee's, meant that pelvic adaptations for walking had as yet few obstetric consequences. Lovejoy (1988) notes that the *afarensis* birth canal was elliptical, wide, but short from front to back; the birth process was only slightly more complex than that of a chimpanzee, where the infant's cranium passes straight through. With *A. afarensis*, one, sideways turn would have been needed, followed by a tilt, whereas humans require two such rotations. Geschwind and Gala- burda (1987, Ch. 19) observe, incidentally, that most infants exit the birth canal in the left-occiput-anterior (LOA) position. Although there is no evidence that the birth canal is asymmetric, the infant's head, like that of an adult, exhibits a counterclockwise torque. When the head rotates into the LOA position, the longest diameter of the head lies in the axis of the birth canal with the narrowest diameter at right angles to this axis, thus offering the least resistance to movement. (A LOA position would incidentally have been associated, in utero, with constriction of the left arm of the fetus against the mother's back, thus perhaps favoring the use of, and later even preference for, the freer right hand. Alternatively, asymmetrical brain damage from pelvic promontories during delivery could also possibly account for dextrality in humans (see e.g., Ehr- lichman, Zoccolotti, and Owen 1982).

Rak (1991) is particularly struck by the extreme width of the pelvic inlet of the *A. afarensis* specimen, Lucy, especially in relationship to body size, and in light of the small fetal head size of offspring (see Figure 7.6). He

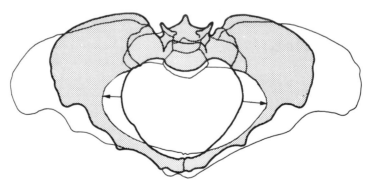

Figure 7.6 Pelvis of *Australopithecus afarensis* (Lucy) superimposed on that of a modern *Homo sapiens* female, both scaled to body weight. Although the anteroposterior length of the two pelvic inlets is similar, the width of Lucy's (indicated by the arrowheads) is considerably greater. Note that the *external* boundary of the shaded area marks the extent of the (externally smaller) *human* pelvis, and the *internal* boundary of the shaded area designates *maximal* inlet dimensions in *either* species. From Y. Rak, Lucy's pelvic anatomy: Its role in bipedal gait. *Journal of Human Evolution*, 1991, 20:283–290, with permission.

believes that this adaptation, with an associated increase in distance between the hip joints and a short femur, constituted an essential element that, when combined with the horizontal rotation of the pelvis during bipedal walking, minimized vertical displacement of the center of mass at each step. Thus part of the step length is now economically achieved via *pelvic rotation.* He concludes that *A. afarensis* was clearly bipedal and highly terrestrial, and was not simply an intermediate stage between a chimpanzee-like hominoid and *H. sapiens;* the solution to the problem of achieving a bipedal gait from the starting point of a chimpanzee-like hominoid was entirely unique. We could further ask whether this widened pelvic inlet, seen here as a locomotor adaptation, served in fact as a *preadaptation* for subsequent enlargement of the fetal head. We shall shortly return to the question of pelvic diameter and its obstetric consequences in the context of thermoregulation in *H. erectus* (Ruff 1991).

Leakey and Hay (1979) reported trackways with numerous footprints of three bipedal hominids preserved in a hardened ashfall at Laetoli, Tanzania, and dated to 3.6 mya (see Figure 7.7). Tuttle (1990) notes that they indicate fully fledged bipedalism (at least 1 million years before the advent of stone tools), and presumes they can be attributed to *A. afarensis.* He studied the feet and footprints of modern Peruvian Indians (who are used to walking without footwear, an important consideration since shoes can deform or alter the shape of the foot), and found that the Laetoli prints and strides were well within the modern range.

Homo habilis: The First True Human and User of Tools?

While *Homo habilis* is known to have continued to the time of and overlapped with *H. erectus* at 1.5 mya (see e.g., Toth 1987), and to have coexisted with *A. boisei* (Klein 1989), it first appeared around 2.3 mya (Harrison et al. 1988). Again, debate continues as to whether there is a single, very sexually dimorphic species with respect to brain volume (one group clusters around 700 cm^3, and another around 500 cm^3) and body size, or two separate species. Harrison et al. (1988) opt for the latter, *H. habilis* proper and a smaller-brained species superficially resembling *A. africanus*, and note that otherwise the degree of sexual dimorphism would exceed that of any living species. This view however is by no means universal. Indeed, Miller (1991) is adamant that variability in brain volume does not provide evidence for more than one species.

Be that as it may, Lewin (1989a) notes that this first species of *Homo* has a larger brain than any australopithecine, though smaller than that

Figure 7.7   Footprints of an early hominid preserved in hardened ash from Laetoli, Tanzania, c. 3.6 mya. Photograph by John Reader, courtesy of Mary Leakey.

of *H. erectus,* with a reduced face relative to the size of the cranial vault, a thinner cranial bone, tooth rows tucked under the face, a less massive jaw and dentition (associated perhaps with a new reliance on tools), smaller cheek teeth, and narrower premolars, although with similar wear patterns to those of the australopithecines. This last feature would indicate that it continued as a generalized frugivore (although see tool evidence discussed later), as there was no major change until the time of the partly carnivorous *H. erectus. H. habilis* appears to have retained any residual tree-climbing abilities (see previous section for dispute on this matter) of the earlier *A. afarensis,* with a hand, while essentially modern (with greater vascularity and nerve supply to the finger ends, indicating greater sensitivity and manipulative skill associated with tool behavior), nevertheless still retaining some curvature of the finger bones. This is indicative of a powerful grip and (with the creature's long arms and short legs) of a possibly retained arboreal capacity. Its pelvic and leg bones differed from homologues of *Australopithecus,* and resembled those of *H. erectus,* being similar to if more robust than those of modern humans (Harrison et al. 1988).

With respect to its brain, *H. habilis* possessed a rounded cranium resembling an enlarged, allometrically scaled version of *A. africanus.* Since a larger body needs to be driven by a larger brain, we must allometrically compare *relative* rather than *absolute* brain sizes (Jerison 1973); body size can be estimated from the postcranial skeleton. As Tobias (1987) notes, bones from the latter are usually scattered, compounding the problem that a species tends to be identified predominantly on cranial and dental evidence, rather than from postcranial bones. *H. habilis* represents the first real increase, allometrically, in brain size from the apes, attaining around 50% of the level of *H. sapiens,* that is, less than the 70–80% level of the later *H. erectus.* (We shall discuss such questions as whether *H. erectus* did evolve from *H. habilis,* and was the progenitor of *H. sapiens.*) The expansion of the habiline brain was largely one of broadening and, to some extent, of an increase in height, without noticeable lengthening (Tobias 1987). Increase was especially marked in the frontal and parietal lobes, both regions which we shall later see are intimately concerned with our own essential "humanness."

Tobias (1987) notes that gyral and sulcal impressions from the anterior surface of the developing brain are prominently imprinted on the interior of the braincase and are maximally evident in juveniles and young adults; resorption with increasing age tends to weaken such impressions. Fortunately for paleoneurology, many hominid fossils are of anatomically immature individuals. As Tobias observes, the presence of an impression can be highly informative, whereas its absence may be of dubious morphological significance. The sulcal and gyral patterns on the frontal lobe of *H. habilis* seem very similar to our own, and distinct from those of the pongids. A prominent Broca's motor speech area in the posterior part of the inferior frontal convolution of the left hemisphere is distinctly humanlike, and may indeed foreshadow speech. The parietal lobe also was well developed: The superior parietal lobule was especially prominent in the left hemisphere, as too was the inferior parietal lobule, especially in the region of Wernicke's receptive speech area, the supramarginal and angular gyri, despite current debates on their cytoarchitectonic and functional limits. Apart from the usual pattern of a higher-terminating Sylvian fissure on the right hemisphere, there is as yet, according to Tobias, no other evidence of left hemisphere–right hemisphere asymmetries, due only to poor preservation of the relevant dimensions (although see Holloway, 1983, for evidence of left-occipital and right-frontal petalias). Articulate speech, though perhaps very rudimentary, remains therefore a distinct possibility for *H. habilis*.

Tool use, too, seems to have been a first for *H. habilis*, appearing initially around 2.5 mya (Klein 1989; Lewin 1989a), or a little before (Harrison et al. 1988), the so-called Oldowan industry of Olduvai in east Africa (see Figure 7.8). Sharp flakes are found, together with stone cores, carried several kilometers from known sources; these cores may be both the source of flake tools, and tools in their own right. Although they were clumsily made by direct percussion, microscopic wear analysis of some flakes suggests that they were used for cutting plant material, meat (indicative of some degree of carnivory), and wood—which would suggest their use for manufacturing *new* tools, for example, digging sticks (Harrison et al. 1988). All these early tools were found, both scattered and aggregated, near lakes and streams, with and without animal bones which may display cut marks from stone tools, as in the camp sites of modern hunter–gatherers. Such evidence is suggestive of food-sharing, division of labor, cooperation, and reciprocity, all of which as we shall see may be important triggers of early evolution in *Homo*. Disagreement continues as to whether meat was obtained by scavenging, hunting, or both.

As Klein (1989) observes, the archaeological record of tools would be largely invisible if hominids had not the habit of returning regularly to the same site, and stone tools are both far more indestructible than bones and less likely to be accumulated by nonhuman forces. However, the accumulation of stones and bones around such natural, shady campsites as streams and rivers could still be due to natural causes—sweeping and sorting by water flow, carnivore kills (which also may have attracted human scavengers), natural death, and scavenging by animals, with

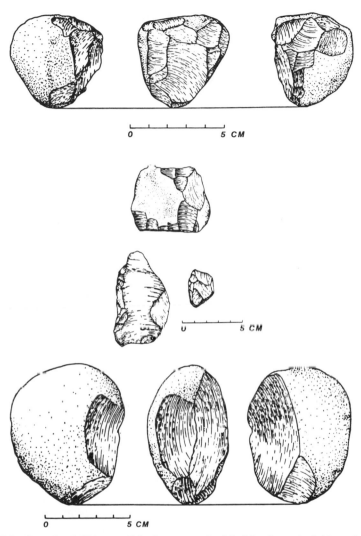

Figure 7.8 Sample of Oldowan flaked stone tools. Modifications via flaking of the core were apparently designed merely to provide a cutting edge; there is little apparent attempt to achieve an overall shape. From T. Wynn and W.C. McGrew, An ape's view of the Oldowan. *Man,* 1989, 24:383–393, with permission.

associated tooth marks often hard to distinguish from tool cuts (Binford 1985; Klein 1989). Moreover, while a tool is clearly an artifact, the problem is that an artifact is not always readily and clearly distinguished from the result of natural breakage (Ackerman 1989; Loy 1990–1991); the further you go back in time, the more primitive and natural-looking even *real* tools become. One cannot subject all fractured stone material to electron microscopy for traces of marks left by cutting and scraping, especially if very old or eroded. Ackerman notes that sites should ideally meet five criteria to conclusively establish human presence: There should be clear dating, unambiguous human artifacts, a faunal context, other traces of human activity (e.g., fire or cut-marked bones apart from the artifacts), and, finally, actual human remains; all five rarely appear together.

Despite these caveats, Potts (1988) concludes that hunting *did* occur at these habiline sites, from an analysis of bone accumulations and cut marks on bones apparently caused by stone flakes. These sites may not, however, have been home bases, but rather perhaps stone caches for raw materials for making artifacts. It could be more efficient to bring carcasses for butchering to these sites, together with stones from sources up to 11 km away, and to make and use tools there on-site. (According to Boesch and Boesch, 1984, chimpanzees in the wild may behave similarly, collecting at a convenient site both hard edible nuts and hammer stones with which to crack them.)

The Oldowan industry is said to continue on at 1.6 mya, still with *H. habilis*, into the Acheulian stone tool tradition, itself otherwise the hallmark of later *H. erectus*. According to Tattersall and Eldredge (1977), in light of the fact that *H. erectus* possesses certain cranial features out of line with a conventional *habilis*-to-*erectus*-to-*sapiens* lineage, a direct descent from *habilis* to *sapiens* has been proposed. (A direct *africanus*-to-*erectus* lineage, without an intervening *habilis* has also beenproposed; see Lewin 1989a.) Similarly, Clarke (1990) describes and reconstructs a crushed hominid cranium from an Acheulian horizon at Lake Ndutu, Tanzania, of 200–500 kya. He assigns it to archaic *H. sapiens* on morphological grounds, claiming that this taxon evolved from an African species usually classified as *H. erectus*, but here argued to be a separate species that evolved 1.5 mya from *H. habilis*, and which continued the Acheulian hand-axe tradition that began with *H. habilis* 1.6 mya. *H. erectus* is instead said to be a species originating in Asia east of Bangladesh, possibly from a habiline population isolated to that part of the world by mountains, rainforests, and other geographical and climatic features. *H. erectus*, confined and isolated thereto, would not have been cognizant of the later Acheulian hand-axe technology, and therefore was not ancestral to *H. sapiens;* the real ancestor, *H. leakeyi,* itself deriving from an African *H. habilis*. Certainly *H. erectus* is plentiful in China and Java, where, however, Acheulian industry is absent despite the presence of Oldowan artifacts. (Of course alternative reasons for this absence, e.g., lack of suitable stone coupled with an abundance of hard bamboo as a suitable alternative raw material, cannot be discounted; Rightmire 1990.)

According to this novel scenario, the habiline populations in southeast Asia and China then evolved into classic *H. erectus,* whereas African *H. habilis* evolved into an *erectus*-like species, although with sapient-like affinities. While 500–600 kya is the commonly accepted dating for Europe's oldest hominid fossils, Ackerman (1989) and Delson (1989) separately report on controversial conference claims by Eugène Bonifay that simple quartz tools from St.-Eble in south-central France are dated to 2.2–2.5 mya. For many reasons the dating is secure, but once again the question is whether they really are tools or are naturally broken rocks. If confirmed, we would have to revise drastically the conventional wisdom that *H. habilis,* evolving considerably over 2 mya in Africa, never left that continent; it is normally accepted that migrations out of Africa were first effected by *H. erectus,* which appeared in Africa some time before 1.6 mya and reached other places in the Old World around 1 mya. *H. erectus* evolved into archaic *H. sapiens* around 0.5 mya, and gave rise (in Africa) to anatomically modern *H. sapiens* around 200 kya, which then replaced other *Homo* populations elsewhere. The revised formulation would instead have *H. habilis* inhabiting western Europe 2.5 mya. However, for the moment we must follow conventional wisdom—and *H. erectus*—on the path to *H. sapiens.*

Homo erectus: A True Species, a Grade of *Homo sapiens,* or an Evolutionary Cul-De-Sac?

Even if the last word on *H. erectus* is not due to Rightmire (1988, 1990), his reviews are surely the most comprehensive and lengthy. First described a century ago by Eugene Dubois, with the oldest specimens commencing perhaps between 1.6 and 1.8 mya at the Plio-Pleistocene boundary, *H. erectus* is known from Indonesia, China, south, east, and northwest Africa, and almost certainly Europe. It exhibits remarkable similarities over a very wide area of the globe (although see the last section) and for a very long period, down to perhaps 150 kya (400 kya, according to Klein 1989). As always, however, there are gaps and scarcities in the fossil record, problems with chronology, and questions as to whether it is truly a separate species or merely a grade in an evolutionary continuum between *H. habilis* and *H. sapiens*—or even quite separate from that lineage.

H. erectus is characterized by a long, low cranium with an endocranial capacity of around 1000 cm^3, that is still considerably below ours. (There is debate as to whether or not there was a gradual increase during its long history of more than 1 million years; those like Wolpoff 1989, who see *H. sapiens* evolving locally from separate populations of *H. erectus* claim to see such an increase, which is denied by proponents of the alternative "replacement" theory, see next section.) It possessed a robust face projecting in its lower regions (prognathism, perhaps indicative of the use of the front teeth for gripping and tearing) with heavy brows, a flattened frontal squama often with a midline keel, marked postorbital narrowing of the cranium, sharply flexed rear of the vault, a blunt or shelflike transverse torus which projects most prominently near the midline, a flattened base of the skull (unlike our modern flexed configuration), a large and robust mandible, and little indication of a bony chin.

Rightmire (1990) also lists other features which distinguish *H. erectus* from *H. sapiens*. Since many of these features are plesiomorphic, having been also found in earlier *Homo* or even *Australopithecus*, we need derived (apomorphic) features for a true species diagnosis. Such features include a thickened brow backed by a flattened supratoral shelf, frontal keeling, expression of a parietal angular torus, an angled occiput, distinctive morphology of the transverse torus, robust bones of the vault, and increased volume of the upper nose. The latter perhaps permitted retrieval of moisture from the exhaled breath, thus allowing the habitation of arid environments. Rightmire concludes that it is indeed a real species, rather than just an arbitrary grade or stage in the evolution of a lineage, which perhaps only in Africa evolved into *H. sapiens*, eventually replacing all other more archaic forms of *Homo* everywhere else. In Europe, *H. erectus* may have been eradicated by hominids, represented by finds from Petralona (Greece) and Arago Cave (France), which had bigger brains and perhaps developed into Neanderthals, themselves to be wiped out later by African-derived anatomically modern *H. sapiens*. Rightmire believes that hominid evolution is essentially mosaic, with brain enlargement, the differential development of different brain areas, and the reorganization of the face and vault all occurring at different times and at different speeds over long periods. This evolution would probably operate via continuous phyletic transformation rather than as a result of episodic change.

To these observations we might add that marked sexual dimorphism was present in face and brow-ridge thickness (Harrison et al. 1988), whereas height for the sexes was in the modern range, suggesting perhaps increased cooperation between the sexes and maybe language. The robust skeleton was paired with a narrow pelvis that could have had obstetric consequences, resulting perhaps in the evolution of relative immaturity in the newborn. The neck of the femur was long as in australopithecines, whereas the femoral head was large as in anatomically modern skeletons. Generally, the postcranial skeleton indicated a terrestrial positional repertoire, with tool making, carrying, and long-distance running and walking. Simons (1989) adds that the lower cervical and upper thoracic vertebrae had nearly horizontal spines arranged as is typical in both African apes and Neanderthals, rather than inclining downwards as is more characteristic of ourselves. The shape of the thorax was conical like that of African apes, rather than barrel-shaped as in ourselves.

We have already discussed thermoregulation in the context of the origins of bipedalism (Falk 1990), and the large (relative to body size) pelvis of *A. afarensis* (Rak 1991). Ruff (1991) notes that both temperature and humidity seem to determine body size and shape in modern human populations in terms of adjustment of the ratio of surface area to body mass in the dissipation or conservation of heat in different environments. He shows that to keep a constant ratio of surface area to body mass while increasing (for whatever reason) body mass, body breadth, as indexed by pelvic dimensions, must increase. He therefore argues that the marked decrease in maximum pelvic breadth, relative to stature, from early small australopithecines to later larger hominids in Africa, may have largely stemmed from thermoregulatory constraints. These would have entailed the need to maintain a reasonably large surface area to body mass ratio in a tropical hominid by limiting increases in absolute body breadth while height increased. Ruff further argues that this constraint on pelvic breadth, together with obstetric and biomechanical factors, may have contributed to the evolution of the rotational birth mechanism and secondary altriciality, otherwise invoked in large-bodied African *H. erectus;* this taxon may therefore have been adapted for open, dry conditions and for running, thereby filling a niche as a diurnal endurance predator. Its pelvis would have effected a compromise between obstetric, thermoregulatory, and locomotor–biomechanical factors.

Harrison et al. (1988) note that in Africa, shortly after the appearance of *H. erectus,* some new tools appear, especially the symmetrical and larger biface (teardrop) forms, in an industry known as the Acheulian (see Figure 7.9), which also retained many of the earlier Oldowan features. (In China the Acheulian is absent, as we saw, and few stone tools are found in Java, perhaps because hard, otherwise suitable but nevertheless perishable bamboo artifacts were constructed.) The Acheulian industry was not replaced by the Mousterian (typically Neanderthal) in Europe until 200 kya or later. Nevertheless, the Acheulian culture of *H. erectus* coincides with the first migration to colder climates outside of Africa (itself indicative of greater adaptive capacities), the first appearance of systematic hunting, the first true home bases, the first indication of an extended childhood, and the first secure use of fire 500 kya in Zhoukoudian cave, northern China, although there is no evidence of elaborate

100 millimetres

Figure 7.9 Two Acheulian bifaces (handaxes), from between 500 and 300 kya, from Rajasthan on the left, and from central Sahara on the right. Courtesy of Martin Williams.

cooking. (Later we shall present a detailed analysis of the use of fire.) However, this taxon kept out of truly cold climates (unlike the later Neanderthals), possibly constructing shelters, though these could have been natural phenomena associated with the clumping of rock piles, such as frost heaves (Klein 1989). The stone tools do not seem to have been transported any great distance from their source, despite their more sophisticated use to cut, whittle, scrape, shred, and butcher (as shown by microscopic edge-wear analysis) on bone, antler, meat, hide, wood, and plant tissue. Interestingly, there was little evidence of the use of bone for tools, despite the fact that bone can preserve well. There is no unequivocal evidence of planned or coordinated hunting, trapping, or ambushing, or transporting large quantities of meat over a distance, despite the presence of bones from large game animals. The growing levels of carnivory, reduced jaws, and enlarged brain probably all led to an increasing reliance on tools, the use of which may itself have fed back to a further increase in brain size, though the basic toolkit changed relatively little over a very long period.

Holloway (1981a,b) discusses cortical asymmetries in *H. erectus* and Neanderthal endocasts. He finds that a combination of left-occipital and right-frontal asymmetries in both anterior–posterior and lateral dimensions is particularly strong in both fossil hominids, though such petalial asymmetries extend back to the australopithecines.

Archaic *Homo sapiens* and the Neanderthals

Archaic *Homo sapiens* emerged (from whatever provenance, see previously) in Africa and Europe between 400 and 700 kya (Harrison et al. 1988; Klein 1989; although Bräuer, 1989, proposes a later date of around 300 kya). It possessed a larger braincase, reduced skull thickness, expanded parietals, and a more rounded occiput, though it retained the brow ridges and low flat frontal features of *H. erectus*. It merged progressively into the Neanderthals in Europe alone; no such trend was apparent in Africa. The Neanderthals (*Homo sapiens neanderthalensis*) appeared in west, central, and southeast Europe and west Asia some time before 130 kya (Klein 1989)—200 kya, according to Harrison et al. (1988), and more than 250 kya in Eurasia according to Lewin (1989a)—and lasted until 50–35 kya, depending on the locality. Indeed, they may have been present in western Europe as late as 34–31 kya at Saint-Césaire (Foley 1989). This duration is short for evolution to occur. It was the last truly "primitive" human group, though it was more sophisticated than *H. erectus* and resembled anatomically modern *H. sapiens sapiens* in many ways. Neanderthals even shared the same toolkit in the Middle Paleolithic of 50–100 kya in Israel and probably elsewhere (Simons 1989). These two kinds of late hominids probably differentiated allopatrically, the Neanderthals having arisen in the north and modern *H. sapiens sapiens* having entered Eurasia relatively recently from Africa.

Like earlier *H. erectus*, which they resembled although with additional muscular hypertrophy, the Neanderthals had extremely dense skeletal bones, short forearms and lower legs (an adaptation to a cold climate?), thick skulls, and projecting brow ridges, both sexes being extremely muscular with powerful grips. The nose was large (to warm cold air?) and the face jutted forward prognathously with powerful jaws and large front teeth which from wear patterns seem to have been used routinely as a clamp or vice. Nevertheless, the very large brain (another adaptation to a cold climate? compare modern Eskimos, according to Klein 1989) was comparable in size to our own or even exceeded it. The braincase was lower and longer than ours, and globelike in cross section. The chest was barrel-shaped, with very thick ribs and general hypertrophy of the scapular region for strength. (The modern human upper-skeletal morphology is perhaps more adapted for throwing behavior and projectile use, the ease of shoulder dislocation under such circumstances might suggest incomplete adaptation in that direction.) The massive Neanderthal skeleton, implying high energetic requirements, suggests less efficient cultural adaptations than the superceding anatomically modern people, who with their more tropical limb proportions must have been better able to control environmental and climatic stresses (Harrison et al. 1988). The more primitive and flatter basicranium, even more than in other archaic forms of *H. sapiens,* might suggest a high laryngeal position and undeveloped pharyngeal anatomy, with consequently a more restricted repertoire of sound production (but see the following discussion on Neanderthal speech capacities). In any case, as Harrison et al. (1988) observe, there could be a range of bony anatomies for a particular pharyngeal configuration, the flatter Neanderthal basicranium being optimal for the long cranium and projecting face. As previously discussed, petalial asymmetries and convexities in speech-related areas resemble our own. The birth canal was wider than in modern females, suggesting either larger babies (like the adults), or a longer period of gestation with more independent precocial offspring—which again could be a cold-climate adaptation. In the latter instance this would slow fecundity and population growth, and lead to a further, slight selective disadvantage compared to incoming anatomically modern people. Finally, despite its large brain, the Neanderthal's relatively smaller frontal lobes might have resulted in reduced behavioral flexibility and an inability to engage in long-term planning, a further competitive disadvantage compared to *H. sapiens sapiens.*

Big game hunting was now well established, to judge by the evidence of killing and butchering sites (Kozlowski 1990). (Dorozynski and Anderson, 1991, review recent claims by Marrioti that Neanderthals in fact ate mostly meat; an analysis of C^{13} and N^{15} isotopes of collagen extracted from 40,000-year-old skeletal remains enabled determination of food sources that provided the protein building blocks.) Indeed, microwear analysis of Middle-Paleolithic Mousterian tools indicates butchery, working of hides, and of wood (Anderson-Gerfaud 1990), suggesting an "invisible" (i.e., lost to the archaeological record) technology in the manufacture of wooden artifacts. The lithic industry consisted largely of flakes, the size and number of bifaces decreasing compared to Acheulian industries. Knives, cutting tools, and stabbing spears often show evidence for the first time of hafting. Development as in the Acheulian was slow (although in the terminal-Mousterian Châtelperronian period, when anatomically modern people arrived in western Europe and briefly coexisted with the Neanderthals, there may have been borrowing of new traditions). There was little carving or polishing of bone, antler, ivory, or shell, little firm evidence of art (although see later), and little evidence of substantial dwellings (Klein 1989). (Kozlowski, 1990, however, describes dwellings constructed from mammoth bones at Molodova in Russia and elsewhere.) Fire was used for cooking, warmth, and perhaps exclusion of carnivores, and the dead were buried either to physically remove the body or for spiritual purposes. (We shall return to this issue later.)

Reasons for their demise are unknown, though competition with incoming modern people seems likely. However, as we shall see, the two groups, though coexisting for a thousand or more years during the Châtelperronian (around 35 kya) of western Europe, may in southwest Asia have coexisted (or alternated?) for very much longer, from 90–40 kya (Tillier 1989) where both groups had a similar simple toolkit and little art. Why were they able to coexist for so long if one group was, as is so often claimed, so markedly inferior to the other? Alternatively, why did they not hybridize if their cultures were so similar and their common origin so genetically recent? Were they in fact separate species (Stringer 1990a,b)? Alternatively, did they possess such fundamentally different adaptations that a form of *behavioral* isolation resulted, throughout their joint ranges, just as *we* do not to wish to interbreed with chimpanzees, even though technically we perhaps could?

While not everyone excludes the possibility (or evidence) of such intermingling, we may also note that the extinction of the Neanderthals need not have been catastrophic (Lewin 1989a); it could have occurred locally, over perhaps 1000 years or 30 or more generations, and this scenario would need only a slight competitive advantage (perhaps around 1%) by *H. sapiens sapiens*. Language superiority might be one such possibility; another might be birth rate. A third possibility (see Lewin 1991a) relates to pelvic shape and running. For the first time, a Neanderthal pelvis has been fully reconstructed from fragments in the Kebara cave. Sockets for the hip joints were found to be backwards, relative to our own. Thus the line of the center of gravity of the trunk passed through the hip joints and down the legs in a single, straight vertical line. In us, the line forms a zigzag in the pelvis; the line through the trunk hits the pelvis behind the position of the hip joints, moves horizontally forward within the pelvis, then vertically down through the hip joints and legs. Not only does this configuration result in a cantilevered shock-absorber, but it permits a bounce-back effect during progression in some ways analogous to the recoil–spring progression of the Australian kangaroo. With the Neanderthals, themselves massive, the hip joint takes the strain. As Lewin (1991a) asks: Which configuration is unique, ours or the Neanderthal? Unfortunately, no *H. erectus* pelvis is available for detailed comparison. Indeed, Hale (1991) notes, as an orthopaedist, that the Neanderthal hip joints may be superior to our own with respect to resultant back alignment (we must hyperextend at the lower lumbar joints to stand fully erect by increasing the lumbar curvature), back pain, and hip disease. Nevertheless, our energy-efficient adaptation may be superior for long-distance walking, jogging, or running, with its more striding gait.

Anatomically Modern *Homo sapiens sapiens*: Out of Africa or Locally Evolved?

While we cannot yet exclude the possibility that anatomically modern humans appeared in a number of regions more or less simultaneously, and more or less independently (Wolpoff 1989), the consensus currently emerging is that they first appeared in Africa, like so many earlier species, at least 100 kya when Neanderthals were the sole inhabitants of Europe (Bräuer 1989; Klein 1989). Despite earlier doubts about excavation and dating, the taxon has now been securely dated to more than 90 kya at the Klasies River Mouth and Border caves in South Africa (Stringer 1989, 1990a,b); Rightmire and Deacon (1991) give a figure of around 100 kya. Foley (1989) provides a date of 120 kya for Laetoli LH-18 and 100 kya for Omo. Rightmire and Deacon note that these remains of early Late Pleistocene individuals show all the indications of modernity.

   In southwest Asia, dates for very similar individuals have been given of at least 95 kya (possibly 115 kya) for Qafzeh and Skhul caves in Israel (Rak 1990; Stringer 1990a,b). Stringer et al. (1989) note that these two caves also contain Middle Paleolithic artifacts (Levallois-Mousterian) similar to those *later* used by the morphologically distinct Neanderthals found elsewhere locally (e.g., Tabun). This would imply that Neanderthals and early modern humans either coexisted for tens of millennia or alternated, with perhaps the latter group only episodically present due perhaps to the extreme climatic fluctuations at that time, to which the tropical newcomers may not have adapted well. Either way the latter were to be found in southwest Asia long before their occurrence in Europe, and at about the same time as their estimated first appearance in Africa. Thus they are recorded in southern Europe around 50 kya, central Europe around 35 kya, and western Europe 5,000 years later; a sequence of dates at least compatible with a continuing north-westerly migration from southwest Asia and Africa. Indeed Rapacz et al. (1991) determined the genetic distances between various extant populations according to apolipoprotein B allelic frequencies (see Figure 7.10). They found a vectored genocline from central Africa (Bantus) northward to the Middle East (Turks), with a branch westward to Europe (Swiss) and eastward to India (Indians and Tamils), China (Tibetans and Chinese), Indonesia (Indonesians) and Australia (Australian Aborigines).

Harrison et al. (1988) note that with body proportions now fully modern, the skeleton (e.g., long bone shape and thickness) and muscle mass (e.g., depth and extent of insertions) was less bulky. The skull now was rounded, the face flatter, front teeth were smaller (because they were now perhaps less often used as a vice or tool), the cranial base was now fully flexed, and while the brain volume was largely unchanged, there were subtle changes in proportions. The higher, shorter, thinner-walled cranium was in some ways more like *H. habilis* than *H. erectus* with

Figure 7.10 Genetic distances determined between various groups, according to apolipoprotein B allelic frequencies, indicate a vectored genocline from central Africa (Bantus) northwards to the Middle East (Turks), with a branch westward to Europe (Swiss) and eastward to India (Indians, Tamils), China (Tibetans, Chinese), Indonesia, and Australia. Redrawn from J. Rapacz, L. Chen, E. Butler-Brunner, M.-J. Wu, J. U. Hasler-Rapacz, R. Butler, and V. N. Schumaker, Identification of the ancestral haplotype for apolipoprotein B suggests an African origin of *Homo sapiens* and traces their subsequent migration to Europe and the Pacific. *Proceedings of the National Academy of Science,* 1991, 88:1403–1406.

its long, low-vaulted, thick-walled cranium and sharply angled occiput (Klein 1989), suggesting to some authorities that *H. erectus* should be removed from a direct *habilis-sapiens lineage*; alternatively, the distinct cranial features of *H. erectus* might merely reflect jaw function. Either way, we should again note the mosaic nature of human evolution, with different components changing at different times. Stringer and Andrews (1988) summarize these derived characteristics of anatomically modern *H. sapiens* as specifically incorporating a gracile skeleton, voluminous and remodeled cranium, reduction or absence of a supraorbital torus and external cranial buttressing, reduced dentition and supporting architecture, an orthognathous face, and the presence of a mental eminence on the mandible. Other anatomical changes associated with decreased arm muscularity include concomitant changes in shoulder, elbow, wrist, and hand, due perhaps to more efficient limb and weapon use, and a more successful dependence on manufactured tools (Simons 1989). Thus, with higher populations and greater technological competence, the stage was set for a major exploitation of all kinds of mammals, birds, and fish, even though fully modern behavior may not have been achieved until the Upper Paleolithic cultural explosion of around 35 kya in western Europe. At this time, tools for killing at a distance appear; for example, spear throwers, bows and arrows, net sinkers, fish hooks, and so on. Harrison et al. (1988) review the evidence for longer individual survival rates as the result of a more efficient society and system of tool manufacture. The evidence includes new and better blades of various kinds, now displaying considerable regional and temporal diversity, the bow and arrow, spear thrower, nets and snares, food storage, improved hearths and clothing, and better dwellings, differentiated (as now) into areas for living, food preparation, and sleeping.

A modern hunter–gatherer lifestyle seems to have been adopted, which included more elaborate burials and grave goods, implying complex belief systems. Elaborate personal decoration now appeared, together with rock art and cave paintings after 35 kya, encoding information on hunting, critical life transitions (e.g., puberty) and religion. Thus information could now be culturally transmitted across generations. Behavior generally required less energy, helped by longer legs and narrower pelvises which led to more efficient walking and running. However, the resultant narrower birth canal necessitated the birth of less mature children after a shorter pregnancy; this required cultural advances to maintain altricial offspring, including the development of language—all aspects of which will be discussed more extensively later on.

Not everyone accepts that anatomically modern humans first appeared in Africa and spread to the rest of the Old World, sooner or later replacing preexisting populations. According to the multiregional evolution model (otherwise known as regional continuity), there was an Early and Middle Pleistocene radiation of *H. erectus* from Africa; thereafter local differentiation led to the establishment of regional populations which successively evolved, through a series of evolutionary grades, to produce anatomically modern humans in different parts of the Old World. Thus there was regional evolution from Neanderthals to early modern humans in Europe (except perhaps in the west where some form of replacement almost certainly occurred), and from *H. erectus* or archaic *H. sapiens* to anatomically modern *H. sapiens sapiens* elsewhere in Eurasia (Wolpoff 1989).

The appearance of *H. sapiens sapiens* was therefore primarily the result of long-term trends in human evolution, occurring mainly through the resorting of the same genetic material in similar fashions in different regions, rather than by evolution and radiation of novel genetic material and morphologies (Stringer and Andrews 1988). Speciation events consequently were probably absent, according to this model, during the last 1.5 million years, and *H. erectus* could be designated a grade of *H. sapiens*. This model suggests that modern populations have very deep roots, having been separated from each other, with perhaps only minimal opportunities for horizontal gene flow, for maybe 1 million years.

The crucial question is the evidence (said by its proponents to be strongest in Australasia and east Asia) of regional continuity. Thus as we shall see morphological continuities are claimed between Javan *H. erectus* and some Australian fossils, and between Chinese *H. erectus* and some anatomically modern Chinese specimens. There is general agreement that there is some evidence of a diachronic trend from *H. erectus* to *archaic H. sapiens* in China; however, the evidence is far less convincing of a similar trend between archaic and early modern *H. sapiens*, there or anywhere but in Africa (Smith et al. 1989). Indeed, anatomical characteristics apparently indicative of such regional continuity are hotly disputed with different interpretations possible (Lewin 1989a,b); for example, a genuine transition, or simply regional variation, or even the consequence of hybridization. Moreover, little can be inferred from skeletal material about behavioral differences, color, and so on—all important isolating mechanisms in living species. Thus the various species of *Cercopithecus* monkeys can hardly be differentiated simply on the basis of skeletal material. Anatomical characteristics which may seem to indicate regional continuity between archaic and modern humans are also generally primitive, and so cannot be used to link specific geographic populations through time (Stringers and Andrews 1988). Nevertheless Pope, (1991), from analyses of the midfacial region of fossil and recent material, concludes that at least some traits associated with the emergence of modern humans resulted not from a single African origin, but from admixture and gene flow between different regions of the Old World. Thus he claims that fossil Asian faces contrast markedly with many of their European and African contemporaries, and in comparison with African and European populations all these features are also found in higher frequencies in extant Mongoloid populations.

In any case, let us assume that the Neanderthals could not or did not interbreed with contemporary populations of anatomically modern immigrants at the time of the Upper Paleolithic in western Europe. This is the predominant view (although see Bräuer and Rimback (1990) and Stringer et al. (1989) for the possibility of some robust features in the otherwise anatomically modern material at Skhul and Qafzeh). It would then seem strange that if all modern populations evolved separately, over a period of maybe 1 million years, they could so readily interbreed, given that their common origins would be even more distant than those of the Neanderthals, with their 300 thousand years or more of European evolution. Of course, as noted by Davidson (1990), species identifications by paleoanthropologists are classifications of the fossils and do not have the same status as zoological classifications of living species; we can never know whether particular morphologies of fossils could have generated between them fertile offspring. Indeed, archaic *H. sapiens*, from which the specialized Neanderthals derived, had a general morphology which was closer to our own than to typical Neanderthals. Evolution from Neanderthals to anatomically modern Europeans would require an impossibly rapid rate of change (Vandermeerch 1989). Generally, evidence of regional continuity between archaic and fully modern forms seems to be restricted to Africa, with abrupt transitional replacements elsewhere; this is consistent with the alternative, "Out of Africa" replacement model. It is interesting to note that the two alternative models of human evolution have their counterparts in the broader debate in biology and evolution, between gradualism and punctuated equilibrium (Gould and Eldredge 1977), corresponding respectively to multiregional evolution and replacement theory.

Bräuer and Rimbach (1990) studied the craniometric affinities among Neanderthals, Upper Paleolithic Europeans, early anatomically modern southwest Asians, and archaic and modern Africans, with uni- and multidimensional methods to assess the two competing accounts. They found that Neanderthals and Upper Paleolithic (Cro-Magnon) Europeans showed the greatest morphological differences; there were close similarities between late archaic Africans and Upper Paleolithic Europeans, and between early modern humans from Africa and Europe. They also found that the Skhul and Qafzeh finds are essentially modern, though the hint of some Neanderthal or at least robust affinities leaves open the possibility of some gene flow. (Smith, Falsetti, and Donnelly, 1989, suggest that any archaic traits could be simply primitive retentions in the incoming African populations, or homoplasies.) Thus all the evidence supported the hypothesis of anatomically modern humans originating in Africa, probably as a new *species*, in the absence of any significant admixture. This new species then spread to the Near East and then Europe, replacing (outliving, outcompeting, outlasting, or exterminating?) the Neanderthals, and to the rest of the Old World, similarly replacing relict archaic populations.

We are left with the question of why such a migration occurred out of Africa. Population pressure is unlikely, and an explanation in terms of environmental change, though easily offered, lacks empirical support according to many writers. However, Foley (1989) notes that there were major glaciations and retreats after 150 kya, and savannah, the understorey of which was grass-dominated, alternated with rainforest. Such mosaic and isolating habitats favor evolution, in contrast to the extreme stability of the southeast Asian rainforests and the extreme harshness of Eurasia (and see Foley 1991a). On the other hand, a move by tropical peoples into more northerly and rigorous climates, where adequate clothing and shelter would have acquired a new significance, might have been driven by nothing more fundamental than an urge for exploration, which we still see today.

Bräuer (1989) accepts the current consensus view of an African origin, via replacement, for anatomically modern humans. However he rejects the notion that we are therefore the result of a biological speciation event. He argues that expanding (and replacing) modern populations assimilated some archaic genes into their gene pools via limited hybridization, thus accounting for the temporary retention of some possibly archaic features (e.g., as we saw in southwest Asia). In particular, Bräuer proposes that developed *H. erectus* spread from Africa around 800 kya; the European branch led to a locally developed form of *H. erectus*, and then on to "Ante-Neanderthals" and classic Neanderthals. The African branch led on to early and then late archaic *H. sapiens*, and then to early, followed by fully, anatomically modern people. This group *also* spread via the perennially hospitable Nile valley from Africa (perhaps 100 kya, although he suggests 70 kya) to replace European Neanderthals. At this point he accepts the possibility of some gene flow, for example, between late Neanderthals and the Cro-Magnon people of European Upper Paleolithic. Thus the Châtelperronian period may not simply have been a cultural mix whereby some innovative traditions from anatomically modern invaders were grafted onto an essentially Neanderthal society; it may also have involved a *gene* mix. After all, the time scale of mutual geographical isolation of the two *sapiens* groups until then is not likely to have preempted interbreeding, even though behavioral and cultural factors might have been sufficient to lead to voluntary isolation.

Interestingly, Bräuer rejects regional continuity in Australia, as proposed by Wolpoff (1989). Thus there is currently some debate on the significance of robust hominid material at Kow Swamp and Cohuna in southern Australia. Some see it as evidence of an earlier *H. erectus* colonization (followed by later modern immigrations merging with the first inhabitants) which may have left its mark, the "mark of ancient Java" (Solo or Nyandong) on the brow ridges of the modern Australian aboriginal peoples (see Macintosh, 1965, cited by Habgood 1989). Others see it merely as a local or regional divergence. A major paradox of the Australian scene is the presence at a very early date, at least as early as in Europe, of complex behaviors involving use of ochre, art, burial, and cremation, accompanied by a very simple (some would say depauperate) stone tool industry of the Middle Paleolithic type (Mellars 1989a,b). Whereas some might argue that this was an adaptation to the local environment and poor basic materials, as Australians who know the country well, we can reject such a proposition. Alternatively, the colonization of Australia by anatomically modern people may have occurred before the technological revolution which took place at the commencement of the European Upper Paleolithic; this revolution of course happened well *after* the acquisition of a fully modern morphology.

Until very recently, the oldest secure date for the occupation of Greater Australia (i.e., the land mass, including the current continent, together with Tasmania and the islands to the north, which were all interconnected at times of lowered sea levels caused by glaciation) was 40 kya from eastern Papua, New Guinea. However, Roberts, Jones, and Smith (1990) report thermoluminescence dates indicating the arrival of people between 50 and 60 kya in northern Australia, at the Malakunanja II and Mulangangerr sites on the western escarpment of the Arnhem land plateau. Stone artifacts were present in the thousands, together with red and yellow ochres and hematite, indicating cultural activities. The authors note that these dates are not only the oldest yet proposed for aboriginal occupation, but may in fact mark the time of initial human arrival. Thus sediments devoid of human material continue uninterruptedly below the layers containing the artifacts, which commence abruptly. Since Java (Solo) forms of *H. erectus* were present immediately to the north of Australia between 110 and 55 kya, the possibility of an origin from that source, together with a continued genetic influence, cannot be discounted (although see the next section on contrary evidence from mitochondrial and nuclear DNA studies). Indeed, between 60 and 70 kya there was a cold spell when the seas were 60 or 70 m below their present level; this might have been when these first Australian settlers arrived, a period for which there is as yet no evidence of the presence of anatomically modern people in southeast Asia (although see Turner II 1989, 1990).

Relatively few stone tools have in fact been found in the Java or Sundaland (Asian) area, suggesting that perhaps more perishable materials (e.g., bamboo) may have been used (Jones 1989). Separating the Sahul (Australian) and Sunda (Asian) continental shelves are deep ocean troughs and island arcs. During the southern hemisphere monsoons, onshore north-westerlies could have helped immigration to Australia from, for example, Timor, to be completed in around 7 days, perhaps on rafts of bamboo or palm.

Jones (1989) considers the minimum number of both sexes needed in a group (or series of successive groups) to found a colony, taking note of the problems of incest taboos and recessive genes, and notes that previous speculations have inadequately addressed such issues. He observes that aborigines have long fired the land systematically in connection with foraging, with detectable consequences on pollen and geomorphic sequences and sediment cores, with a transitional event of this sort noted at 60 kya. This date, therefore, for a number of reasons, may well mark the earliest human arrival in Australia. Successive immigrant waves could have occurred from different sources, some with gracile and some with robust skeletal morphology, thus accounting for the confused archaeological picture in Australia (and see Habgood 1989). We have good evidence for multiple arrival times from, for example, the absence in Tasmania of the dingo, a partly domesticated dog similar to a type present still in India. The island of Tasmania was finally isolated from the Australian mainland by rising sea levels around 12 kya. Indeed, firm dates for a human presence in Tasmania go back now to over 30 kya and possibly 37 kya (Cosgrove 1989; Jones 1989).

Turner (1989, 1990), from a dental analysis, attempts to document the prehistoric migrations that peopled Asia, the Pacific basin, and the Americas, and concludes in favor of a two-fold split: Sundadonty (now including the peoples of southeast Asia, Indonesia, and Polynesia) and Sinodonty (China, Japan, Siberia, Mongolia, and the Americas, although with an *original* center in southeast Asia). He concludes that anatomically modern humans arrived (presumably from Africa) in southeast Asia before 50 kya, with a generalized dental pattern. This pattern is still present in the Australian aborigine, who therefore migrated *before* the evolution of Sundadonty. Sinadonty then evolved (in northeast Asia) around 20 kya, with descendents crossing to America more than 12 kya. Conventional wisdom has the first settlers reaching America from Siberia, across the then dry Bering Straits, with the Clovis industry *after* 12 kya; the immigrants traveled down an ice-free, if forbidding, corridor running through Alaska, the Yukon, and the Northwest Territories. Much disputed claims put pre-Clovis dates as early as 40 to 32 kya at Pedra Furada in Brazil (see Guidon and Delibrias 1986, and Bednarik 1989a). Equally disputed dates, though as late as 13 kya, have been offered for sites in Patagonia, Chile, and Venezuela. However, just as the Australian megafauna seems to have disappeared coincidently with the putative arrival of the aborigine, so too did the American megafauna vanish around the conventionally held time of the arrival of humans, around 11.5 kya. (Gould, 1991b, notes, however, that massive extinctions of *invertebrates* also occurred around 11 kya, arguing against the hypothesis of human overkill of megafauna in Eurasia, America, and Australia; he argues instead for a climatic–thermal factor influencing food supplies.)

We are left with the evidence that Australian cranial morphology retains a number of apparently archaic robust features (although with increasing gracility over the last 10,000 years) in face and cranium, with flatter foreheads and strong reinforcing ridges across the rear of the skull. A recent proponent of morphological continuity of at least 1 million years between Javan hominids and modern Australian aboriginal peoples is Kramer (1991). He compared hominid mandibular remains from Sangiran, central Java (1 mya), and from modern aboriginal peoples in Australia and Africa. Many more features linked Sangiran to modern Australian aborigines than linked the African and Australian groups. Stringer (1990a,b) asks whether features such as these stem from one or more migrations; for example, a robust form from the west originating perhaps in Java, and a more gracile form from the north, coming perhaps from China via New Guinea, which maintained separate identities for tens of thousands of years before finally merging—or, alternatively, did they evolve locally from a single, more or less homogeneous population which varied over time, gradually becoming more gracile? If these archaicisms *are* partly Solo derived, apart from the possible ethical implications, there are taxonomic consequences. Thus *H. erectus* would have to be incorporated within the taxon *H. sapiens,* as hybridization can only occur at a subspecies level, and *not* between species. As we shall now see, nuclear and mitochondrial DNA evidence suggests that we *all* originated within Africa within the last 200,000 years, implying that aboriginal specializations must be late; indeed, in the two best documented regions, western France and the Middle East, evidence seems decisively in favor of the replacement model (Mellars 1989a,b). Nevertheless, were we to characterize the nature of such replacement, we might still have to place a question mark over Australia (see Figure 7.11).

Molecular Evidence for an African Origin of Anatomically Modern Humans

So far, we have considered paleontological evidence which bears upon modern human origins. Complementary to such a study of *phenotypes* is modern molecular research, the domain of molecular anthropologists and evolutionary geneticists, who attempt phylogenetic reconstruction by focusing on the *genotype* (Cann 1988). Genes are units that replicate, with or without modification, and are less ephemeral than phenotypes. Phenotypes may be subject to multigene modifying influences and to the environment itself, whereas genes, if nuclear, are subject to genetic recombination through mating in each generation. Cann (1988) reviews how evolutionary trees can be inductively constructed for ancestral populations with data from variable gene states in modern populations.

Until now, molecular biology as a major source of quantitative information about our evolutionary past has had to rely largely on comparisons of genes within the nucleus. Mutations accumulate slowly in nuclear genes and, since they are inherited from both parents, mix (recombine) in each generation, thereby obscuring individual histories. Mitochondrial DNA (mtDNA), however, gives a magnified view of the diversity present in the human gene pool since mutations accumulate in this DNA several times faster (2-4% per million years in primates) than in nuclear DNA (nDNA) (Cann, Stoneking, and Wilson 1987). Moreover, since mtDNA is haploid and is clonally inherited maternally, without recombination, it provides a much more suitable procedure than nDNA for determining the relatedness of individuals to each other and for estimating the histo-

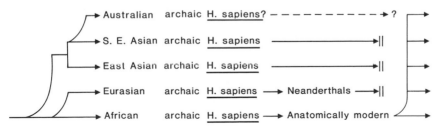

Figure 7.11   Continuity or replacement of modern *H. sapiens* in the major regions of the world.

ries of female lineages. To follow the males of a species, the nuclear genes or their products still need to be examined.

Mitochondria, about the size of bacteria, are cytoplasmic organelles involved in, among other things, converting energy from food sources into the energy useful for cellular activity (Spuhler 1988). They are quite independent of nDNA and may have originated (before the Cambrian) from free-living bacteria-like forms that established a symbiotic relationship with the host cells, in much the same way as chloroplasts in plants. Each mitochondrion contains a few copies of a small circular DNA molecule that codes for several of the enzymes needed for energy production (see Figure 7.12). Use of mtDNA to determine lineages assumes that mutations occur at a constant rate, are not repaired, and are not culled or favored by natural selection; "junk" DNA fulfills such requirements as it seems not to affect the phenotype (Stringer 1990a,b). If we know *independently* (e.g., from archaeological or physical dating techniques) when two branches diverged, we can use this knowledge to calibrate the molecular clock and determine when other divergences may have occurred. However, as we shall see, there are in practice numerous disagreements on the regularity and rate of change and the number

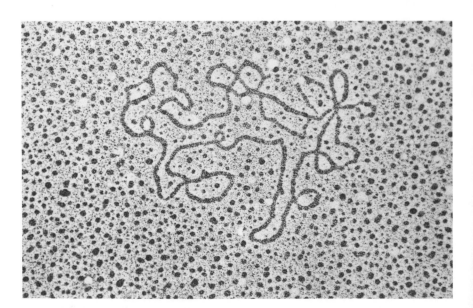

Figure 7.12   An electron micrograph of intact mitochondrial DNA, from a species of lizard. From C. Moritz, Coping without sex in the outback. *Australian Natural History*, 1991, 23:643–649, with permission.

of mutations per million years, and known divergence times for the calibration. Moreover, it is by no means impossible that leaks of paternal mtDNA may occasionally contribute to a maternal lineage (Avise 1991), although, despite possible expectations to the contrary, evidence so far for such mtDNA recombination in higher animals comes only from inter-specific crosses (hybridization) of *Mus musculus* and *Mus domesticus*.

Cann et al. (1987; and see also Stoneking and Cann 1989) sought to assess the variation between mtDNA sequences among different individuals drawn from various present-day geographic groups, assuming that the degree of variation reflected the time for which the populations have remained separated. The DNA was cut with restriction enzymes, each cleaving the DNA chain at discrete combinations of nucleotides on the chain (Lewin 1989a) to map about 9% of the mtDNA. If the mt genomes for all subjects were identical in sequence, the same pattern of fragments would always result. Any differences are usable in measuring sequence variability between individuals and populations. They studied 147 individuals from 5 geographic locations—Africa, Asia, Europe, Australia, and New Guinea—generating 133 different fragment patterns, within which there was only 0.57% overall sequence variation. The 133 different types were then arranged into a genealogical tree in the most parsimonious fashion; that is, in terms of the smallest number of changes that could link the different patterns and eventually trace back to a common ancestor. Many near-minimal solutions were possible, but all in fact produced a split between Africans alone, on one branch, and all the rest plus some Africans on the other branch (see Figure 7.13). This was taken to suggest a common African ancestor. Moreover, the African branch displayed the greatest variability, an observation compatible with the greatest antiquity.

Cann et al. (1987) made certain assumptions on the arrival times of settlers in Australia, New Guinea, and the New World to calibrate the molecular clock, and on the constancy and rate of mutations, all assumptions with which there may be disagreement (see e.g., Wolpoff 1989). They calculated that the common ancestor of all modern populations lived between 140 and 280 kya in Africa, that is, around 200 kya, and that the ancestral stock of modern Eurasians diverged from African stock between 90 and 180 kya. Indeed, if the resident populations of archaic *H. sapiens* or *H. erectus* had contributed mtDNA types to modern populations, Stoneking and Cann (1989) observe that non-African mtDNA types would be expected with an estimated divergence date around 1 mya. Since there is no such evidence, the dispersing African population seems to have replaced the prior inhabitants *without* interbreeding—contrary to Bräuer's (1989) views (discussed previously). Very recently Vigilant et

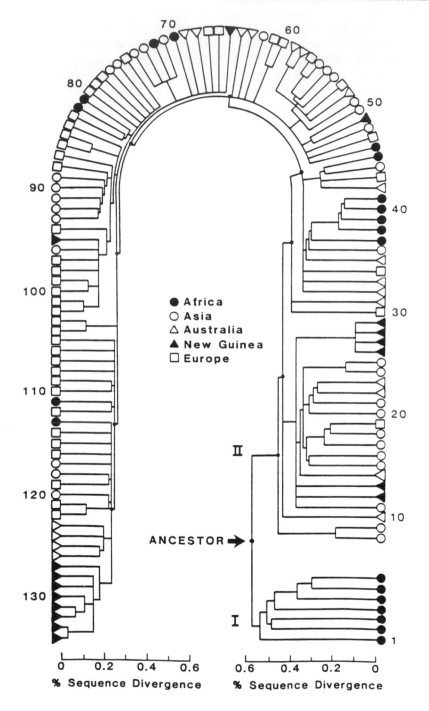

% Sequence Divergence

al. (1991) replicated and extended their earlier mtDNA study, narrowing the upper and lower limits to 166 and 249 kya for a common African origin. However, it should be noted that at press time, the validity of the genealogical tree computations has been called into question (Templeton 1992), and Stoneking himself (Hedges et al. 1992) has accepted these criticisms. Parsimonious trees with non-African roots can indeed be constructed. It is currently unclear exactly what are the implications of these recent developments for the African origin hypothesis.

If there was a population bottleneck associated with such an African speciation event—a breeding population being reduced to only a few individuals—then the "mitochondrial Eve" of 200 kya would represent the very beginning of the modern human lineage. Speciation, of course, could have taken place much earlier, with a severe population bottleneck (i.e., loss of many lineages) occurring around 200 kya to produce the same mtDNA pattern as currently observed. Note, however, that at such a population bottleneck, loss could occur of all mtDNA variability, while some nDNA variability could still be retained. Note also that the single mitochondrial Eve, representing the single, specific mtDNA type, could in fact have belonged to a population of very many others who due to chance stochastic processes *down the line* happened to leave no latter-day female descendents (Spuhler 1988). While the majority of the Africans in the sample were African Americans who would probably bear many non-African *nuclear* genes, perhaps, some might assume, being contributed in large part by Caucasian males rather than vice versa, those males would have introduced little if any *mitochondrial* DNA to the African American population.

Wolpoff (1989) questions whether the main source of mtDNA variation is simply mutations without subsequent selective pressures; whether the rate of mutation is constant everywhere, with a linear accumulation over time; and whether the colonization events used in determining (calibrating) mutation rates are accurately dated, singular, and unidirectional. As we have already seen, Australia received its first colonists perhaps a further 30 thousand years before the date assumed by Cann et al. The 2–4% divergence rate derived by Cann et al. from such figures, although the authors say it matches that of a wide range of vertebrates,

---

Figure 7.13 Phylogenetic tree of 134 mtDNA types from 148 people whose maternal ancestors came from 5 geographical regions shown by symbols in the center of the tree. Two branches, labeled I and II, indicate the two major subdivisions found. Drawn distances from bifurcations reflect degree of sequence divergence. From R. L. Cann, DNA and human origins. *Annual Review of Anthropology*, 1988, 17:127–143, with permission, © 1988 by Annual Reviews Inc.

if applied to the human–chimpanzee divergence produces a date of only 2-3 mya according to Wolpoff. Wolpoff claims that the average rate of base pair mutations is much slower, around 0.71% per million years, giving the human–chimpanzee split at 6.6 mya, a date matching paleontological and *nuclear* DNA hybridization data (see previously). This 0.71% value, according to Wolpoff, gives a divergence time for the ancestors of modern populations of around 850 kya, when only *H. erectus* was present, probably in Africa alone, with no indigenous local populations either to mix with or to isolate from.

While she does not specifically address such criticisms, in an earlier paper, Cann (1988) notes that the transmission genetics of mtDNA have been consistently misunderstood by both professional geneticists and biological anthropologists. To claim that all living humans can trace their mt genomes back to a single female founder is not to say that we all *come* from a single female ancestor, or that we necessarily experienced a severe population bottleneck early in the founding of *H. sapiens sapiens*. We should take account of the likelihood of lineage extinction, although in an expanding emigrant population, according to Smith et al. (1989), this effect is less likely. The phenomenon can be simulated by the fact that in an isolated island population the number of surnames—which conventionally of course happen to be *paternally* inherited—progressively reduces over the generations (see e.g., Avise, Neigel, and Arnold 1984). We should note the near certainty that mt variation will eventually reduce to a single maternal lineage in a species, simply because of chance stochastic processes. This constriction of human lineages to a few ancestral individuals, due to small population size, *cousin mating,* and high mortality rates before reproductive age, is known as pedigree collapse. It accounts for the well-known item of "trivia" that all English people, of native stock, are likely to be related to King Alfred. Thus the occurrence of cousin marriage results in one person assuming multiple places in a pedigree. Otherwise, as Cann observes, anyone starting with two parents and stepping back through the generations, and at each point multiplying the number of ancestors by two, would in 1000 years arrive at an impossibly large number of ancestors, more than the world's population at that time. Such effects cause pedigrees to assume a diamond shape, starting small, expanding, and then contracting.

Studies of the Y chromosome (i.e., involving *nuclear* DNA) have suggested that "Adam" (our genetic "father") might have been an African whose closest modern counterpart is an Aka pygmy in the Central African Republic (see Dorozynski, 1991, for comment). The Y chromosome is of course limited to men. Only a small portion (the "pseudoautosomal" region on the short arm) is subject to recombination with the X chromosome; consequently most of the Y chromosome does not change through generations, except for rare mutations, and so should be an accurate tracer of an ancient patrilineage. However as yet the claims are to be considered at best controversial.

Cavalli-Sforza et al. 1988) analyzed genetic information from a very large collection of gene frequencies for "classical" (non-DNA, i.e., amino acid, sequences of enzymes and proteins) polymorphisms of the world's aboriginal peoples. The data were grouped in 42 populations studied for 120 alleles. The first split in the phylogenetic tree separated the Africans from the non-Africans, just as in the study of Cann et al. (1987), and can perhaps be represented as in Figure 7.14. They found that the average genetic distances between the most important clusters was proportional to the archaeological separation times, except that the procedure predicted a peopling of the Americas at 35 kya rather than as conventionally at 11–12 kya; however, as we saw such an earlier date has in fact at least once been proposed on archaeological grounds. A second, remarkable, finding was the amount of overlap in the mapping of *linguistic* data upon the genetic tree. Thus linguistic families corresponded very closely to population groupings, indicating a very considerable parallelism between genetic and linguistic evolution, and the conclusion that the geographic isolation that led to different genetic lines also resulted in the development of different linguistic groups.

We can conclude with Mellars (1989a,b), although noting some very recent problems, that mtDNA, nDNA, protein analysis and blood group studies all converge in favour of the replacement model, probably from Africa, via a major population bottleneck. The African origin may be

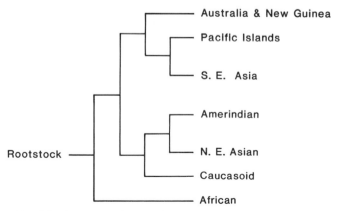

Figure 7.14 Possible clustering of extant human races in terms of average genetic distances.

dated at around 200 kya, though this figure can be challenged on a number of grounds (recent reanalyses, mutation rate, and calibration dates), as can the adaptive neutrality of mutations in mtDNA. Alternative estimates would then push back the date of Eurasian colonization by *H. erectus* to 850 kya. However, the archaeology of southeast Asia and the Middle East, especially Israel, is more compatible with the sudden entry of anatomically modern humans at a moment somewhat in excess of 100 kya.

Summary and Conclusions

Phylogenetically, we are another species of ape, closer perhaps to the chimpanzee than that species is to the gorilla. It is surprising that we exhibit so many superficial physical and behavioral differences given the comparative recency of our common divergence, and our near identity genetically. The australopithecines were our immediate ancestors and they walked upright more or less like us and had a more or less modern grasp, despite possessing brains, scaled for body size, only slighty larger than chimpanzees'. It is not entirely clear how the australopithecine and the human lineages developed; there are alternative nodes and branches along which our trajectory may have traveled, and we do not yet know the details of speciation. Nevertheless, we do know that an upright bipedal posture and an ability to grasp, make and use stone tools, and possibly to fashion other secondary tools (e.g., of wood) by means of such primary tools, long antedated major growth in cranial capacity. We do not know exactly why our ancestors adopted an upright bipedal mode of locomotion, though a number of plausible scenarios have been offered. We do know now that it is not necessarily inefficient, as had previously been thought, and that, as so often in the context of evolutionary change, more than one selective pressure may have operated. Brachiation, reaching up for fruits, and extractive foraging may all have preadapted our ancestors to make the necessary changes for upright walking. This in turn had pelvic and obstetric consequences, together perhaps with resultant changes in cranial blood flow, which directly or indirectly may have led to or permitted subsequent development of the hominid brain. Again we would emphasize multifactorial causality in evolution, with input from, for example, maturational and social factors. Just as an ecosystem is a complex feedback-driven web of elements all mutually interacting in space, so do the factors that drive evolutionary change all interact over time.

Whatever the exact australopithecine trajectory that led to *Homo habilis*, that taxon was probably the first to construct and employ stone tools on a regular basis. (Chimpanzees may naturally wield stone hammers to crack nuts on anvils, and can even learn, in captivity, to flake stone cutting implements to the level of Oldowan *H. habilis*; they do not however in the wild habitually construct and use tools in this fashion.) We do not know for sure whether what we call *H. habilis* was indeed a single, sexually dimorphic species; nor do we know for sure that we are descended from that taxon, but both are probably the case. (According to Hartwig-Scherer and Martin, 1991, the postcranial skeleton of *H. habilis* resembles that of African apes more closely than that of *H. afarensis*, which would in fact argue against it occupying an intermediate position between *A. afarensis* and later *Homo*.) Whereas *H. habilis* probably could not talk, as we do, its brain was larger than that of the australopithecines, and the neurological changes associated with human speech and language were already becoming apparent in brain endocasts. Similar taxonomic questions apply to *H. erectus*; that taxon, with its enlarged brain and Acheulian technology, was the first hominid equipped intellectually and behaviorally to leave the warmth of tropical Africa. It probably, but by no means certainly, was our direct ancestor, as well as of our recently extinct cognate, the Neanderthal.

A big question is whether anatomically modern humans evolved in Africa and migrated to the rest of the world, replacing earlier archaic populations because of a competitive advantage arising from, for example, superior speech and articulatory mechanisms. Alternatively, is some version of the minority view correct, that populations in many regions evolved in parallel, with or without horizontal gene flow, into modern peoples? The latter viewpoint, which we believe is incompatible with archaeological and molecular evidence, would suggest that differences between human populations have a long evolutionary history. Nevertheless, a question to be faced by replacement theorists is why there seems to have been little or no interbreeding or hybridization between Neanderthals and invaders. Behavioral isolation having to do with language has been proposed as a mechanism. However, in the Châtelperronian period, at the commencement of the European Upper Paleolithic, not only did the two taxa appear to coexist (indeed they may also have done so for considerably longer, beforehand, in the Middle East), but there seems to have been a two-way cultural flow. Thus the declining Neanderthals seem to have acquired aspects of Aurignacian culture and technology, and, according to Marshack, the invading Cro-Magnons may also have borrowed certain Mousterian techniques.

The dates of arrival of humans in Australia and the Americas are being continually pushed back. For Australia, a figure in excess of 50 kya seems secure, considerably before the cultural flowering of the European Upper Paleolithic; however more than one migration certainly ocurred.

The complexity and antiquity of Australian rock art is only just being realized. In the Americas, evidence is growing of human arrivals considerably prior to 12 kya, the generally accepted baseline. Indeed, just as the first Australians had to use boats to cross a series of stretches of open water tens of kilometers wide, even at sea-level minima, so too we must not exclude the possibility that boats may have been used to speed the spread of human arrivals in the Americas. Did the extinction of the megafauna, the very large animals which were widespread throughout the world in the Pleistocene, coincide with the arrival of anatomically modern humans in Europe, Australia, and the Americas, or were both phenomena dependent on a third factor, climatic change?

Further Reading

Cann, R. L. DNA and human origins. *Annual Review of Anthropology*, 1988, 17:127–143.
Falk, D. Brain evolution in *Homo*: The radiator theory. *Behavioral and Brain Sciences*, 1990, 13:333–381.
Harrison, G. A.; Tanner, J. M.; Pilbeam, D. R.; and Baker, P. T. Human Biology: An Introduction to Human Evolution, Variation, Growth and Adaptability. 3rd Edition. Oxford: Oxford University Press, 1988.
Klein, R. G. The Human Career: Human Biological and Cultural Origins. Chicago: University of Chicago Press, 1989.
Latimer, B., and Lovejoy, C. O. The calcaneus of *Australopithecus afarensis* and its implications for the evolution of bipedality. *American Journal of Physical Anthropology*, 1989, 78:369–386.
Lewin, R. Human Evolution: An Illustrated Introduction. Oxford: Blackwell Scientific Publications, 1989.
Lovejoy, C. O. Evolution of human walking. *Scientific American*, 1988, 11:82–89.
Mellars, P., and Stringer, C. The Human Revolution: Behavioral and Biological Perspectives on the Origins of Modern Humans. Princeton, N.J.: Princeton University Press, 1989.
Rightmire, G. P. The Evolution of *Homo erectus*: Comparative Anatomical Studies of an Extinct Human Species. Cambridge: Cambridge University Press, 1990.
Ruff, C. B. Climate and body shape in hominid evolution. *Journal of Human Evolution*, 1991, 21:81–105.
Smith, F. H.; Falsetti, A. B.; and Donnelly, S. M. Modern human origins. *Yearbook of Physical Anthropology*, 1989, 32:35–68.

Culture, Tool Use, and Art

8

Introduction

Human evolution is characterized in its later stages by the appearance and development of such capacities and activities as tool manufacture and use, using fire for various purposes, burial of the dead, language, art and culture, and other higher intellectual faculties including conscious self-awareness. Not all of these of course are directly observable in the archaeological record—indeed conscious self-awareness can only be inferred, with difficulty, in a living creature which is available for direct study—and alternative explanations are possible, for example, for the apparent use of fire or what seem to have been deliberate burials by, for example, *Homo erectus*. Moreover, tool behaviors and symbolic communication are by no means unique to our species; both can be studied in wild and captive chimpanzee populations.

This chapter starts by examining how evolution has shaped the human hand for grasping and manipulatory functions, essential for tool behaviors. Tool use is not restricted to humans or to even primates. We shall review recent studies which indicate that chimpanzees provide a model for many of the tool-using and manufacturing behaviors currently ascribed to early hominids. We shall also note the very slow development of changing patterns of tool behavior in the archaeological record, at least until the time of the European Upper Paleolithic, and at the same time review the evidence for very early dextrality. The question of art, symbolism, ritual, and burial remains vexed until the cultural and artistic flowering of the Upper Paleolithic, although fire seems to have been in use well before then. Finally, we shall examine in detail the cascade of developments in art-related behaviors which coincided with the arrival of anatomically modern people in Europe, even though these same peoples prior to reaching western Europe seem not to have displayed high levels of innovative creativity. Indeed, we shall see that a case can be made that they built upon preexisting indigenous (i.e., Neanderthal) developments.

Object Manipulation: Evolutionary Developments

The subset of terrestrial vertebrates known as the Tetrapoda (two pairs of limbs) includes amphibians, reptiles, birds, and mammals. Some fly, some swim, some slither, some crawl, and some run. The archetypal tetrapod is traditionally viewed as having possessed four limbs, each with five digits—the pentadactyl (five-fingered) limb. However, not all extant species possess this canonical number of digits; some possess more and some less (Gould 1991c). Had the hominid lineage reduced this complement, our manipulative capacity and tool-use might not have developed, along with our other cultural achievements. This is not to say that manipulative capacity drove an increase in either brain size or culture; whales and elephants both possess large brains, even after allometric correction for body size. Elephants possess some limited manipulative abilities with their trunks, but apart from feeding, they perhaps put them to use only in greeting, when bathing, or to wield sticks (a *tool*) in order to scratch themselves (Beck 1980). Whales may have a complex acoustically based culture, but are quite incapable of modifying their environment, and seem unable to learn how to combat predation by *H. sapiens*. Recently, even the canonical pentadactyly of the oldest tetrapods has been called into question (Gould 1991c). Newly collected material of *Ichthyostega* and *Acanthostega* of the Devonian period (390–340 mya) possess, respectively, seven digits on the hindlimb and eight on the forelimb (Coates and Clack 1990). See Figure 8.1.

Shubin and Alberch (1986) describe the embryogenesis of the tetrapod limb as the outcome of interactions among branching (making two series from one), segmentation (producing further elements in a single series), and condensation (union between elements). The forelimb builds from the body out, shoulder to fingers, with the humerus extending from the trunk, then branching to produce the next two paired elements (radius and ulna). Further branching involving the wrist bones sets the distinctive pattern that eventually forms the fingers (Gould 1991c). The radius in fact ceases to branch, yielding but a single row of segments as the limb continues to develop, whereas the ulna serves as a focus for all subsequent multiplication of elements, including the actual digits. (Similar processes produce the leg and foot bones, and toes.) The five fingers then form sequentially, the penultimate digit first, and the thumb last. The thumb is the first to be lost when tetrapod lineages reduce the number of digits from five to some smaller number, followed by the digit which is normally next-to-last to form in a normal pentadactyl animal, and so on. Amphibia, in fact, typically produce four digits on their front limbs. Five fingers, perhaps an optimal number for human manipulation,

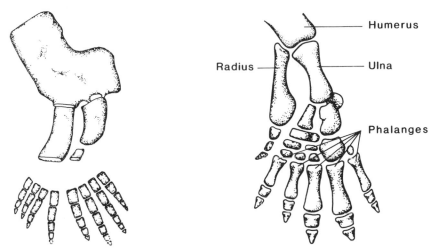

Figure 8.1 Forelimb of early tetrapod *Acanthostega* (left) has eight digits; the embryological development of a modern tetrapod hand (right) proceeds generally along the ulna-derived phalanges from left to right, as shown. Partly redrawn from M. I. Coates and J. A. Clack, Polydactyly in the earliest known tetrapod limbs, *Nature*, 1990, 347:66–69; and from S. J. Gould, Eight (or fewer) little piggies, *Natural History*, 1991, January, 22–29, with permissions.

may in fact be "a happenstance, not a necessity" for tetrapods (Gould 1991c). As Gould observes, had we, like the Amphibia, only four fingers on each hand, at the very least we would probably employ an arithmetic to base eight and would find that keyboard music and typing would now pose a rather different set of requirements.

Grasping is not of course limited to primates; various marsupial species and the squirrels all can hold and manipulate food with varying degrees of efficiency (Megirian et al. 1977), as indeed can such birds as the parrot, which we have seen may also display behavioral asymmetries in this respect. Grasping is, however, a mark of the primates, and this ability of a limb extremity to surround an object is made possible by the evolution of the prehensile hand. It was probably an early adaptation for arboreal locomotion among prosimians, although their possessing frontally located eyes, which are not necessary for arboreal locomotion per se, suggests that grasping may also have been an adaptation for visually directed predation, for example, on insects (MacNeilage 1990). Prosimians are restricted to a whole-hand power grip. Lorisoids reach by spreading their digits widely apart, touching the object with the distal part of the palm, and then close the fingers around it. Lemuroids, in contrast, reach with their finger tips more or less parallel and touch with their fingers first, then they hook their fingers around the object, pressing it against the palm. With the higher primates the precision grip develops, the object being held between the terminal phalanx of the finger, or fingers and the thumb. This process is optimized by the development of a fully opposable thumb in humans and in some of the more terrestrial Old World monkeys. Thus the carpometacarpal joint permits rotation of the thumb so that it can sweep across the palm, and its pulp (tip of the volar surface) can directly oppose the pulp of any of the digits. Precision grips can to varying extents also be attained without fully opposable thumbs; thus pseudo-opposability occurs in some arboreal taxa, where the thumb is opposed to the side of the index finger. In the scissor grip, an object may be held between two fingers.

While the precision grip is of course a sine qua non for efficient tool manipulation, the latter almost certainly was not the major evolutionary driving force for the former. Thus as we saw, non-tool-using australopithecines were already well on their way in this particular evolutionary respect, although a precision grip doubtlessly served as an essential preadaptation for tool behavior and further cognitive development. MacNeilage (1990) in fact opts for extractive or invasive foraging as a driving force in the evolution of fine hand control in primates (and see also Parker and Gibson 1979). Thus, as discussed earlier, a left-hand preference is typically manifested in prosimians for reaching, with a right-hand preference for postural support (the power grip) leading to fine manipulation, by the right hand, of objects grasped by the left. A variety of fruits and seeds have to be cracked, peeled, shelled, unravelled, and so on, by frugivorous extractive foragers (Gibson 1990) via a wide range of techniques, to say nothing about cutting through animal hide to obtain meat for carnivores.

Parker and Gibson (1979) see prosimians, Old World monkeys, great apes, and humans as constituting a series of grade levels of terminal intellectual abilities corresponding to the Piagettian sequence of stages of intellectual development in human infants and children. (It should be noted that over the last decade Piaget's stages of intellectual development in children have been increasingly criticized (see e.g., commentaries to Chevalier-Skolnikoff 1989) both from the standpoint of theory, and from an inability to confirm empirically the predictions made. The account is, however, useful for classification purposes and for making comparisons between taxa with respect to behavioral acts differing in complexity; it also attempts to specify the cognitive processes underlying complex behaviors like tool use.) Parker and Gibson try to match these four evolutionary grades with the four stages of human sensorimotor development—although it is not clear why the four evolutionary taxa *should* exactly match Piaget's four sensorimotor stages. Thus Parker and Gibson propose an ontogenetic recapitulation of evolutionary stages, though as observed in the Commentary to their paper, the concept of ontogeny recapitulating phylogeny is also to be viewed with considerable skepticism.

We can of course compare object manipulation by humans and nonhuman primates without invoking Piagettian stages of cognitive development, or ontogenetic recapitulation of phylogeny. Vauclair (in press) notes that adult apes rarely act on objects with the apparent intention of engaging an infant's attention, unlike human caregivers, especially mothers, who will regularly and repeatedly do so to attract and sustain the infant's attention. (However, see Boesch, 1991, for an example of deliberate coaching in tool use by chimpanzee mothers.) Similarly, infant apes attend much less to the actions of their caregivers, although in the presence of human caregivers they may act much more like human infants. Moreover, human infants transfer objects between hands more so than do infant apes. Role differentiation between hands, for example, one hand holding an object and turning it appropriately, while the other hand operates on it, processes which in humans are usually the domain of the nonpreferred and the preferred hand respectively, may be largely a human phenomenon. (For an exception, see termite dipping by chimpanzees, discussed later, where one hand wields the twig and the other sweeps off adhering termites.) Such role differentiation between hands is said to appear in the second half of the first year of a child's life (Vauclair, in press). The left hand may thus delineate the *frame* into which the right hand inserts the *contents*, a concept with overtones of a right-hemisphere global–holistic and a left-hemisphere serial–analytic dichotomy (Bradshaw and Nettleton 1981).

This concept also finds parallels with the view of *language* as a dualistic and serial structure (MacNeilage, Studdert-Kennedy, and Lindblom 1984). According to this view, at a phonological level, consonant and vowel elements are inserted into syllabic frames, whereas at a morphological level, words (contents) are inserted into syntactic frames. We shall discuss later the idea that the syntax of language may have its counterpart in the syntax of manual praxis involved in making and using tools, and the fact that the (motor) language centers in the brain are closely adjacent, in the same left hemisphere, to the centers that control manual praxis (Kimura 1982). We note in passing that the syntax of language involves hierarchical organization and the embedding of subsequences within larger sequences, whereas the praxis of tool manufacture (though not necessarily of use) instead tends to require simple chaining or sequencing, in the manner of beads on a string. That said, Vauclair (in press) speculates whether complex object manipulation is linked to the emergence of *H. sapiens* and language, and whether there is a common software (a capacity to understand means–end relationships) which underlies both object manipulation and language. Thus both processes, language and complex object manipulation, use intermediaries to achieve a goal; those intermediaries are, respectively, other individuals and tools.

Nonhuman Tool Use

Tool use, language, and upright bipedal locomotion are commonly held to characterize humans in some kind of unique fashion. However, none of these are in fact totally unique to us. Depending on how we define language (see below), apes may or may not be said to be capable of learning to use a form of language. Other species (bears, dogs, and chimpanzees) can briefly adopt a bipedal posture, but only we do so habitually, and even then this posture may possibly not have been fully and permanently achieved until after the demise of the australopithecines. Tool use in one way or another is certainly not our prerogative (Beck 1980), although again problems of adequate definition have to be addressed first. If we call a tool something that is detached and separate from the user's or operator's own anatomy, which is not attached to any substrate, and which is used to change the state of another object either directly through action of the tool on the object or at a distance as in aimed throwing (Parker and Gibson 1977), we can eliminate beaks, claws, or stings, but include such things as the anvil stones held by Californian sea otters on their chests for cracking shellfish while floating supinely, or the pieces of dead grass carried by the Imported Fire Ant (*Solenopsis invicta*) to soak up honey to carry back to their nests (Barber et al. 1989).

In the avian context, Pepperberg (1989) describes how three parrots from two quite different species, an Australian and an Amazonian, used bottle caps, empty nut shells, and teaspoons to bail water and seeds from containers the levels of which were too low for access otherwise. jays shredded newspaper bedding to rake in otherwise inaccessible fo· and to mop up wet, adhering food so that it could be eaten. A roo.. plugged a hole in its aviary floor so that water could accumulate for drinking and bathing. Finally, Pepperberg describes how a pair of ravens flipped stones onto human intruders to protect their nest.

Hauser (1988) reports how an old female vervet monkey (*Cercopithecus aethiops pygerythrus*) *discovered* in the wild how to use the seed pods of *Acacia tortilis* to soak up and extract nutritional exudate from recesses in the tree; she deliberately used old dry pods and left them to soak for about 2 minutes before removing and consuming them. Subsequently, other members of the troop observed and learned to initiate this useful tool behavior. Japanese macaques in one locality have also been reported (Kawai 1965) discovering how to separate grain from sand by dropping the mixture in water; grain floats, unlike sand. This tradition was then passed on. Superficially similar is the habit of tits in the United Kingdom "learning" to steal cream from milk bottles by piercing the tops (Hinde and Fisher 1951), a habit that spread nationwide. However, beaks are not tools, and the birds have a natural probing behavior in their repertoire, for example, for seeking grubs under bark. Nevertheless, macaques are known to hammer open oyster shells with stones, and to use them for crushing scorpions and for grooming (Falk 1989).

Indeed, the New World capuchin (*Cebus capucinus imitator*), a genus whose cerebral cortex closely resembles the Old World macaque's in sulcal patterns (Falk 1989), and which is renowned for its manipulative dexterity, seems to possess a repertoire as complex and diverse as chimpanzees with respect to tool use (Chevalier-Skolnikoff 1990), despite the popular belief that New World monkeys are less "advanced" than those of the Old World. Thus wild individuals have been observed breaking off and dropping branches on one another, flailing and throwing sticks at another, probing holes with twigs, and bashing two nuts together to break them. Captive individuals (Chevalier-Skolnikoff 1989) used objects to rake in inaccessible food and to knock down suspended food, made ladders out of sticks to reach food, stacked and climbed boxes for the same purpose, pushed food out of tubes with sticks, and used sticks to pry lids off boxes, hammered nuts with stones, used a container to collect food and water, and retrieved objects out of sight to use as such tools. The behaviors all appeared deliberate and insightful. The same applies to tool use in the orangutan (Galdikas 1982); semiwild animals seem to employ a wider range of tool uses than completely wild animals, but are still far behind chimpanzees in this respect.

In the wild, chimpanzees strip twigs to fashion probes for termite "fishing," first sweeping off the leaves, holding the stick with one hand, probing, and then wiping off the insects with the other hand (see Figure 8.2). Chimpanzees (like Egyptian vultures) use stone hammers to crack nuts, and chew leaves to make sponges to collect water from otherwise inaccessible crannies (Goodall 1968). They use sticks as weapons, rakes, and ladders. Charles et al. (1991) report that a captive chimpanzee learned how to strike sharp flakes off a cobble held in one hand with a hammer stone wielded by the other, and to use the flakes to cut string around a box containing food. Wild chimpanzees may bring several tools to bear on a task, applying them in a precise sequence, like a *toolkit*, to

Figure 8.2 The chimpanzee, Flint, "fishing" for termites. From J. H. van Lawick Goodall, The Chimpanzees of Gombe: Patterns of Behavior. Cambridge, Mass.: Harvard University Press, 1986, with permission.

get access to a honey source (Brewer and McGrew 1990). Thus a large branch with a sharp end may be used as a chisel to jab at the wax coating of a bees' nest. This may be followed by a shorter thinner stick with a better point for more accurate working on the hole made so far. Next a green branch around 1 cm in diameter and trimmed to around 30 cm long may be pushed into the hole to puncture the seal. Finally, a green vine dipped through the hole may be employed to extract the dripping honey. This sequential use of a range of tools by animals in the wild is in principle almost indistinguishable from our own tool-using behavior.

Wild chimpanzee mothers have even been reported as actively *coaching* their offspring in tool use to crack nuts (Boesch 1991), at some cost to themselves in terms of time, tools, and nuts. Thus they leave nuts and a "hammer" near an appropriate "anvil," and intervene when the young chimpanzee runs into difficulties, demonstrating how to hold the hammer and how to reposition the nut. The mother presumably understands what the pupil does *not* know, recognizing the extent to which its performance falls short of optimal, so that she can best correct it.

Kortlandt (1989) noted that wild chimpanzees cracking oil palm nuts use the same motor patterns and lithic technology (a stone hammer) as the surrounding villagers, suggesting imitation or learning. Conversely, chimpanzees cracking nuts from forest trees use heavy hammer stones (or wooden clubs) with stone mortars or tree-root anvils, with the direction of the hammer blow following the perpendicular to the flat surface through its center of gravity; however, no human stone tools, even from the Oldowan culture of *Homo habilis*, apparently show this characteristic wear pattern. Moreover, to date, there is no evidence that chimpanzees make tools for more than one purpose (as we do), or to fashion *other* tools, for example, digging sticks. Edge-wear analysis of stone tools from the Oldowan Koobi-Fora formations suggests that in some cases they may have been used to cut wood, implying that they may have been used to fashion (perishable) wooden implements. However, according to Wynn and McGrew (1989), all behavior that can be inferred from Oldowan tools and sites falls within the range of the apes' adaptive grade, with nothing exclusively humanlike in the oldest known archaeological evidence. However, they note that the Oldowan did include two behavior patterns that point in the direction of adaptations found later in hominid evolution: carrying tools and food for considerable distances (although this may also be to some extent a characteristic of chimpanzees), and competing with large carnivores for prey.

Finally, though it is not strictly tool use per se, we should mention recent reports (see Newton and Nishida 1990) of chimpanzees ingesting herbs for their medicinal properties. Several species of plant are reported to be used, both by the chimpanzees and local villagers, for a range of health conditions, the leaves having been shown to contain various pharmacologically active substances. In particular, the animals were observed not to swallow the leaves but to hold them between the tongue and the buccal surface, massaging them without chewing for a length of time. The authors note that this technique is extraordinarily similar to buccal and sublingual drug administration used in human medicine when the stomach route would inactivate the active substance. Indeed, the active substances in these herbs were such as to be optimally absorbed via the buccal mucosa. The authors do not speculate on how nonhuman primates could have evolved these behaviors.

Human Tool Use

The first stone tools associated with the hominids date back to around 2.5 mya—the Oldowan industry, associated presumptively with *H. habilis*, and consisting of crude scrapers, choppers, and flakes produced by a few blows with a hammer stone (Lewin 1989a). (Although such *wooden* implements could of course never have survived, Bahn, 1989a, concludes from an analysis of characteristic dental grooves that *H. habilis* was a habitual user of tooth picks.) There was very little change in the lithic record until 1.5 mya (*H. erectus*) when the Acheulian culture gradually supervened, with teardrop-shaped handaxes appearing alongside the Oldowan for half a million years. In South Africa, toward the end of the Acheulian, the Levallois technique for making flakes arose around 200 kya. There was much more intensive core preparation, and complete flakes were struck off at a single blow. Around 150 kya, the Middle Paleolithic Mousterian culture arose, when an apparent sense of stylistic order first appeared with the Neanderthal populations; however, available technology and materials, and the need for resharpening, were constraints that necessarily imposed shape and form, and to some extent the end product was an inevitable result of such forces. This tradition, with its increasing use of bone, antler, and ivory lasted until around 35 kya, when the Upper Paleolithic revolution of the Aurignacian culture appeared and the anatomically modern Cro-Magnon peoples rapidly replaced the archaic forms of *H. sapiens* in Eurasia.

The Aurignacian and subsequent cultures were marked by ever more rapid changes and displays of technical virtuosity (for review, see Chase and Dibble 1987; Lewin 1989a). It is only with the Aurignacian that we get a feeling of art for art's sake, of shear technical virtuosity and bravura, and an inevitably accompanying linguistic tradition. While many authors have invoked cognitive categorization and linguistic representation as guiding forces in tool making before the Upper Paleolithic (see Chase and Dibble 1987), appealing to common syntactic processes in both domains, it is far from clear that the transmission of knapping techniques required verbalization. Indeed, even today such skills are far better taught by imitation and demonstration. Nevertheless, tool making in the Middle Paleolithic was clearly learned; it was purposive and involved planning despite being modified by local conditions such as available raw materials, end-use requirements, and repeated utilization and re-shaping.

In western Europe, the Châtelperronian, technically part of the Mousterian tradition, overlapped and alternated with the early Aurignacian in several sites for some centuries (Mellars 1989). Either the two groups of people alternated (compare similar proposals, previously discussed, with respect to the same two groups of humans around 70,000 years earlier in the region of Israel when *H. sapiens sapiens* and Neanderthals seem to have alternated for a very much longer period), or they were sufficiently adapted to coexist and not compete. Either way, the Neanderthals seem to have acquired Aurignacian technology and forms of personal adornment.

Wynn (1989) adopts Piagettian principles in an attempt to assess spatial thinking and the evolution of spatial intelligence, as manifested in the nature and evolution of lithic industries. As mentioned earlier, not everyone accepts Piagettian theory or that such principles can be applied comparatively to phylogeny, especially with respect to extinct species whose behaviors can fossilize only to the extent that they leave behind tools and artifacts; nevertheless, potentially illuminating comparisons and conclusions can be drawn. He concludes that the spatial repertoire of hominids in Oldowan times was very limited; the chopper and scraper required only the concepts of proximity, separation, and order in positioning successive blows to create the implement. With the advent of the Acheulian culture, standardization and symmetry indicated that the artisan had a clear idea of what was being aimed at, together with perhaps an aesthetic sense. The final stage (late Acheulian, around 300 kya) saw the discovery of perspective, according to Wynn, and essentially modern ideas involving three-dimensional symmetry, straight edges, regular cross sections viewed from any direction, and parallel lines—all far beyond trial-and-error knapping. He regards these step changes as reflecting quantal advances in spatial intellect, at the very least, analogous to those experienced by the developing child.

However, Davidson and Noble (1989) note that apparent standardization in lithic cultures can result from the mechanical and practical constraints of working with a medium, and need not reflect cultural tradition or even topological concepts or "mind's-eye" models. Moreover, they observe that it is not necessarily the case that the final form of flaked artifacts, as found by the archaeologist, need necessarily represent an intended shape; we may tend to discover mostly failures or castaways, and who is to know whether apparent core tools really *were* tools or just the residual from which useful flakes had been struck until no more could be produced? These authors even advance the radical speculation that the apparently sophisticated, symmetrical, and aesthetically pleasing Acheulian biface "handaxes" were merely what was left after extracting all useful flakes.

Early Evidence of Dextrality in Tool Use and Art

We have already reviewed evidence for handedness in nonhuman primates, and for morphological asymmetries in the brains of monkeys, apes, and fossil hominids. At the behavioral level, evidence is mounting for hand asymmetries comparable to our own dextrality from the time of *H. habilis*. (It is not clear how much reliance we can place on an early report by Dart, 1949, that the skulls of baboons found with the remains of australopithecines belonged to animals killed by the application of "dextrous force"; he claimed that radiating fractures from the impact of hard objects appeared more often on the left side of the animal's skull than on the right, thus seeming to indicate dextrality in the attacker.)

According to Toth (1985), an analysis of prehistoric stone artifacts from Lower Pleistocene sites Koobi Fora in Kenya (*H. habilis*) and the Acheulian (*H. erectus*) horizons at Ambrona in Spain, reveals a preferential clockwise rotation of stone cores during flaking, consistent with that produced by dextral toolmakers (see Figure 8.3). Toth notes that when flakes are struck from a core, some flakes will normally exhibit areas of weathered outer surface (cortex). When such flakes are oriented with their striking platforms (butts) upwards and the outer dorsal surface towards the viewer, those with the cortex only on the right side of the flake suggest that the hammer stone struck to the right of the previously removed flakes. Then, the noncortical left side of the flake usually represents the scar or scars of previous flake removals. To remove flakes efficiently, one should use the acute angles on the edges of the cores and follow ridges or bulges on the cores when detaching the flakes. When a sequence of flakes is removed, and the core is held in the left hand and the hammer stone is wielded by the right, the left hand usually turns clockwise (viewing the core from above) to achieve fine muscular control of the flexors and supinators. The resultant core flakes will then exhibit

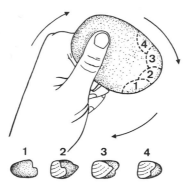

Figure 8.3   The pattern of successively removed flakes from stone cores indicates that the earliest tool-using hominids may have been dextral. Dextrals hold cores in the passive left hand and rotate them clockwise, leaving a sequential trace of characteristically superimposed scars. Redrawn from N. Toth, Archaeological evidence for preferential right handedness in the Lower and Middle Pleistocene, and its possible implications. *Journal of Human Evolution*, 1985, 14:607–614.

a rightward orientation. At Koobi Fora, Toth obtained a right-to-left percentage ratio of 57:43 flakes, and at Ambrona a ratio of 61:39.

Microwear analysis of flake tools from Acheulian (Lower Paleolithic of 200 kya) horizons at Clacton, England, which were used for boring, indicates a clockwise direction of rotation at the time of applying downward pressure (Keeley 1979). Similarly, according to Oakley (1972), Middle Paleolithic tools associated with *H. erectus* from Choukoutien (China) appear mostly to have been chipped by dextrals. Cornford (1986) likewise provides a detailed analysis of dextrality by individuals intermediate between *H. erectus* and Neanderthals excavated at La Cotte de St. Brelade (France), who employed the characteristic *coup du tranchet* technique; this was done to resharpen or modify the edges of existing flint tools, probably in response to diminishing supplies of suitable stone. The technique was first recognized through its numerous byproducts. These consisted of two types of flake, one having been struck longitudinally (long-sharpening flakes, LSFs), and the other laterally (transverse-sharpening flakes, TSFs). By removing these flakes a new edge was created, of the greatest possible length and sharpness, on the parenttool. Cornford notes that the essential asymmetry of the mode of production, as mirrored in the byproducts, makes it possible to observe a strong bias to one side in the relative position of the new and old bulbar surfaces, which seems to be linked to the maker's use of a preferred hand, usually the right. Indeed, from a sample of 1590 flakes the dextral pattern occurs somewhere between 71% and 84%, depending on criteria adopted. Similar conclusions also apply to the Neolithic (Spennemann 1984b); an analysis of grinding striations on bone and antler implements from southern Europe of 6 kya indicated that sinistrality ranged around 6.3% in Switzerland and about three times that figure in southern Germany.

Teeth, too, provide evidence of an ancient dextrality. Bermúdez de Castro, Bromage, and Jalvo (1988) review reports of striations on the buccal surfaces of upper permanent anterior teeth of Neanderthals and Anteneanderthals from sites in France and Spain (see Figure 8.4). A microscopic analysis indicated that the individuals cut materials, for example, meat held between the maxillary and mandibular incisors, with stone tools which inadvertently scratched the enamel in a characteristic configuration. The authors also performed experiments with false teeth held in place while meat was cut as it was gripped between the false teeth, using the left or right hand. Analysis of the location and orientation of the striations, modern and fossil, indicated that the latter were predominantly made by dextrals, suggesting that the individuals held meat (or possibly hide) between the teeth and cut it with a (stone) knife wielded by the right hand.

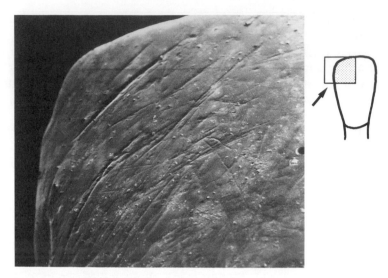

Figure 8.4   A scanning electron micrograph of a lower central incisor of an anteNeander-thal from Spain. The striated wear patterns on such anterior teeth are compatible with abductive cutting of meat by a stone, wielded by the right hand, held in the front of the mouth. Courtesy of T.G. Bromage.

In this context, an interesting aside is the recent debate concerning grooves found in the teeth of humans, including Paleolithic, Neander-thal, archaic *H. sapiens, H. erectus,* and *H. habilis* (Bahn 1989a; Puech and Cianfarani 1988). These grooves occurred along the junction between the cementum and the enamel, mostly on the upper and lower molars and premolars. Possessing a semicircular or trough-like shape with longitudi-nal and parallel striations, the grooves were often bordered by ridges of reactive cementum, and seemed to have been produced as a response to irritation caused by some implement. Root caries, gritty saliva forced between the teeth, and fiber processing were dismissed by Bahn; he notes that the constant location of the grooves, their longitudinal polish, and striations suggest a prolonged back and forth movement of an inflex-ible probe, and by reference to similar grooves in, for example, Amerindi-ans and Australian aborigines, use of toothpicks is invoked. Ekhardt (1990) and Brown and Molnar (1990) instead opt for an explanation involving the stripping of animal sinews, although one wonders whether that would have fallen within the behavioral repertoire of *H. habilis.*

An analysis of art may assist in the elucidation of ancient dextrality–sinistrality ratios. We tend to draw left- rather than right-facing profiles if right handed, because the activation pattern of adductors and abductors in the wrist and arm is easier when a dextral performs a counterclockwise loop, and profiles tend to be drawn from the head down rather than vice versa (Alter 1989). However, with respect to the Upper Paleolithic cave paintings of western Europe, an analysis of around 300 examples from Lascaux and associated sites indicated an approximately equal incidence of left- and right-facing paintings (Perello 1970; though see Wilson 1885). This counterintuitive finding from such comparatively recent art (around 16 kya) might possibly reflect the nature and size of the medium. In fact, Cro-Magnon hand silhouettes are usually of the left hand (i.e., drawn by the right around the left, or with the right hand supplying pigment, e.g., to be blown via the mouth; see e.g. Uhrbrock 1973). Similarly, microscopic analysis of finely engraved lines (illuminated from the side) of Upper Paleolithic material indicates that they were mostly produced by dextrals.

Coming now to historical times, Spennemann (1984a) examined the monumental structure of Borobudur in central Java, which was constructed just over 1000 years ago and is lavishly decorated with scenes drawn from life. Of over 1000 depictions of individuals performing skilled unimanual actions (playing musical instruments, serving food, writing, using tools, beating, and whipping), around 9% involved the left hand; with respect to unskilled actions (leaning on a stick, using a bell, holding an object) the figure rose to around 38%. Essentially similar conclusions were drawn from a monumental analysis of historical and late prehistoric art by Coren and Porac (1977). They assessed more than 12,000 photographs and reproductions of drawings, paintings, sculptures, tools, and weapons produced over the last 5,000 years; out of 1,180 scorable instances, they found that the right hand was used in 93% of cases, regardless of historical era or geographical location. Of course, the artist may not be able or wish to mimic contemporary patterns of hand preferences, nor may the artists themselves necessarily be dextral or sinistral in proportion to the rest of the population (Mebert and Michel 1980). Moreover, will an artist paint dextrals and sinistrals in proportion to their real-life incidence, or might he or she represent *all* subjects in accordance with the hand preference of the majority of the population?

Art and Symbolism before the Upper Paleolithic

Marshack (1989a) notes that when the Upper Paleolithic "creative explosion" of rock art and artifacts occurs between 35 and 40 kya with the appearance of anatomically modern humans in western Europe, and the production of material of considerable technical sophistication, it does so suddenly and only in regions of prior Neanderthal habitation and Mousterian culture. There is no evidence of comparable activity among similar anatomically modern humans in central Europe, the Near East, Africa, or the Mediterranean. He concludes that the sudden west-European phenomenon could not simply have been a genetically controlled increase in symboling capacity (the rate of cultural change was too fast for that), or just the intrusion of an *already* highly developed symbol culture from elsewhere, although this certainly did occur. Rather, this activity was built on local Neanderthal Mousterian precedents (i.e., in this case a flow from archaic to modern peoples, rather than vice versa), which he argues was far more complex than previously thought. In particular, he claims that the use of animal bones, animal teeth, sea shells, and fossils as items of symbolic value or personal decoration in the European Upper Paleolithic of *H. sapiens sapiens* all had origins in the Neanderthal Mousterian traditions of drilling beads and boring decorative objects, and so on. Thus he rejects the common view of Neanderthals as a regional, temporary, specialized subspecies of reduced cognitive capacity, subhuman and lacking the intellect and "two-handed vision-centered neurology" of anatomically modern humans.

Chase and Dibble (1987) reject the idea of *symbolic* activity occurring before the appearance of anatomically modern people in the Upper Paleolithic. They accept that there is evidence of an *aesthetic* sense in the Lower Paleolithic, as evidenced, for example, in the pleasing symmetry of biface tools, the use of hematite and ochre, the various manuports, objets trouvés, and simple items of personal adornment. By contrast, they see Upper Paleolithic art as rich, complex, and obviously symbolic. Lindly and Clark (1990), however,, stress the fact that the development of symbolism was itself only apparent considerably *after* the first appearance of anatomically modern people in Africa, at least 50,000 years later, in fact. Thus, like Marshack in this respect, they do not see the sudden developments across the Middle–Upper Paleolithic transition as being susceptible to neurological or biological explanations, but as to be understood solely in cultural evolutionary terms—perhaps even as the result of the development of some sort of "critical social mass" (Bradshaw 1988).

What then is the evidence for artistic, aesthetic, or even symbolic precedents in the earlier cultures of archaic *H. sapiens* or *H. erectus*? Marshack (1989b) reviews the doubtful evidence from the Acheulian that *H. erectus* may possibly have crudely carved pebbles, or instead perhaps collected stones which seemed naturally to resemble real objects. Oakley (1981) describes Acheulian flint handaxes constructed around fossils (a shell and a sea urchin) which seem to have been preserved during the making in a prominent and aesthetically pleasing location. See Figure 8.5a,b). (Less attractive but possibly more accurate explanations could appeal to the likelihood of this happening by chance in highly fossiliferous flint, or of the need to knap the stone in this way *around* the fossil to produce a viable tool.) Ochre seems to have been collected (and therefore used perhaps for personal or object decoration) in the Acheulian, for example 300 kya at Terra Amata, France, and Becov, Czechoslovakia (see, e.g., Marshack 1989a). Similarly, Bednarik (1990) reports on the discovery of an Acheulian hematite pebble with striations from Hungsi (southern India); the abrasion marks "are clearly not recent . . . and from the position of the striated facet, from the pebble's shape and from the oblique direction of the striations, we can perhaps infer that the most likely direction in which it was moved over the surface being marked was from the upper left to the lower right. This movement is more comfortably made with the right arm than the left, which would suggest that the artist was right handed" (Bednarik 1990, p. 75).

Between 300 and 250 kya at Clacton (England) and Lehringen (Germany), beautifully shaped and carved spears of fire-hardened yew wood were made during the Acheulian (see Figure 8.6). At Pech de l'Aze (France), a rib carved with a series of connected, festooned double

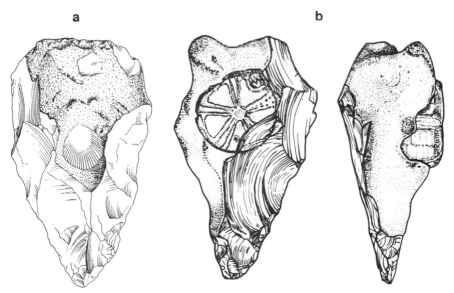

a   **b**

Figure 8.5   Pointed Acheulian flint hand axes, viewed from above (a), and above and from side (b), constructed around Cretaceous fossils preserved in a prominent and aesthetically pleasing location. From K. P. Oakley, Emergence of higher thought, 3.0-0.2 MaB.P. *Philosophical Transactions of the Royal Society of London B*, 1981, 292:205–211, with permission.

Figure 8.6   Point of carefully shaped spear of yew wood, Lehringen, Germany. Acheulian, c. 250,000 years before present. Courtesy of A. Marshack.

arcs, apparently engraved intentionally, was reported from deposits of 300 kya. Bones with purposeful geometric designs of a similar date were found at Bilzingsleben, East Germany (see e.g., Bahn and Vertut 1988, and Lewin 1989a; although see Davidson 1990, 1991, who believes that the marks were probably made during meat stripping). Marshack (1991) also reports that microwear studies of tools from the European Acheulian showed that they had been used to work skins and fibrous plants, and for boring and reaming. Thus, a bifacial tool from Clacton had wood polish from a rotary boring motion, which in fact indicated clockwise (i.e., right-handed) movement with downward pressure.

Turning now to the Neanderthal Mousterian, the archaeological record becomes much richer and more evocative. Not only, for example, do we now find evidence of the building of huts or windbreaks from mammoth bones at Molodova (Ukraine) 40 kya or more, but at the same site was found a shoulder blade with pits, color patches, and notches forming complex cruciform and rectangular figures (Bahn and Vertut 1988). At Nahr Ibrahim (Lebanon), bones of fallow deer (*Dama mesopotamia*) were found piled up and topped with a skull cap surrounded by ochre, suggestive of a symbolic ritual. At Tata (Hungary), a carved and polished (from much use?) segment of mammoth molar was found from 100 kya, together with a fossil nummulite (a disk-like shell) showing a line engraved at right angles to a natural crack and thus forming a cross (Bahn and Vertut 1988; Marshack 1989a). Similar Mousterian objects include a double arc incised on bone (Riparo Tagliente, Italy), a bovid shoulder blade with long, fine parallel lines at La Quina, together with pendants of a bored reindeer phalange and a fox tooth (see Figure 8.7), and nonutilitarian round stone balls and disks. A bone with a series of fine intentionally notched marks constituting a geometric pattern was found at La Ferrassie (Dordogne), together with carved, paired, and randomly placed cup-shaped marks apparently associated with a burial (Marshack 1989a); and at Bacho Kiro (Bulgaria) a zigzag motif on bone was described. Early Mousterian (or Micoquian) beads made from a bored wolf foot bone and a vertebra of 111 kya have been reported from Bocksteinschmiede (Germany). See, for example, Bahn and Vertut (1988), and Marshack (1989a) who notes that the use of the "hole" seems to start

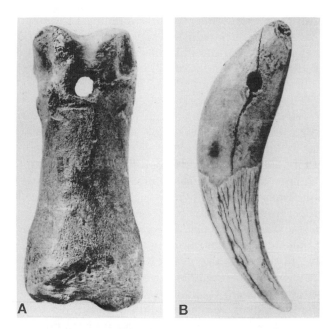

Figure 8.7 Phalange of deer and canine tooth of fox, both bored. La Quina, possibly Mousterian. Courtesy of A. Marshack. From A. Marshack, Evolution of the human capacity; The symbolic evidence, *Yearbook of Physical Anthropology*, 1989, 32:1–34. Copyright © of A. Marshack. Reprinted with permission of Wiley-Liss, a division of John Wiley & Sons Inc.

at this point. Thus, Marshack (1991) describes an awl made from a skull bone, perhaps for piercing skins, from the Bocksteinschmiede level.

Davidson (1990, 1991), however, remains skeptical concerning the origin and significance of many of these holes and lines, appealing to natural causes and reiterating "that the documented early objects with deliberate marks [are] too scarce to show any evidence of shared meaning . . . a sorry catalogue of wishful thinking" (1990, p. 55). He predicts that no tradition of early nondepictive marking will be found from which may be inferred a convention or code through which meaning could be recognized. Of course, marked objects may still have been important in the eventual evolution of marking behavior, given, as we have already stressed, that selection takes place from a range of existing features (preadaptation). "Thus an evolutionary argument about the emergence of marking requires that marks were being made before they were made to show meaning . . . accidentally or incidentally" (Davidson 1990, p. 56). Marshack, however, sees a mosaic evolution of potentially variable capacities that favored visually mediated problem solving, categorizing, mapping, and modeling, which led to symboling and imaging systems, and on to language and increasing social and cultural complexity. In these respects he sees Neanderthal and Cro-Magnon capacities as comparable, if not exactly equal.

Burial

The way a creature responds to the death of conspecifics may tell us something about its cognitive processes and possible belief systems, although a simple, and otherwise uninformative, emotional response to a situation generally associated with danger can never be excluded. Thus there are apocryphal and poorly documented accounts in the animal literature of chimpanzees displaying expressions of horror at the sight of a dead conspecific, of undergoing a state of apparent depression and grief resulting in emaciation and even death on the loss of a close relative (Goodall 1989), and of elephants covering their dead with earth and boughs and appearing to display behaviors reminiscent of mourning (Douglas-Hamilton and Douglas-Hamilton 1975, and see also Joubert 1991). While no one doubts the occurrence and significance of burials and grave goods in sites containing anatomically modern humans, the archaeological literature contains much poorly founded speculation and badly drawn conclusions and interpretations from apparent burial sites before the Upper Paleolithic (Chase and Dibble 1987; Gargett 1989).

Thus we read emotively about Neanderthal burials at Shanidar (Iraq), with the body placed upon a bed of pine boughs and ritually covered with flowers, of a bison leg being found with the Neanderthal skeleton of La Chapelle-aux-Saints (Corrèze), of a ring of stones surrounding the skull of tightly flexed skeletons, aligned east–west, at Monte Circeo, and of a Neanderthal child at La Ferrassie covered with a huge limestone block carved with cupules or dots gouged often in pairs and randomly distributed. Our feelings and conclusions may well be correct, but Gargett (1989) offers a salutary and critical evaluation and reexamination of the evidence for purposeful burial of the dead, and other inferences of ritual behavior, in the Middle Paleolithic, which might suggest that the Neanderthals had a concept of the afterlife or an emotional capacity comparable to our own. He notes the importance of considering the sedimentological context via the new science of taphonomy; caves and rock shelters can be places of rapid deposition and preservation may be enhanced there by chemical and atmospheric factors. Indeed, as Dettwyler (1991) notes, behavior does not fossilize, and attempts at reconstructing scenarios of behavioral evolution are at best imaginative and at worst subjective.

The archaeology of Neanderthal burial began in France in the early twentieth century. Gargett (1989) observes that Neanderthal skeletons, sometimes in anomalous contexts, convinced excavators that they had been intentionally buried, who then inferred the existence of accompanying ritual. Recently, more conservative interpretations have appeared. Thus, instead of invoking ritual cannibalism to explain splintered bones and missing basicrania, an awareness of cave taphonomy has explained, in terms of a rain of rubble from the roof, the selective decay of weak areas of bone. Apparent cyst burials can be due to the washing or drifting of bones into natural depressions. Burial with flowers (Shanidar) may just be wind or animal transport of plant material. Apparent stone arrangements may really reflect the organizing power of the observer's visual system. Bones of course accumulate rapidly in caves because animals (and humans) retreat there with prey or to die. Remains which happen to find their way to side walls or to niches in rock falls may be selectively preserved. The fact that whole skeletons are more likely to be preserved from later than from earlier periods does not imply a later development of ideas about burial and the afterlife merely that more recent remains are less likely to have been disturbed by accumulation of chance events.

Gargett (1989) sets a number of criteria for the evidence of intentional burial (compare the criteria in the following discussion of the intentional use of fire): a flexed posture (though death during sleep can naturally lead to flexion mimicking deliberate postmortem flexion of a corpse); a dug grave (although as just noted natural forces can mimic this); protection of the corpse by rocks (which are not just the consequence of a roof fall); food or other grave offerings (which have not been dragged there by animals); magic or ritual manifestations (like flowers, but flowers which have not been brought in as bedding by, e.g., rodents). Gargett concludes that in fact the evidence so far for deliberate Neanderthal burials is at best tenuous, and that "the removal of mortuary ritual from the behavioral repertoire of the Neanderthal may make the observed disconformity in material culture at the Middle/Upper Paleolithic boundary a little easier to understand" (Gargett 1989, p.177). Indeed, where burials *were* deliberate, they may merely have been undertaken to remove the body, and a tightly flexed fetal position could simply result from reasons of digging economy. The fact that so many Neanderthal skeletons are of old individuals with many signs of chronic or healed damage is perhaps the best indicator that Neanderthal society cared for the infirm, even if their intellect may not have been sufficient to save them from unnecessary wear and tear (Chase and Dibble 1987; Klein 1989). However, Dettwyler (1991) is critical of claims that the prolonged survival of injured, crippled, blind, toothless, malformed (a Neanderthal spina bifida infant skeleton was found), aged, or dwarf individuals proves the existence of compassion. He notes that the young, and pregnant mothers, have always needed shelter, food, and support, and that damaged individuals in a human society are not necessarily unproductive.

Whatever the true situation might have been in Mousterian times, the incidence of indisputable grave goods, decoration, and ritual increases enormously, and suddenly, in the Upper Paleolithic.

Fire

The adoption of fire by hominids with a tropical origin was a major factor permitting colonization of colder, higher latitudes during periods of global cooling. Its importance is reflected in its prominence in creation myths around the world; for example, the bringing of fire by Prometheus to the Greek's earliest ancestors and his punishment by the gods for such presumption. Not only did it provide warmth and light in dark places (thereby extending the length of waking hours and encouraging social interaction, mental and spiritual development, and language via story telling), and protect the sleeper from predators and enemies, but it also modified natural ecosystems in potentially useful directions, provided a channel for signaling, and above all permitted cooking.

Cooking—its origin, significance, and reasons for adoption—seems to have been largely ignored in the archaeological literature. It may well of course have had an accidental origin: attractive olfactory stimuli from a corpse after a bushfire has swept an area, followed by equally attractive gustatory sensations. It might have been adopted originally as a way of removing hair and feathers, or softening hide. It certainly gets rid of bad taste from partially decomposing game, kills parasites, and under certain circumstances can aid preservation (although, generally, a new and more dangerous series of bacteria can attack cooked as opposed to raw meat). It is an excellent way of detoxifying alkaloids, saponins, and various agents that plants adopt as defensive measures against grazing. Indeed, the complex preparations, often lasting days, that nonliterate peoples undertake to achieve such detoxification of, for example, sago and the fruit of *Macrozamia* in Australia testifies to the insight and skills that aborigines have developed over the millennia. Finally, of course, cooking aids digestion, and consumption by the toothless. If we make the distinction between cooking and warmth, we could say that the presence of charred bones is evidence of cooking, and their absence is evidence of the use of fire for, perhaps, warmth. However, campers know the tendency to dump bones (cooked or otherwise) after eating in a fire for disposal—not necessarily just to avoid attracting the attention of scavengers or predators.

Early on, natural ground fires (which are common after lightning strikes in savannah country, as in Australia and Africa) may have been opportunistically captured. Subsequently, attempts would have been made to store it, and probably much later to produce it (James 1989). The presence of a hearth is, of course, the best critical evidence for intentional use. Sillen and Brain (1990) review the evidence for early use of fire. The Zhoukoudian cave (China) is perhaps the earliest fairly convincing case, dating around 400 kya (*H. erectus*). Burned bones of deer, charcoal, burned stone tools, and thick layers of ash suggest a series of ancient hearths. At Yuanmou (China) there is ash with two *H. erectus* incisor teeth, dating to around 500 kya (1.7 mya, according to James 1989), though the evidence for deliberate use is by no means conclusive. The earliest European findings (Vértesszölös cave, Hungary) date to 400 kya, and includes hominid remains and burnt animal bones. In southeast Asia, Pope (1991) reports a circular arrangement of fire-cracked basalt cobbles of 1 mya, and in South Africa, Sillen and Brain (1990) describe 270 pieces of burnt or blackened antelope bone from Swartkrans cave (1 to 1.3 mya); while no hearths were present, the authors discount natural fires because of the high temperatures apparently achieved in the bones and their distribution within a cave known to have been inhabited by *H. erectus*.

James (1989) critically reviews the nature of the archeological evidence for deliberate use of fire by early hominids in the Lower and Middle Pleistocene in Africa, Asia, and Europe. He notes that early hominid sites tend to be often revisited by later hominids who will obscure and "overwrite" the evidence. On the other hand, *natural* redispositional processes are more likely to *destroy* than to spuriously *create* evidence of deliberate fire use. Moreover, the less-structured hearths of early people will provide poorer and more ambiguous evidence even if they survive intact. He itemizes the criteria for accepting that fires were deliberately made and kept: burned layers of soil, ash, fire-cracked rocks, the charred remains of bones, shells and tools, and a hearth (the *best* evidence) consisting of a heap of stones to confine or enhance the fire. No actual hearths are in fact found in Neanderthal times at the end of the Middle Pleistocene (128 kya). Natural processes (burned tree stumps in a clay soil from a bushfire and local mineral deposition) can often simulate hearths. James concludes that Zhoukoudian (400 kya) is still the best candidate for the earliest deliberate use of fire, and one accepted by most archaeologists, though not unreservedly, with a good case for unprepared Mousterian hearths in Europe at Terra Amata (300 kya) and St. Estève-Janson. The strongest African evidence is at Kalambo Falls (northern Zambia, 180 kya) with charred logs, charcoal, reddened areas, carbonized grass stems, and (possibly fire-hardened) wooden implements. By the Middle Stone Age, deliberate fire use was well-established, for example, at Klasies River Mouth (around 130 kya).

The Art of the European Upper Paleolithic

As we shall see, while anatomically modern people eventually replaced earlier archaic forms throughout their range, with a new technology and society, it was only in western Europe that the transition was both rapid and dramatic. Not only was the period of probable coexistence between Neanderthals and *H. sapiens sapiens* likely to have been shorter there (though by no means inconsiderable, even then), but above all, art and technology explosively took off, in a fashion suggesting the operation of internal societal factors rather than sudden evolutionary changes or mutations. After all, anatomically modern people had been present elsewhere, for example in the Middle East, at least sporadically, for 70,000 years or more, with only modest advances in art and technology. The mosaic ecology of western Europe at that time, during the last great Pleistocene glaciation between 75 and 10 kya, may have been at least partly responsible (Klein 1989)—indeed, the glacial maximum at 18 kya coincides with a high point in art. As Marshack (1988a,1989a) observes, seasonally migrating herds of horses, bison, aurochs, reindeer, ibex, woolly mammoths, and rhinoceros along a fertile network of richly grassed river valleys which provided shelter, access, passage, and food for bands of hunters, would have been hospitable places in a largely hostile world. In addition to the herbivores, migratory birds and fish would have been seasonally abundant.

Exact dates in western Europe for the commencement of the Upper Paleolithic are still under debate. Traditionally, the Châtelperronian, the end of the Mousterian tradition and a period of influence (at least in western and central France) from contemporary Cro-Magnon invaders, is put at between 35 and 30 kya; some suggest a narrower range of dates within those limits, whereas Lewin (1989a) gives a broader range of 40 to 30 kya, although presumably at any one geographical locus the range may have been considerably less. The fact of such influence from an arguably superior culture suggests that the Neanderthals were at least intellectually equipped to learn by example. Indeed, the continued coexistence of the two groups without apparent evidence of hybridization indicates that they were adapted to two rather different ecologies and did not interact or, despite competition, managed to hold their own for many generations. Thus the Neanderthals, with their Middle Paleolithic Mousterian culture, seem to have borrowed technology from the anatomically modern Cro-Magnon invaders and survived for some generations into the Upper Paleolithic (Mercier et al. 1991; and see Stringer and Grün, 1991, for further commentary).

The Aurignacian is the earliest cultural period of the Cro-Magnon Upper Paleolithic, and extended across the whole of western, central, and eastern Europe and the Middle East. It is typically given a similar (i.e., overlapping) range to the (extremely localized) Châtelperronian, that is, 34 to 30 kya (to 27 kya, according to Klein 1989), although it reached western France considerably later than other southern and eastern regions. Four other cultural periods are traditionally recognized: Gravettian (30–22 kya), Solutrean (22–18 kya), Magdalenian (the richest period of cave art, coinciding with the terminal ice age, 18–11 kya), and the Azilian (11–9 kya, when sea levels again rose). There followed the periods of Mesolithic hunter–gatherers and Neolithic agriculture. Klein (1989) generally gives slightly later dates throughout. However, in southern and eastern Europe, dates as early as 43 kya have been suggested for the commencement of the Aurignacian in Bulgaria (Bacho Kiro), and only a little later in Hungary, Austria, and Spain, a period coinciding with the Hengelo or Cottés interstadial, suggesting a role for climatic factors. Straus (1989) reviews recent reports that reach three chief conclusions: the Aurignacian of western Europe should be dated to at least 40 kya, much older than previously acknowledged for that region; the Aurignacian technology may have spread from Eastern Europe much faster than previously thought; and, finally, the transition from Neanderthal Mousterian to Aurignacian, at least in Catalonia, was extremely rapid, even though the two human forms may have elsewhere coexisted side by side for some time.

The (admittedly localized) phenomenon of the Châtelperronian seems to indicate that the Neanderthals may have learned something from the more advanced invaders; Marshack's (1989a) thesis, that the Upper Paleolithic cultural revolution only took root in regions where there were already preexisting Mousterian traditions, may therefore be the other side of the learning equation. Nevertheless, it is undeniable that cultural changes sometime around 35 kya were enormous and explosive. Stone tools were increasingly sophisticated and standardized (as well as often showing idiosyncratic personal styles), and included blade technology, burins, and hafting, techniques rarely employed until then. New levels of expertise and technical competence were attained, often seemingly so for their own sake; for example, the carving of exquisite animal heads on utilitarian objects like a spear-thrower. Objects were cut, sawn, carved, ground, polished, perforated, grooved, and shaped from bone, ivory, and antlers, for punches, points, awls, needles, and so on, together with nonutilitarian objects for personal adornment (Klein 1989; Mellars 1989). Beads for example were made using an almost production-line technique from ivory "rods" (White 1989). Surviving examples from the Upper Paleolithic are of course only the tip of the iceberg, with much perishable material from bark, fibers, feathers, hide, wood, and clay likely to have been lost (Bahn and Vertut 1988).

The presence of music in Aurignacian times is attested to by findings of bone finger flutes, stone bull-roarers, jaw bone rattles, and lithophones (calcite stalactites which emit clear notes when struck). Ritual burial, huts, and proper fireplaces now were commonplace and elaborate. The penetration of Siberia and northwest America during periods of intense cold indicates warm, sewn clothing and weather protection. Planning, forethought, logistics, prediction, and strategic anticipation of events and behavior is now implied by the archaeological record, together with a resultant considerable increase in population. Styles in ornamentation and technology changed extremely rapidly, both between adjacent regions and over short periods of time, contrasting with the previous long periods of stasis and uniformity.

Marshack (1988a,b; 1989a) sees all this as indicating sharper conceptual abilities to categorize, formalize, and structure, along with the growth of symbol systems. The sudden and relatively localized nature of this Upper Paleolithic revolution (for it was nothing short of that) indicates that it was cultural, rather than biological (and see White 1989) or from a shift in actual neurological capacities. We can perhaps see it as society suddenly achieving a "critical mass" (Bradshaw 1988), similar to other more recent historical times in western (and eastern) civilization; there is no reason, for example, to believe that during the Dark Ages in Europe people became less intelligent, or that during the Renaissance intellectual *potential* again increased.

Some of the earliest animal carvings were found at the Aurignacian site of Vogelherd (Lonetal) in Germany, dating to 32 kya: highly sophisticated mammoth ivory carvings of lion, leopard, bison, bear, horse, mammoth, and reindeer (Marshack 1988a), with detailed renditions of eyes, nose, mouth, mane, ears, and hair (see Figures 8.8 and 8.9). Also in the Lonetal (Hohlenstein-Stadel) was found a carved "therianthrope" (half-

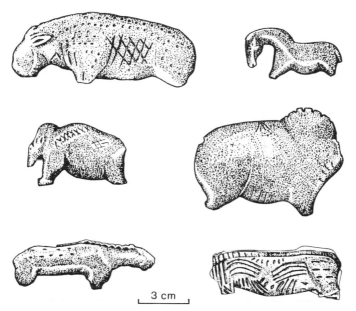

Figure 8.8   Animal figurines carved from mammoth ivory from Vogelherd, Germany, c. 30,000 years before present—Aurignacian. From P. Mellars, Major issues in the emergence of modern humans. *Current Anthropology*, 1989, 30:349–385, with permission.

Figure 8.9   Detail of ivory horse from preceding illustration. Courtesy of A. Marshack.

human, half-animal) figure of mammoth ivory: a standing human male with a feline head, reminiscent of an Egyptian animal-headed god, and perhaps part of a sorcery ritual (see Figure 8.10). Subsequently, almost every imaginable image of mammal, bird, reptile, and fish was faithfully portrayed, with extreme realism and often indicating seasonal or breeding condition. Polish and wear suggest much handling and long use. The properties and shape of the raw materials were often used to great advantage, determining the resultant image. While the composition was sometimes made to fit the material, often the composition appeared to be carefully planned, requiring a search for an appropriately shaped piece of bone or ivory. The fact that drawings were sometimes continued around the side and back of a piece of stone or bone suggests that the concept of the image might in these cases have come first.

"Venus" figures feature prominently in popular conceptions of Ice Age art. They are common in the later Gravettian period, although carved and engraved vulvas and phalluses are found in the Aurignacian itself (Marshack 1988a,b). Bednarik (1989c) describes the recently discovered and highly sophisticated "Venus of Galgenberg" from near Krems (Austria) and dating to the Aurignacian of 30 kya (see Figure 8.11). Carved from serpentine or schist, it is one of the earliest sculptures known. Unlike the later Gravettian statuettes which are sculptured in the round (and emphasize breasts, buttocks, hips, and thighs), the Galgenberg figurine is flat. The stone's brittle physical properties would not permit fashioning of a free-standing limb, especially an arm. To overcome this limitation, two different conventions were employed and are still used by contemporary sculptors. The right arm and the legs are structurally supported (and thus braced) at both ends, while the left arm is shortened to half its anatomical length by being depicted in a folded-back position. As Bednarik suggests, this implies a lengthy prior tradition, perhaps on perishable material. The whole appearance is naturalistic, without obesity, steatopygia, or exaggerated female characteristics. One can only wonder why realism was lost in later Gravettian sculptures, with their exaggerated sexuality and symmetry and abstract schematization (Marshack 1985). (Bednarik also mentions an Acheulian female figurine from Berekhat Ram, Israel, and dating to between 800 and 230 kya; it is a small scoria pebble whose shape *naturally* resembles a Venus, but may have been modified by several additional grooves. Its presence at the site suggests that the occupants recognized its iconic properties as an *objet trouvé*, and were interested in the female form, whether or not they actually modified the find.)

Bahn and Vertut (1988) note that not all of the Gravettian Venus figures glorify maternity and fertility; these are of varying apparent female ages

Figure 8.10 Lion-headed anthropomorphic figure ("therianthrope") from Hohlenstein-Stadel, Germany, c. 32,000 years before present—Aurignacian. From A. Marshack, Evolution of the human capacity: The symbolic evidence. *Yearbook of Physical Anthropology*, 1989, 32:1–34. Copyright © of A. Marshack. Reprinted with permission of Wiley-Liss, a division of John Wiley & Sons Inc.

and may simply be "about womanhood," displaying as they often do elaborate coiffures and bracelet ornamentation (e.g., the Venus of Willendorf; see Figure 8.12). Bahn and Vertut comment upon our own obsession with sexual interpretations, for example, seeing vulvas in innocent cracks or faults. They note that very few of the Venus figurines mark the pubic triangle or show the median cleft. Thus the exaggerated

Figure 8.11 "Venus" of Galgenberg from near Krems, Austria, c. 32,000 years before present—Aurignacian. Courtesy of R. Bednarik.

steatopygia may represent fecundity rather than sex; as ancient erotica they were a failure. Bednarik (1989b) takes such considerations further. He notes that there are now around 140 known female sculptures from the Upper Paleolithic, spanning a period of up to 22,000 years, which were produced by people of diverse material cultures from the Atlantic to Lake Baïkal. Collective statistical treatment could help determine their purpose or meaning, if these sculptures possessed (which is debatable) many common characteristics. Some seem pregnant, most are not; some are obese, most are not; a few are steatopygous, a few indicate vulvas, many but not a majority have large breasts, while some lack breasts, and some are not clearly male or female. Many are naked, some suggest clothing; some seem deliberately broken for ritual or for perhaps prognosticatory reasons (compare the deliberately exploded, kiln-fired Gravettian terracotta images at Dolni Věstonice; see Vandiver et al. 1989). They are big or small, pendant (even requiring wearing upside down) or free-standing. He concludes that there may be no one single universal purpose or meaning. Many of the features stem from difficulties in working with fragile materials, so that, for example, limbs are often only vaguely indicated on the surface. He warns us against imposing our own preconceptions, biases, value judgements, and knowledge in trying to

Figure 8.12 ''Venus'' of Willendorf, c. 28,000 years before present—Gravettian. Courtesy of A. Marshack.

interpret their use and significance (a caveat which should be extended well beyond this particular domain in human prehistory), and doubts whether we should in fact lump together material from such a huge geographical and chronological range.

These sculptures seem to have been more than merely representational, and to have had a meaning and a significance to the artists and viewers—to have told a story. The fact that so many seem to have been deliberately broken (or even exploded by kiln-firing with wet material) indeed suggests that they told a story. They may have meant more in the making than in the viewing, being largely irrelevant on completion, as in fact is often the case today with aboriginal rock art in Australia. Thus the later Magdalenian parietal (cave) art often contained paintings that seem to have been "reused," with new (or extra) tails, legs, or heads added, indicating perhaps selective attention to certain parts, perhaps at certain times. Even today, the act of "touching-up" old paintings has a ritual significance to the Australian aborigine.

Marshack (1988a) reports in detail a serpentine (boustrophedon, i.e., back-and-forth, "like the plowing of an ox") notation engraved upon a piece of bone, a "plaque" otherwise used for retouching stone tools (see Figure 8.13). It comes from an Aurignacian rock shelter at Abri Blanchard, France, and dates to around 30 kya. He describes from microscopic, infra-red, ultra-violet, and fluorescence techniques the accumulation of 69 marks, with 26 changes of point style or pressure of marking, and claims that the notation images the waxing and waning of the moon, encompassing 2.25 lunar months with full moon periods at the left, and crescents and days of invisibility at the right. The marks, he says, were accumulated sequentially over a long period using different tools. While they may have been calendrical—to help plan and schedule periodic social, cultural, and economic activity, including hunting migratory herds in the right season—they were not, he says, arithmetical and were not broken into, for example, fives or tens. True arithmetical and astronomical calendars may not have appeared until the emergence of agriculture. Marshack also describes other similar notational material, for example, on a bone fragment from the later Magdalenian period of the Grotte du Täi (France), which he says is a plaque containing notation for 3.5 years, with the notation dipping down when space becomes short. Thus Marshack sees the early presence of complex conceptual systems integrating divergent processes and periodicities—changing seasons, flood and thaw, animal availability, human reproduction, and cycles of subsistence, movement, and aggregation, and possible celestial and astronomical periodicities—into a mythological and metaphorical frame (Marshack, in press).

Not everyone, however, accepts Marshack's interpretation. White (1989) sees the Abri Blanchard plaque as simply representing natural objects and patterns like the decoration of sea shells. D'Errico (1989) experimentally produced and examined similar marks, comparing them with actual finds attributed to lunar calendars and notation systems. He concludes that the marks were always made by rapid, repeated tool movements rather than by accumulation over time, with the same tool making most of the neighboring marks, and that the idea of Upper Paleolithic calendars is wishful thinking (and see also Bahn 1989b).

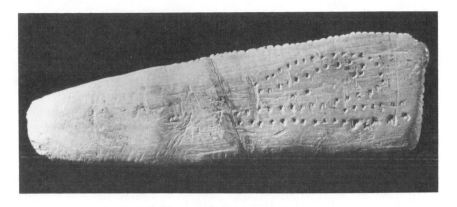

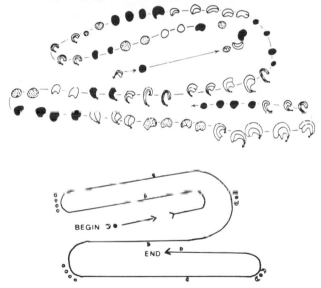

Figure 8.13 Carved bone plaque from Blanchard, France, c. 28,000 years before present. The sequence of marks may be notational and sequential. The schematic rendition indicates the 69 marks made in 24 "sets" and by different tools, forming a serpentine image that may model the phases of the moon. Courtesy of A. Marshack.

Apart from pioneering sophisticated techniques for examining the mode of *production* of engraved lines and marks, Marshack has developed the concept of Upper Paleolithic symbolism, imagery, and visual categorization. Thus he sees the primary factor in hominization not to be language and tool use, but to be, instead, the capacity for abstraction in visual categorization, the modeling of processes and relations in time and space. The use of language and the use of tools, he says, are both referential and depend ultimately upon visual categorization, the ability to recognize and differentiate relevant categories and classes of objects and processes, and to recognize relationships and associations, thereby schematizing processes and forms and syntactically restructuring reality (Marshack, in press). Language is therefore an alternate and later communicatory system that builds upon these cognitive and conceptual foundations. It possesses rules, grammar, and syntax, and can produce a potentially infinite range and variety of sentences. Symbolic imagery, according to Marshack, is similar and potentially no less rich, and is dependent on similar neural structures. (We shall in fact return later to both the origin of intelligence and language, and to the neurological underpinnings of language and perceptual-imaging systems.)

In conclusion, Marshack's personal scenario for human evolution is one of a two-handed, bipedal, and vision-centered creature, with imaging and problem-solving capacities; the system of symbols is not derivable from language but involves conceptualization, abstraction, and symboling, with its own syntax, mode of usage, associations, "vocabulary," and iconography. This system is separate from but complementary to developing language and tool use, and each aspect possesses its own syntax and semantics (Marshack 1985).

Before leaving the subject of art in the Upper Paleolithic, we should briefly mention that aspect which is perhaps most prominent in popular imagination: the parietal Franco-Cantabrian cave art of the Magdalenian at, for example, Lascaux and Altamira. Colors were achieved from a variety of natural minerals—hematite (red), ochres (red and yellow), manganese dioxide and charcoal (black), and kaolinite (white). Many colors could be further modified by heating and were subjected to a variety of binding agents; for example, fish glues, gum arabic, gelatin, egg white, bovine blood, and urine. The subject matter generally consisted of large animals (bison, horse, ibex, and aurochs), with relatively few carnivores, and even fewer plants or people. Painted in startling colors on ceiling and walls, they were often in narrow and inaccessible locations which possibly even required scaffolding, and so probably were not the object of group rituals (Bahn and Vertut 1988; Marshack 1988a,b).

The dark, dangerous, difficult, and uncomfortable ambience of the wet, secret, silent, and claustrophobic caves, and the effect of apparent movement and other "special effects" from flickering hand-held oil lamps, might well have produced strange psychological states in viewers or participants (Bahn and Vertut 1988). The animal images often depicted dimorphic sexual and seasonal characteristics and behaviors, the horse in summer or winter coat, the bison in or out of molt, stags carrying mature antlers or baying during the autumnal rut, deer with springtime antler or horn buds, hinds with fawns, cows with calves, and male salmon with a hook in the lower jaw which happens only in the spawning season. All required sharp observation and an awareness of sequential seasonal changes. Bahn and Vertut (1988) note that large herbivores predominate, carnivores, fish, and birds are fairly rare, while reptiles, insects, and plants are very rare. It all seems not to represent just a random bestiary, but something of significance. Different species predominate in different areas in different times. Human figures are rare and generally less well done, and females tend to predominate, often with some clothing and ornaments and without emphasis of the sexual organs apart from the breasts. Overall, there seem to be no absolute rules for the art, although there is evidence of careful selection of surface, size, technique, color, species, and anatomical detail, with considerable accuracy and some stylization. It does not appear to be simple graffiti or decoration, or esthetics. There has been much subjectivity and wishful thinking in attempted interpretations, for example, hunting or fertility magic, shamanistic ritual, and so on.

With respect to the apparently *nonrepresentational* geometric images, patterns of dots and lines that are also found in caves, on rock walls, and on smaller portable surfaces, there have been attempts to explain them in terms of entoptic phenomena (for review of such claims by, e.g., Lewis-Williams and Dowson, 1988, see Lewin 1991c). These include "phosphenes," the perceptions induced under migraine, drugs, and trance states, and at a cortical level, the building blocks of visual perception. Lewis-Williams and Dowson (1988) argue that altered states of consciousness in recent shamanistic rituals are represented in art and resemble the grid patterns, sets of parallel lines, dots, zigzag lines, curves, and filigrees sometimes found in the art of the European Upper Paleolithic—and indeed in Australia today. It is certainly true that drugs, fatigue, pressure on the eye, and electrical stimulation can result in certain elementary visual patterns (strictly speaking, phosphenes can only result from activity at the level of the *retina*, although Lewis-Williams and Dowson seem to wish to extend the term to include any form of nonveridical perception). Nevertheless, it should be remembered that *both* sets of patterns, subjective and in ancient parietal art, may simply be constrained by the limits that can be placed on dots, lines, and contour; one will end up with patterns like "phosphenes" simply because there are few other simple variations.

Parietal art similar to that of the European Upper Paleolithic is not of course unique to that region. Similar material, if in some respects less dramatic, is to be found in the Americas and southern Africa. However, in Australia, especially in the Kimberley region, Arnhem Land and Cape York, galleries are to be found that are as impressive as anything in Europe, though generally of a much more recent date (see Figure 8.14). It remains to be determined to what extent similar traditions lay behind their creation.

Summary and Conclusions

It may be an evolutionary accident that we have five, and not four or six fingers, but the ability to grasp and, ultimately, to manipulate, was a major factor in our evolution. The example of the cetaceans indicates that a large brain, a complex social organization and system of communication, and the ability to learn complex behaviors are not of themselves sufficient to guarantee the evolution of high levels of intellect and culture. While many species are now known regularly to deploy tool-using behaviors, in some cases even with apparent insight, recent work with chimpanzees and, surprisingly the New World capuchin monkeys, indicates how close their tool-using (and even coaching) behaviors can approximate our own. Indeed, the Oldowan lithic industry of 2.5 mya is not greatly different from that which the chimpanzee can attain. Neverthe-

Figure 8.14 Parietal art from northern Australia.

less the pace of stone-tool evolution was painfully slow, despite enormous increases in brain size, and it was not until the Upper Paleolithic that we see any really rapid advances. Tool behaviors could not therefore have been the only, or perhaps even the major, impetus behind human evolution. An interesting observation, however, is how very early in the archaeological record there is evidence for dextrality—right at the beginning of the hominid trajectory, in fact, with *H. habilis* and the associated Oldowan culture.

There is considerable debate concerning precedents in art before the European Upper Paleolithic, and the extent to which they may have provided a necessary substrate, in an indigenous Neanderthal culture, for invading anatomically modern humans. Evidence for even earlier, Acheulian, artwork is even more controversial. Similar conclusions relate to the question of burial; fire seems to have been in use at least 400 kya. However, the big questions which remain concern the origins and causes of the cultural explosion of the European Upper Paleolithic, particularly as manifested in all forms of art, and the relationship, if any, that these phenomena may have with the appearance of language. The latter issue is addressed in the next chapter.

Further Reading

Bahn, P. G., and Vertut, J. Images of the Ice Age. Leicester: Windward, 1988.

Chase, P. G., and Dibble, H. L. Middle Paleolithic symbolism: A review of current evidence and interpretations. *Journal of Anthropological Archaeology*, 1987, 6:263–296.

Chevalier-Skolnikoff, S. Spontaneous tool use and sensorimotor intelligence in *Cebus* compared with monkeys and apes. *Behavioral and Brain Sciences*, 1989, 12:561–627.

Gargett, R. H. Grave shortcomings: The evidence for Neanderthal burial. *Current Anthropology*, 1989, 30:157–177.

James, S. R. Hominid use of fire in the Lower and Middle Pleistocene. *Current Anthropology*, 1989, 30:1–26.

Klein, R. G. The Human Career: Human Biological and Cultural Origins. Chicago: University of Chicago Press, 1989.

MacNeilage, P. F. Grasping in modern primates: The evolutionary context. In M. A. Goodale (ed.), Vision and Action: The Control of Grasping. Norwood, New Jersey: Ablex Publishing Corporation, 1990, pp. 1–13.

Marshack, A. Evolution of the human capacity: The symbolic evidence. *Yearbook of Physical Anthropology*, 1989, 32:1–34.

Parker, S. T., and Gibson, K. R. A developmental model for the evolution of language and intelligence in early hominids. *Behavioral and Brain Sciences*, 1979, 2:367–408.

Wynn, T. The Evolution of Spatial Competence: Illinois Studies in Anthropology Number 17. Urbana and Chicago: University of Illinois Press, 1989.

Brain, Language, Intellect, and the Evolution of the Self

9

Introduction

There is a continuous tradition of tool-using behaviors back through the early hominids and, judging from the capacities of contemporary chimpanzees and New World monkeys, probably back to our common primate ancestors. Indeed many species of vertebrates and even inverte brates are definable as tool users. Tools may leave traces in the archaeological record; but this is not the case with language, although as we shall see, some would argue that art, which may indeed be "fossilized," can provide us with clues as to the evolution of language. Other clues come from paleoneurology, the examination, for example, in endocasts, for the presence of cortical structures which are known to be related to functions such as speech in living humans. Modern neurology is beginning to tell us much about the underpinnings of those quintessentially human characteristics—speech, language, and art-related behaviors.

Our other quintessentially human characteristics are a high level of intellect, and an awareness of the self in place and time. Just as the evolution of tool use and language are still subject to debate, so too is the nature and origin of our intellectual capacities and to what extent we are unique in our capacity for self-awareness. Increasingly, the boundaries between ourselves and other species are being blurred, particularly with respect to tool behaviors, language, thought, and consciousness. This chapter presents these capacities as being subject to multifactorial and interacting influences in evolution, and as approachable from a number of converging viewpoints, notably experimental psychology, neuropsychology, neurology, archaeology and paleontology, ethology, philosophy, and cognitive neuroscience. A clearer picture is beginning to emerge of the nature and origins of the human being.

Speech and Language: Its Nature and Evolution

As we have seen in the last chapter, a recurrent theme runs through Marshack's writings on Upper Paleolithic art and symbolism: both domains, language and nonverbal symbolic behavior as manifested in Upper Paleolithic art, given the paucity and depauperate nature of comparable material before some point in the trajectory of anatomically modern *H. sapiens sapiens*, involve referential categorical capacities for differentiating, integrating, generalizing, and communicating. As we shall shortly see, it may be no accident therefore that the neurological substrates of language partly involve those cortical areas of the parietal lobe in the left hemisphere which, in the right hemisphere, are given over to spatial processing and are known to contribute to visuoperceptual pattern recognition.

Temporal and frontal cortices on the left, of course, *also* make major contributions to speech and language processes. Indeed, in the event of major developmental breakdown of language, as in childhood autism, it is not uncommon for an outlet to be sought in another communicative modality like drawing, where exceptional ability may be reported as an island in a general sea of disability (Selfe 1978; Treffert 1989). We shall also shortly see that it may be no neurological accident that closely adjacent, anterior left-hemisphere structures mediate speech and praxis, and fine motor control of the forelimb extremities (Kimura 1982). Nevertheless it would be simplistic to look to common evolutionary origins or neural mechanisms in the symbolism, categorization, and referencing of art and language (although compare Davidson and Noble 1989, and Noble and Davidson 1991, to whom we shall later return). Davidson and Noble argue that language is the endpoint of an evolutionary trajectory moving via throwing, pointing or deixis, gesture, and art, which is itself a form of accidentally or deliberately frozen gesture. Language, they say, is parasitic on the capacity to generate and reflect on an image.

It would also be naive to overemphasize the link between tool construction and use, and language. Both, as we saw, involve sequential, syntactic behaviors (Frost 1980), but so too do many other skilled behaviors from hunting to play, and the hierarchical nature of syntax in language may be unique to that domain. Nor should we invoke symbolic language (or even mental images) when observing the levels of standardization to be found in many kinds of stone tools. Within the constraints imposed from working with certain kinds of raw materials, continually resharpening existing tools, and producing something for a particular purpose, there may not be too many degrees of freedom available for alternative forms (Dibble 1989). Indeed, even sophisticated modern languages serve as a poor medium of instruction in tool manufacture or use; imitation and observation are far more effective. Imitative learning has incidentally been proposed as another reason why we are predominantly dextral; certainly two individuals of the same handedness can instruct each other in knot tying far better than a mixed dyad, as Michel and Harkins (1985) demonstrated. However, as we have seen H. habilis was probably already dextral at the dawn of tool use, and anyway its tools were almost certainly too simple to demand dyadic tutoring.

Nor could tool use primarily have driven the evolution of brain, intelligence, and language, because despite continual and considerable increments in brain size, toolkits were very slow to change (Calvin 1987). In a later section we shall review the evidence for social behavior ("Machiavellian intelligence") as a driving force for our increased encephalization; it is tempting now to account for language in similar terms, noting that competitive and cooperative social behavior in a hierarchical society is most developed in the pygmy chimpanzee (Pan paniscus), which with its social, cultural, and communicative abilities is much closer to us than the common chimpanzee, Pan troglodytes (Raijmakers 1990). Language may possibly have been an integral component in the elaboration of our social intelligence, though some might argue that our societies are not intrinsically so very much more complex than those of other primates, whatever might be true of technology. Nevertheless, language was not just discovered, like writing. The latter has to be learned, with effort, whereas a major genetic component ensures the relatively effortless acquisition of the former, even though learning is also involved if appropriate developmental milestones are not exceeded. We can in fact process visual representations even faster than the spoken word; it would be interesting to speculate on our alternative evolutionary and technological history had our ancestors evolved a visual rather than an auditory channel of communication. In any case, whether spoken or written, genetically programmed behaviors can now be subordinated to learned cultural programs.

The general scenario of an upright posture, freeing up the hands and leading to tool use, gesture, and on to language (Hewes 1976; Kimura 1976) is still popular. We readily fall back on gesture; retardates may rely on it, and we can teach a chimpanzee a gestural channel of communication with us. Among primates, only humans and orangutans possess palms that are unpigmented and clearly visible (although the significance of this observation is reduced when we note a similar lack of pigment on the soles of the feet). Gesture may be superior to speech in showing how to do things and there may be syntactic commonalities between gesture and language (Kimura 1982) with gestures closely following speech patterns. However, gesture comes too late in the proposed evolutionary sequence because chimpanzees (and so presumably our common ancestor) share so many gestures with us (Harré and Reynolds 1984), despite not being habitual tool users. Thus chimpanzees bow, bob, crouch, present, kiss, touch, reach hand to hand and hold hands, pat, embrace, prod, beckon, slap, stamp, wave, clap, beg, offer, point, head-shake, and so on. Tomasello (1990) debates the extent to which these gestures are species-specific, shaped in a social environment, or directly imitated. He concludes in favor of a model of conventionalized learning, as when a child raises its arms to facilitate and later to request being picked up. Monkeys even share with us a range of facial expressions, for example, frowns, pouts, threat grimaces, smiles, eyebrow flashes, and expressions of disgust, fear, mirth, surprise, and nausea, all of which are immediately and mutually intelligible. Indeed, increasing tool use might well have *hampered* gesture and instead favored a vocal–auditory channel of communication, as also would have the need to communicate at a distance, in the dark, or among vegetation (Hewes 1977), although vocal compared to gestural communication does have the disadvantage of alerting predators or prey, as well as interfering with breathing and swallowing.

In fact, hunting has been proposed as a single-factor explanation for the development of language (Noble 1990), whereby a weaker carnivore could compete by creating efficient hunting teams via enhanced communication (compare canids and hyaenids). However, wolves and jackals seem able to operate efficiently as hunting teams without vocal signaling. Indeed, except at the planning stage of a hunt, vocal signaling would simply alert prey. Just as unifactorial explanations for the increase in hominid encephalization and behavioral repertoire are both varied and unsatisfactory, so too is any such attempt to describe the origin of language. In Wind's (1989) multifactorial feedback model, notable factors include hand and arm use, changes in function and anatomy of the vocal tract (partly driven by changes in posture and flexion of the basicranium), pre- and postnatal development, a switch from an arboreal to a terrestrial life, increases in brain size, social interrelationships, tool use, diet, number of offspring, neoteny, and changes in facial structure including nose and jaw reduction. While the ultimate origin of human language must remain speculative, hints as to its possible progress may be derived from studies of its ontogenetic development in the child, from its breakdown in the aphasias, and from its appearance *ab initio* in, for example, sign languages for the deaf, pidgin, and creoles.

Duchin (1990) notes that speech, a form of vocal articulation apparently unique to our species, is the product of peripheral anatophysiological events, whereas language includes reasoning within the cerebrum; we should not therefore interchangeably use the terms communication, speech, language, and vocalization, though in practice it is often not easy to maintain these otherwise convenient distinctions. Raijmakers (1990) proposes that communication refers to the ability to influence other individuals via signals; language, however, is a form of communication spontaneously acquired that, via symbols transmitted with gestures and vocalizations, can convey perceptions, intentions, impressions, and abstractions, whereas speech is the manifestation of language. The fact that so much of our brain is somehow involved in language, far more than just the speech areas, suggests that language plays more than just a communicatory role, but may help in the cognitive construction of models of reality (Jerison 1986). Conversely, imagery and nonverbal problem solving may be alternative, and not reducible to language or propositional thinking (Kosslyn 1990); in any case, and despite claims to the contrary, for example, by Eccles (1973), we know that the isolated right hemisphere of split-brain patients, when appropriately interrogated, possesses a fully human mentality despite lack of access to the articulators (see, e.g., Bradshaw 1989).

Of course, when we communicate, we do not just communicate external facts, but also share experiences, enabling others to participate in our current view or model of the external world. Ethological studies of nonhuman communication indicate that commands, sign stimuli, releasers, and fixed action patterns are far more important. The communicatory expression of emotions in lower vertebrates is almost entirely preprogrammed, and largely so in all primates including humans. Many but not all species of birds learn their song from conspecifics; in humans, language has to be learned entirely and only the ability to acquire speech is innate. Indeed, chimpanzees, like humans, pass through a critical age during which they can learn their species-specific communicatory vocalizations.

There are perhaps six components in human speech (Cutting 1990). These include:

Phonology–A largely categorical left-hemisphere aspect involving the speech-related sounds (phonemes, the smallest sound differences that distinguish meaningful utterances). All languages are based on a small fraction of the sounds that we are physically capable of making, and not all languages use the same subset. We form words from rearrangements of such sounds.

Morphology–The denotative significance of individual words or of meaningful units smaller than a word (morphemes).

Syntax–The grammatical rules to combine morphemic strings into uniquely meaningful propositions, another largely left-hemisphere function.

Semantics–The formal meaning of a string of morphemes, to which the left hemisphere may contribute denotative meaning and the right connotative meaning.

Prosodic or suprasegmental aspects–Involving phonological variations within categorical limits to provide extra, interpretative, connotational meaning, via changes in pitch, rhythm, stress, and similar perhaps right-hemisphere emotional aspects.

Pragmatics–The practical use to which the components of language may be put, as with jokes, irony, sarcasm, metaphor, and context—the point of the discourse to which the right hemisphere may contribute.

According to Chomsky's (1980) nativist account, our language ability, which he sees as both species- and task-specific, derives from an innate language-specific neural mechanism or mechanisms. He sees the latter as discrete modules for such separate components as syntax, the lexicon, and so on, with no prior evolutionary history, no prior preadapting counterparts in earlier species, and no biological precedent. Thus a language organ without counterpart in other species suddenly appeared, full blown, like Athena from the head of Zeus. He sees it as manifesting as a byproduct of selection for other abilities, or as a consequence of an as-yet-unknown locus of growth and form, or as an emergent property of a system sufficiently complex as a result of other selective pressures (Pinker and Bloom 1990). This language organ is the basis of universal syntax, an innate set of (transformational) grammatical laws enabling us to generate and comprehend speech. All the world's languages are in principle reducible to deep structures which are largely independent of meaning—"Colorless green ideas sleep greedily"—a semantically nonsensical but syntactically well-formed sentence. This Chomskian universal grammar is said to underlie all known languages, which differ only with respect to surface characteristics. Thus all languages are hierarchically organized and can embed phrases; *any* child can learn *any* language, despite the complexity of the grammars which usually are only *implicitly* known.

This account denies the major premises of modern biological thought, that of an evolutionary continuum from simpler prior systems, with at most a quantitative gulf rather than a qualitative dichotomy between human and animal forms of communication (and see Lieberman 1986). Thus proponents of the Chomskian viewpoint argue that a biological specialization for grammar would have needed more evolutionary time and genomic space than is available, and that, like a partially evolved eye (Catania 1990), would be useless and could not exist in an intermediate form or forms. Of course, there *are* eyes in different species at apparently intermediate levels of evolution, pidgin and creole, as we shall see, are new partly evolved languages, and we can readily comprehend abbreviated ungrammatical telegrams and the utterances of a Broca's aphasic. Likewise, the argument that the single possessor of a beneficial mutation favoring language could do little with it among his or her less-endowed conspecifics, is also invalid; such a mutation would probably have been shared among family members who would *all* have benefited. Homeotic mutations *can* lead to quantum changes in the phenotype, and evolution probably favored increasing automaticity and effortlessness in communication; thus familial sufferers of a language disorder *can* comprehend conversation, though with effort. As noted by Pinker and Bloom (1990), evolution often makes innate and automatic those functions which hitherto have had to be performed by general learning and processing mechanisms at a deliberate, conscious, effortful level, thereby freeing up the organism to undertake another simultaneous task. Indeed, the very fact that language disabilities (inherited or acquired) rarely eliminate *all* language capacities may indicate a long-term, piecemeal evolution of language functions.

As Dingwall (1988) and Raijmakers (1990) observe, it is most unlikely that human communicatory behavior arose in our hominid ancestors by one-shot genetic saltation, and that apes, so close to us genetically, would show no trace of this. Any differences are likely to be quantitative, therefore, rather than qualitative; although as Bickerton (1990) notes, the imperceptible accumulation of quantitative changes can eventually topple over into a qualitative distinction. Human communication probably developed gradually over time via a process of mosaic evolution, with structures subserving various aspects evolving at different rates. The concept of preadaptations is important here; some mutations become advantageous when an aspect of behavior not originally prevailing in the original selective situation becomes involved, or preexisting structures become utilized or adapted for new functions. Indeed, adaptive evolutionary changes cannot occur just *because* they would be advantageous; there must be some prior element that can be adapted or modified, some preexisting infrastructure that can be built upon. Thus the swim bladders of lung fish became lungs, the airway between lungs and mouth was adapted for phonation, and many other peripheral and central structures have been adapted in our species for speech. The concept of preadaptation therefore demands that we differentiate structure from function; a series of small, gradual *structural* changes can lead to abrupt *behavioral* changes, thereby opening up a whole new set of selective forces and leading to new evolutionary potentials (Lieberman 1991).

Of course, we *do* possess certain extremely specialized central and peripheral adaptations for speech and language. Our vocal tracts are tailored to the demands of speech to the detriment of breathing and swallowing. Some language disorders (involving, e.g., delayed speech onset, articulation difficulties, problems with grammar, and even reading difficulties) are genetically transmitted; conversely, the inherited disorder of calcium metabolism, Williams syndrome, leads to profound mental retardation with preservation of even *hyperlinguistic* abilities (Bellugi et al. 1990; Pinker 1991; Yamada 1990). Children effortlessly learn any language without the need for formal instruction, and can invent (as with pidgin and creole) languages more systematic and complex than those they hear. Critical periods for language acquisition are universal in all populations and cultures (Johnson and Newport 1989). Language cannot be easily suppressed by physical or sensory isolation (Poizner, Klima, and Bellugi 1987), although if *total* isolation persists beyond certain critical periods, language may not be properly acquirable thereafter, and lateralization may proceed anomalously (Fromkin et al. 1974). Moreover, human speech is dualistic (MacNeilage 1987), being simultaneously ordered at two syntactic levels: sound and meaning. Each level is organized in a frame–content mode, whereby vowels and consonants are appropriately inserted into syllabic frames, and at the morphological level, lexical items (words) are inserted into frames composed of syntactic morphemes. In this way, the system is open-ended, permitting an infinite number of possible sequences to be generated from a finite set of elements, like molecules from atoms and organisms from genes (Studdert-Kennedy 1990).

There is no evidence of anything even approximating this level of complexity occurring in other primate call systems. Indeed, had the evolution of human speech merely been driven by the need to communicate what is of interest to most predator and prey species, we would probably have evolved a far simpler and more automatic system. Even the sounds of speech are "special" (Lieberman 1985), permitting the transmission of data perhaps ten times faster than that attainable by any other auditory signaling system. Nevertheless, the fact that our *visual* information-processing rate, for example reading even of nonphonological, ideographic symbols, or with Japanese *kanji*, can greatly exceed that of auditory speech perception, suggests that our information-processing capacity evolved before a modern vocal tract morphology. Like other complex yet automatic human and primate motor skills, speech production involves rapid goal-directed responses *toward* target loci, in the absence of innate control mechanisms to cope with every possible starting point. It involves reflexlike motor control mechanisms which can compensate (within 40 msec) for any unexpected, imposed environmental perturbation; for example, the sudden experimental imposition of an unexpected load upon the tongue, lips, or jaw. Nevertheless, nonhuman primates seem to possess neither the oral nor the neural capabilities of producing the full range of human speech sounds that are acoustically distinct and resistant to articulatory perturbation.

All of our putatively unique patterns of behavior have precursors among the higher nonhuman primates (Chomsky's claims notwithstanding), for example, as we have seen, cooperative hunting, food sharing, nest construction, territoriality, bipedal locomotion, tool construction, and use (Dingwall 1988; Goodall 1968). Chimpanzees make and use probes for termite fishing and to obtain honey, they obtain hammers of the correct weight and hardness to pound nuts, they chew leaves to make sponges for mopping water from otherwise inaccessible crannies, and they use leaves to wipe their backsides. At a more cognitive level we can add the capacity for general learning, the ability to recognize oneself in mirrors (see later), symbolic play, some interest in "painting," insightful problem solving and learning, counting (at least up to about five), the ability to categorize, differentiate and generalize, and cross-modal transfer. Apes gesture much like us, beckoning, hugging, begging, offering, withholding, patting, and comforting, and produce very similar facial expressions. As noted by Savage-Rumbaugh and McDonald (1988) the chimpanzee's facial anatomy is very similar to our own except for a large protruding jaw and brow ridges; the nerves and muscles moving the various facial structures are practically the same as ours and, when accustomed to the chimpanzee's slightly different facial architecture, we soon learn to read its expressions accurately, particularly in the case of the pygmy chimpanzee (bonobo). That subspecies has a much smaller brow and jaw and may in fact be more closely related to us.

It is hardly surprising, therefore, that a more or less unconscious recognition of these capacities, while realizing that apes' naturally elicited vocalizations nevertheless do markedly differ from our own, has led people for hundreds of years out of curiosity to try to train apes to communicate with us. The earlier attempts generally used the vocal channel, but it is noteworthy that as early as 1661 Samuel Pepys suggested the employment of manual gestures. In recent decades there have been many successful programs (usefully summarized by Dingwall 1988, and Snowdon 1990), using American Sign Language (ASL) with chimpanzees (the Gardners, Terrace, and Fouts), gorillas (Patterson), and orangutans (Miles), or using plastic symbols or "Yerkish," with chimpanzees by the Premacks and the Rumbaughs, respectively. The relatively large size of the resultant vocabularies (between 60 and 170 words, depending on the study), the ability to generalize the referencing of individual lexical items, their spontaneous use, and so on, all suggest cognitive processing beyond the level of simple, learned conditional associations between symbol and referent; this may be especially so when such behaviors occur under naturalistic social conditions, although other interpretations cannot be totally excluded (Terrace 1984). Thus human language is recursive and open ended, allowing the generation of a more or less infinite number of novel sentences by recombination according to syntactic rules; for example, "The boy who saw the purse on the street where the shop had lost its window which . . . was hungry and . . . ." By contrast, the communications of chimpanzees tend to be concatenations, with each next sequential item determined by a previous element. Moreover, they are seen by some as lacking in spontaneity and to be largely reactive and associative. However, chimpanzees do seem able to use symbols intentionally and conceptually and not just referentially; thus one animal knowing where food is, and the tool needed to obtain it, may request the latter from another animal able to supply it.

In summary, apes can learn words spontaneously, efficiently, referentially (even for absent items), and conceptually from each other, and use them to coordinate joint activities and to inform each other of things not otherwise jointly known. They can learn rules (syntax) for ordering "words," although there is debate as to whether the resultant strings constitute genuine sentences. They spontaneously comment, announce their impressions and intended actions, construct novel sequences, and know what they are doing. They do not simply imitate. Their capacity to use symbols to stand for objects, present or absent, implies an awareness that the symbol now replaces the object, replaces the need to point, and extends beyond the immediate present.

Savage-Rumbaugh et al. (1990) and Savage-Rumbaugh, Romski, Hopkins and Sevcik (1989) report on their recent findings with bonobos (*Pan paniscus*), the gracile and highly communicative species of chimpanzee thought to be very similar to our common ancestor with chimpanzees, and which may have separated from the *Pan troglodytes* lineage only 2.5 mya (de Waal 1989). The two young individuals studied showed much evidence of *naturalistic* (as opposed to trained) acquisition of spoken English, comprehending many quite complex statements the first time, without the need for gestures. Often novel strings of multiword requests were comprehended, in the complete absence of drilling, and when spoken by different people with regional accents and even where mood, stress, intonation, sex, and age information was filtered out via synthesized speech. Early exposure to naturalistic spoken English seemed to be the critical factor, just as with human children, who in some respects scored lower until the age of 3-4 years. Indeed, Lewin (1991b) notes that by 10 years of age, one individual (Kanzi) had acquired a vocabulary of more than 200 individual words, and could cope with syntactically complex sentences like "get the orange that is in the colony room," even when confronted with *another* orange before departing to the colony room. It also appeared to exhibit frustration in trying to *produce* speech. On the other hand, there seems to be no evidence as yet that they spontaneously *teach* each other their new-found skills. Moreover, chimpanzees can learn to sign, to draw, and to communicate via gesture in the laboratory, but cannot *create* or *maintain* these behaviors *naturally*, out of the laboratory. These considerations lead some critics to say that, in the final analysis, rather than tapping language per se, the studies address capacities for categorization, association, problem solving, and communication (and see Terrace 1984). Others might reply to this criticism that this is itself a fair description of language! Chimpanzees may intentionally deceive in their voluntary actions outside of these studies; however, there is as yet no evidence for such prevarication in any of the experimental work.

Why then do chimpanzees not communicate with us vocally? They cannot produce the vocal sounds /i/, /a/, or /u/ or the phonemes /g/ or /k/ due to limitations of their vocal tract anatomy, but even then passable speech should still be attainable; they presumably lack either the central nervous system mechanisms or the volition, preferring as a species visual signaling via facial expressions, gestures, and posture. Indeed, it is only recently that the question of interspecific *vocal* communication by chimpanzees has been raised (Boehm, 1989). They may call to mark territorial boundaries, to signal alarm on the appearance of a predator, to summon help from an ally, or to locate lost individuals. Some calls can occur in monosyllabic isolation or may be combined multisyllabically, with changes in inflection, consisting of hoots, screams, pants, and whispers of varying intensity. Goodall (1989) claims 34 discrete calls, but it is unclear whether they merely involuntarily express the caller's emotional state or also communicate information on the social and physical environment, or even signal deliberate *intentions*. So far, they seem to vary in pitch, tenseness, phrasing, duration, volume, formant frequencies, vowel quality, intonation patterns, and the number of repetitions. They may possibly convey information on age, sex, identity, emotional state, size and makeup of a party, the nature and intention of agonistic situations, interest in being sociable, feeding conditions, and intentions to hunt, patrol, or nest, the presence of neighbors, and the location, distance, and direction of travel. However, as yet this is still hypothetical (Boehm, 1989); there is no evidence that the chimpanzees' communications (natural or acquired) depend in any way on the commonsense, tacit knowledge, or shared assumptions between listener and speaker, which are so much a feature of our discourse, and which are such a problem for machine translation or even speech decoding. Thus consider the sentence "Mary had a little lamb." From the context and prior experience we know whether she owned one, ate some at lunch, or (if a sheep) gave birth to one. Successful language is parasitic on such a network of unconscious associations, which probably are accessed by, if not resident in, the temporal-hippocampal system, and which can only be programmed with great difficulty, if at all, into artificial "expert systems."

We might end this section with the observation:

"If we were to find a [pygmy chimpanzee] that imitated all human
cognitive and linguistic abilities, we should probably ask ourselves what sort
of genetic disorder led him to be so short and hairy."

(Snowdon, 1990, p.239).

If chimpanzees cannot communicate with us vocally, such peripheral limitations are not present in certain avian species prone to mimicry. Thus there are claims that the African grey parrot can speak meaningfully and intentionally. According to Pepperberg and Kozak (1986), a bird observed its trainer hide a nut under a heavy metal mug, which it tried unsuccessfully to move. It had previously learned English labels and phrases to identify, request, refuse, quantify, and categorize various objects. It walked over to its trainer and apparently said "Go pick up cup," a combination of terms that was novel for its repertoire. Pepperberg (1990, 1991) reviews work on this and other avian species and concludes that birds can acquire many concepts and capacities (labeling, object permanence, similarity–differences, color, shape, number, and so on) formerly considered to be the exclusive domain of humans or language-trained primates. She trained her parrot to communicate with its human interlocutor with a vocal code based on the sounds of human speech, taking advantage of the bird's mimetic abilities. It learned to identify, request, refuse, and categorize more than 30 objects, 7 colors, and 5 shapes, and to use the labels 2 to 6 to distinguish between appropriate numbers of objects. It appropriately used a range of stereotyped phrases like "come here," "wanna go x," "what color?", "what shape?" and so on. Such work throws a new light on the primate studies.

The demonstration, in the brains of nonhuman primates, of cytoarchitectonically homologous regions to our own Broca's and Wernicke's speech areas, would be conclusive evidence against Chomsky's (1980) nativist account of a unique species-specific language mechanism in the human brain, without prior evolutionary antecedents. Galaburda and Pandya (1982) claim to have found such homologues in the rhesus monkey, together with the arcuate fasciculus which links the two major areas. However, these homologues may have no equivalent role in the production of sounds since ablation does not affect vocalization (Aitken 1981; Kirzinger and Jürgens 1982), although they may play a role in *receptive* behavior. Moreover, as with humans, damage to the supplementary motor area may increase latencies. Indeed, areas truly homologous to our own Broca's speech area may lie within the inferior limb of the macaque's arcuate sulcus—that is, more ventrally and medially located than ours (Galaburda and Pandya 1982)—such that the ineffective-ablation studies may have targeted the wrong regions. Phonation and vocalization in monkeys is at least partly under the control of limbic centers, such as the anterior cingulate (Sutton, Trachy, and Lindeman 1981), as indeed may also be true of humans (Laplane et al. 1977; Rubens 1975). Thus human emotional signals like crying and laughing seem to employ similar subcortical circuitry to those mediating animal calls, with later evolution of the cortical system.

Deacon (1986a,b) examines the major pathways in nonhuman primates presumed to link language areas in humans. All are present in monkeys, especially the arcuate fasciculus interconnecting our Broca's and Wernicke's areas. He describes the efferent circuits shown by axonal tracer data to be the substrates of primate vocalizations, together with the efferent circuits originating from the primate homologues to our periSylvian language areas; he notes that these two systems comprise parallel pathways for oral–vocal activity. The limbic midbrain circuit is rather less direct, and is predominantly involved in the control of respiration and the laryngeal muscles, while the cortical system largely controls muscles of the face, jaw, and tongue. The programming of movements controlled by the limbic–midbrain system remains intact after cortical periSylvian damage, whereas the motor programming of the neocortical system depends on that system's integrity. Different sectors of the Broca's area homologue (in the monkey) in the arcuate area are reciprocally connected with the anterior cingulate (limbic) vocalization areas, permitting interaction between primitive vocalization systems and more complex articulatory control.

The apparent neural dichotomy in earlier literature between primate calls and human speech may merely reflect differences in the level of cortical involvement and the range of muscle systems to be coordinated;

it may not, as previously thought, reflect a total shift of vocal functions from limbic and midbrain areas in nonhuman primates, to neocortical centers in humans. As Deacon (1989) notes, our own, older, limbic vocalization circuits are still important in initial speech activation and in prosodic and emotional expression; indeed, they activate our few remaining species-specific calls; for example, laughter, crying, shrieks, sighs, groaning, and sobbing. Generally, cortico–cortical connections in the monkey's periSylvian regions predict the spatial pattern of functionally connected areas in human temporal, parietal, and frontal lobes. This suggests that language functions, during evolution, have recruited cortical circuits that were already present, perhaps for different purposes, in our primate ancestors. Quantitative changes in particular populations of neurons may be more common than novel rerouting of neural connections or the *de novo* creation of new brain structures in vertebrate brain evolution. Deacon again emphasizes that there is no evidence for a species-specific language organ, as proposed by Chomsky, or the *de novo* appearance of speech areas in the brains of early hominids; any differences between us and our ancestors are likely to be quantitative rather than qualitative.

None of these physiological conclusions are necessarily incompatible with cognitive studies (see e.g., Fowler and Rosenblum, 1990, for review) indicating that speech perception undergoes processing that is somehow distinct or additional to that of the general auditory system. Listeners have little conscious awareness of speech processing, in that they cannot describe speech as sounding like anything else but speech, in contrast for example to Morse code which can be heard both as a meaningful message and as a series of meaningless dots and dashes. Thus speech perception is "cognitively impenetrable," and phonology is typically transparent to semantics—we perceive the meaning directly rather than the individual component speech sounds. Moreover, speech processing is "informationally encapsulated"; listeners hearing synthetic speech and even highly impoverished sine-wave speech still process these as speech, even though aware that such signals were not produced by a human vocal tract. Furthermore, whereas objects of nonspeech auditory perception generally correspond fairly closely to acoustic structure, dimensions of the percept often fail to correspond to the obvious structural dimensions in the acoustic speech signal; they correspond more to the patterns of vocal tract activity that might have produced the signal. Thus, in isolation, a frequency glide experimentally extracted from a speech signal sounds just like a pitch glide; integrated, however, within an acoustic signal for a consonant–vowel signal, it contributes to the perception of a consonant of which the pitch glide is not experienced as a part.

Different frequency glides that sound quite distinct in isolation may sound like indistinguishable tokens of a given consonant when they are integrated into different syllabic frames. They sound alike when signaling the same consonantal gesture of the vocal tract. The acoustic signals for the same consonant are different in different vowel contexts (i.e., in different syllables) because of the coarticulatory overlap of consonant and vowel production; each modifies the other, vowel and consonant, making the acoustic cues of phonetic features highly dependent on context.

Such findings are compatible with the widely held (and disputed) motor theory of speech perception; according to this theory the listener attempts to recreate the likely articulatory gestures of the speaker from the speech signal, rather than hearing directly the actual moment-to-moment acoustic input (Liberman et al. 1967). The phenomenon of categorical perception also seems to support the motor theory. We perceive the acoustic properties underlying phonetic features in an absolute, all-or-nothing fashion; we do not hear the different variants of the phoneme /b/ in different vowel environments. All are heard as /b/ even though on an acoustic continuum some are much closer to /p/. The relevant continuum here is the voice-onset time, VOT, the time between the sound burst caused by lips opening and the start of phonation, labial stop consonants with a VOT between 0 and 25 msec are all uniformly heard as /b/, whereas those with longer intervals are categorized as /p/. Thus at 25 msec VOT, a threshold is crossed and the listener "flips" to hearing /p/, and additional acoustic variations along the same continuum make no further difference to the perception of /p/. Consequently, phonetic stimuli from different categories are highly discriminable, whereas stimuli from the same phonetic category, though physically very different, are in fact virtually indiscriminable.

The problem, however, for the motor theory of speech perception is that infants (tested with conditioned discrimination techniques), macaques, and even chinchillas seem to divide the human phonetic continuum in the same way as the human adult; they, too, perceive human speech contrasts categorically, apparently hearing an abrupt change at precisely the same acoustic locations that humans do in, for example, separating /ba/ and /pa/, or /da/ and /ta/ categories (Kuhl 1988b; Maurus, Barclay, and Streit, 1988). Thus not only is the ability to produce articulate speech unnecessary to perform complex speech-related discriminations, but so too may be the human brain. At one stroke, both the motor theory and Chomsky's unique language processor come under serious doubt.

It is however far from clear what is the functional significance of such abilities in nonhuman species, unless it is somehow related to their own modes of communication. One possibility is that *our* speech system has converged toward employing these ancient category boundaries; that is, it has *adapted* to preexisiting interspecific and innate acoustic phenomena. Thus characteristics of the mammalian auditory system that led to non-uniform discriminability along certain acoustic dimensions may have resulted in the selection of sounds differing along such dimensions for inclusion in the communicative repertoire. Sounds drawn from within regions of low discriminability would have been perceived in natural categories, in which some degree of variability could be tolerated without changing the overall perception. Conversely, sounds that contrasted with the more easily discriminable regions of the continuum would have formed perceptually different categories (Moody, Stebbins, and May 1990). It is even possible that such acoustic categories, in turn, stem from the shared possession in higher mammals of an essentially similar articulatory apparatus—lips and tongue, for example—which would tend to impose commonalities upon the sound stream.

Kuhl (1988b) notes that infants are born with certain predispositions; they prefer looking at faces than at equally complex visual stimuli (Fantz and Fagan 1975), they prefer viewing faces judged attractive by adults, indicating that facial attractiveness is innately assessed (Langlois et al. 1991), they imitate (within 72 hours of birth) facial expressions or actions presented to them (Meltzoff and Moore 1983), and they prefer to listen to "Motherese" than to "adult" speech. Kuhl notes that "Motherese" is a highly melodic speech signal used by adults the world over when addressing infants; it seems not to be the syntax or semantics that appeals and holds the infant's attention, but the suprasegmental or prosodic features like higher pitch, slower tempo, and expanded intonation contours that are universal in every language and culture. Indeed, the human phonetic inventory is restricted similarly across all the world's language; far fewer sounds are used communicatively than could be produced by the human vocal tract, and the subset tends to be largely common to all the world's languages.

The near-universal existence of "Motherese" shows that adults do come prepared to "instruct" infants in language, though less formally, perhaps, than in manual skills, by providing salient and useful models for the child to imitate, and subsequently to be corrected. Thus the child masters grammar via a form of concept formation; even though a teacher–parent's grammar is only correct or consistent say 60% of the time, the child improves the odds by then more consistently (say 90% of the time) using the slightly more common (and therefore probably locally correct) form. Erroneous generalizations (e.g., "tooths" as a plural for "teeth" by analogy with "boots") are gradually eliminated. This answers the Chomskian problem of how children struggle to learn their first language via a garbled mixture of correct and syntactically wrong (or nonstandard) utterances; there is no need to invoke an innate language acquisition device. It also highlights two contrasting theories of language mediation (Pinker 1991): the associationist approach, and the approach which emphasizes the part played by rules and representations. The former sees the brain as a homogeneous network of interconnected units modified by a learning mechanism that records correlations between frequently co-occuring input patterns; in particular it is comfortable with the teacher–parent give-and-take just described to successively "shape" an infant's grammar. The latter approach sees the brain as a computational device wherein rules and principles operate; it accounts well for the tendency of children to overgeneralize (compare "tooths" for "teeth" just described). As Pinker (1991) observes, neither approach on its own can satisfactorily account for all aspects of language acquisition and use.

Since infants' and nonhuman animals' "auditory" boundaries coincide, as we saw, with humans' *phonetic* ones, even compensating as we do for variations in speech rate (context effects), Kuhl (1988b) argues that there may be no special speech-specific mechanisms, and that speech production evolved to match the properties of the ear. This contrasts with the motor theory of speech perception, which proposes that a special perceptual mechanism evolved in us to detect articulatory (i.e., motor) invariants. According to Kuhl, the auditory system may have played a key role in shaping the acoustics of language and the mechanisms that produce it. Thus even if early hominids could not produce a full range of "our" kind of speech sounds, and this is itself a disputed proposition, they may still have been able to perceive and discriminate a full range. If Kuhl is proven right, we may have to conclude that the uniqueness of human speech lies not at the phonetic level, but rather at the levels of syntax and semantics.

We have disposed of the strongly held view that whereas our speech is largely cortical and adaptive, that of our other primate relatives is essentially limbic and reactive. Recent work is also showing that primate (and even nonprimate) call systems, though developmentally fairly fixed and unmodifiable with little evidence of either vocal learning or syntax, can code a wide range of environmental happenings and transmit them to conspecifics in an informationally useful fashion. Thus Cheney and Seyfarth (1985, 1990) report that vervet monkeys have at least six acoustically different alarm calls for predators such as leopards, smaller cats, martial eagles, pythons, and baboons. Each type of alarm call evokes a different and appropriate escape response. The loud bark-like leopard alarm call causes them to run up trees, while the more gruntlike eagle alarm call causes them to look up in the air. Vervets also recognize the alarm calls of other species. The common superb starling (*Spreo superbus*), too, has several distinct calls; for example, one for terrestrial predators and the other for eagles and hawks. When these calls are played the vervets respond appropriately, looking down or up respectively. In contrast to their skill at recognizing the alarm calls of their own and other species, vervets are very poor at identifying the various *visual* cues left by predators, for example, tracks and carcasses in trees. The authors speculate that in social interactions ("Machiavellian intelligence," discussed later), monkeys use the auditory rather than the visual channel to signal social information, even though the specialization of temporal lobe structures for facial recognition seems, to us, to imply the opposite. Apes are in fact better than vervets at recognizing visual cues left by other animals, perhaps possessing less constrained, less context-specific intelligence.

Lest we believe that calls varying in frequency, amplitude, or duration and which are able to influence conspecifics' succeeding behavior are the exclusive prerogative of primates, and as such are precursors to our own speech and language, we should note a report by Beynon and Rasa (1989) concerning the dwarf mongoose (*Helogale undulata*). In their colonies, one animal stands watch on an elevated position, giving warning calls on sighting a predator. The authors distinguished between a number of different warning calls coding the dangerousness of the potential predator, its type, distance, and elevation. These calls were combined and modified as the situation changed, and the responses of the other group members was modulated accordingly. Calls elicited by different predator species, even of the same class, differed significantly, indicating a high level of predator discrimination. Generally, pulsed calls encoded distance while frequency modulation encoded elevation and predator type, with degree of dangerousness being encoded by base and carrier frequency, and frequency modulation range. The authors note that the recombination of this encoded information, into a vocal repertoire associated with specific threat situations, fulfills all the generally accepted criteria for language. It also appears unique among nonhuman mammals, though the dance of the honey bee, conveying distance and direction of a food source from the hive, has obvious features in common (Von Frisch 1967).

Bird song, of course has long been held up as exhibiting close parallels with human speech. The young must learn from adults, with dialect differences resulting from that period of learning. There is a critical period during which vocal learning is maximized, and usually only species-specific songs are acquired. Hearing is needed both to acquire an acoustic template of adult song and to monitor progress in acquiring vocal control; "babbling" (or its equivalent) occurs as a form of practice in both species. Lateralization to the left side of the brain is also common to both (see Chapter 2). Recently (and see review by Snowdon 1990) this picture has been challenged; new songs can sometimes be learned after the critical period, even across species, especially in naturalistic and social circumstances compared to laboratory circumstances, and the left lateralization may not be as strong as we thought. We too can learn second languages in adulthood, though with difficulty and probably via a different process (Bradshaw 1989).

The Supralaryngeal Tract in Humans and Earlier Hominids

At the peripheral level, it is commonly believed that an obstacle for the chimpanzee—and perhaps fossil hominids prior to anatomically modern *H. sapiens sapiens*—in producing intelligible human speech is the configuration of the articulatory organs. In this respect, the larynx is notable, although the glottis, pharynx, mouth, tongue, and lips all contribute. As summarized by Lieberman (1988), in the human vocal tract, the trachea connects the lungs (driven by their own system of musculature) to the larynx (voice box), where the esophagus (connecting mouth to stomach) joins the system, followed by the oral and nasal cavities. The velum, the soft flexible extension of the palate, closes the nose to the mouth; the tongue and lips complete the system of interconnecting cavities which can be independently closed off. The vocal cords vibrate when air is driven past from the lungs, producing the fundamental sound frequency of phonation, itself determining the pitch of the speaker's voice. Overtones, harmonics of the fundamental frequency, are determined by the width of the throat, the location of the velum and the tongue, and the position of the mouth, all acting as acoustic filters by letting maximum energy through at certain frequencies and suppressing other harmonics. In this way we produce discrete speech sounds.

Apes and human infants less than three months old lack the physical apparatus to produce the full range of formant frequencies because the larynx is too high. When the child is older, in a process not fully complete until adolescence, the larynx descends to produce a vocal tract that is essentially two tubes, the oral passage and the throat, joined at right angles behind the tongue. This descent of the larynx is at the expense of being able to simultaneously breathe and swallow, and the danger of choking, although Wind (1989, personal communication) nevertheless observes that with a population exceeding 5 billion, our throats must nevertheless still function reasonably efficiently. In all other species (and the suckling human infant) the high larynx can lock into the nasal opening, the nasopharynx, by apposing the velum and epiglottis, thus permitting simultaneous breathing and swallowing. Fluid or food moves to either side of the raised larynx which resembles a raised periscope protruding through the oral cavity, connecting lungs with nose (see Figure 9.1). With human adults, food and drink must make the potentially hazardous passage over the tracheal opening. Not only has our larynx descended, resulting in our ability to make certain important vowel sounds like /a/ and /i/, but the tongue has also grown, intruding into the larynx and becoming more mobile. Such changes are all at the expense of eating; however, they do enable us to perform the single most complex motor activity that mammals are capable of: speech. Our consequent ability to seal off the nose from the rest of the airway during speech reduces nasality and the chance of misidentification or confusion by the listener.

Chimpanzees, according to Lieberman, were they to possess adequate brain mechanisms, should be capable of producing only a nasalized subset of our phonemes and not /a/, /i/, /u/, /g/, and /k/, thereby leading to increased likelihood of misidentification. We can thus vocally transmit phonetic segments (approximated by the letters of the alphabet), via parallel articulatory control, at a very rapid rate, even up to around 20 per second. This lies within the constraints of our very brief (around 3 seconds) echoic memory system, which holds the incoming speech signal just long enough for it to be integrated with the current phrase- or sentence-length sample (Lieberman 1989). Nonspeech sounds cannot be identified faster than 7 to 9 per second. The high transmission rate, resulting from our species-specific supralaryngeal tract and matching neural mechanisms, notably Broca's area, is achieved through the generation of formant frequency patterns and rapid temporal and spectral cues. Formant frequencies are the maximum acoustic energy that can get through a given configuration of the supralaryngeal airway, which acts like a pipe organ in letting through the acoustic energy at certain filtered frequencies. Thus, via the parallel movements of the lips, tongue, velum, larynx, and mandible we continuously change these configurations and the resultant formant frequencies.

Lieberman (1989) observes that the *listener* must also normalize the vocal tract of the currently attended speaker; the listener has to estimate the probable length of a speaker's supralaryngeal tract to assign a particular formant frequency pattern to speech sounds. Conversely, a child with a shorter supralaryngeal tract cannot imitate, for example, his mother's formant frequencies, but produces frequencies proportional to the ratio between the length of his and her tracts. Listeners normalize analo-

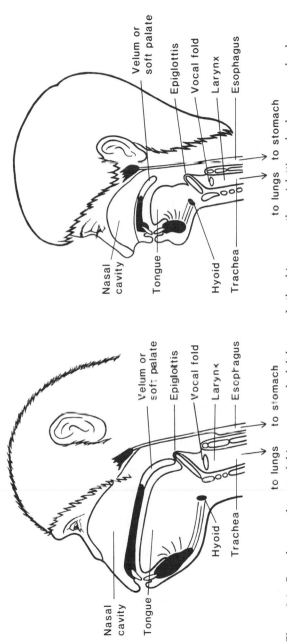

Figure 9.1 Supralaryngeal tracts of chimpanzee and adult human. In the chimpanzee the epiglottis and velum can seal, whereas in the adult human, the lower position of the larynx makes it impossible to lock into the nose. Drawn from illustrations in Corballis (1991); Dingwall (1988); and Lieberman (1991).

gously, the vowel /i/ being an optimal cue for this process. Lieberman argues that chimpanzees and hominids other than anatomically modern *H. sapiens sapiens*, with their reduced range of nasalized vowels, would be at a disadvantage in this respect.

Duchin (1990) notes that whereas the extrinsic tongue and suprahyoid muscles, which help determine the character of speech sounds, have similar configurations and identical innervations in chimpanzees and humans, in chimpanzees the muscles have different angles of insertion than in their homologues in humans, thereby restricting the chimpanzees' potential to articulate consonants. (The muscle attachments to the hyoid bone play an important role in tongue control and movement of the larynx. The hyoid lies between the root of the tongue, which it supports, and the larynx. See, e.g., Marshall 1989.) Duchin compared the inclination of each of the various speech muscles in the oral cavity, using a referent horizontal plane at the level of the mandible, and assessed the relative advantage over chimpanzees with respect to tongue mobility and hence its ability to contact target areas during articulation. Separately, she also measured radiologically the length of the palate and mandible and their position with respect to the hyoid bone and therefore the tongue. Using discriminant function analysis, she characterized the articulatory dimensions of the human (97 adults) and chimpanzee (75) oral cavity and was also able to include estimates from two specimens each of *H. erectus* and Neanderthal humans. (While it is not easy to reconstruct soft tissue vocal tracts from hard fossilized material, the flexion of the basicranium, the direction of the styloid process, the distance between the basion and the palate, and the positioning of the palate and mandible all have been used by, for example, Lieberman to estimate tongue position and shape of the supralaryngeal tract. See Wind 1989). Duchin found that the fossil hominids lay at the extreme edge of the normal human distribution, but well apart from of the chimpanzees. She concluded that both fossils, but not the chimpanzee, had the capacity to successfully articulate the vowels and consonants of human speech.

Dingwall (1988) and Wind (1989) separately list other criticisms of Lieberman's belief that nonanatomically modern hominids were at a selective articulatory disadvantage. Infants, primates, and even nonprimates, as we just saw, can perceive speech in a linguistic mode and discriminate between categories of speech sounds despite having no prior knowledge of the articulations needed to produce a full range of speech sounds. There have also been many major criticisms of Lieberman's reconstructions of fossil hominid vocal tracts; people with grossly deformed vocal tracts (e.g., partial laryngectomies, damage to epiglottis, tongue, palate, teeth, and lips) and birds with quite different vocal tracts can produce quite comprehensible speech sounds reasonably efficiently because of the built-in redundancy of phonatory systems. Finally, Lieberman plays down syntax in his claim that prior to *H. sapiens* the shape of the supralaryngeal tract would have been incompatible with a full range of speech sounds (Fletcher 1991). He argues that the neural and articulatory mechanisms that evolved to subserve rapid articulation and phonology were preadapted for subsequently mediating rule-governed syntax. However, the organizational patterns of sound and syntax are very different. Indeed, they are differently localized and can break down independently during aphasia.

It was the discovery of a Neanderthal fossil hyoid bone from the Mousterian deposits of 60 kya at Kebara cave (Israel) by Arensburg et al. (1989, and see also Arensburg et al. 1990), which really reopened the question of Neanderthal speech capacities (see Figure 9.2). As just noted, previous accounts, for example those of Lieberman, had focused largely on the basicranium as an indicator of speech capabilities. Arensburg and colleagues criticize such methods and instead present the anatomical relationships of the hyoid and adjacent structures as a basis for understanding the form of the vocal tract. In all respects, the size and morphol-

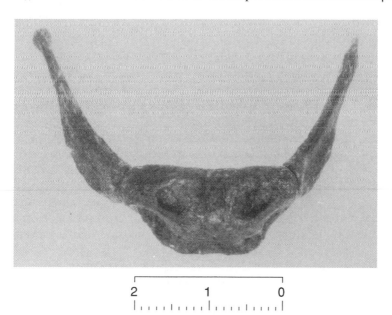

2   1   0

Figure 9.2   The Neanderthal hyoid bone from Kebara 2, c. 60,000 years before present. From B. Arensburg, A. M. Tillier, B. Vandermeersch, H. Duday, L. Schepartz, and Y. Rak, A Middle Paleolithic human hyoid bone. *Nature*, 1989, 338:758–760, with permission from NATURE vol. 338, pp. 758–760. Copyright © 1989 MacMillan Magazines Ltd.

ogy of the Kebara 2 hyoid, its relationship to other morphological components, and by inference its supralaryngeal space, were found to be entirely modern. This contrasted markedly with what is found with the great apes, even though the mandible differed considerably from modern morphology in terms of bicondylar breadth, bigonial breadth, bimental foramen breadth, and maximum body length. They conclude that the morphological basis of human speech was fully developed in Middle Paleolithic Neanderthals (although see Lieberman et al., 1989, for criticisms of the anatomical and functional role of the hyoid in speech production and its relevance to the debate on how Neanderthals might have talked). Of course, a Neanderthal hyoid does not prove that that taxon spoke—only that it *could,* and as Davidson and Noble (1989) observe, in an evolutionary context structures must emerge before the functions for which they can be used. It would however seem odd that an anatomically modern hyoid and a large and appropriately structured brain was incompatible with speech of some sort.

The Neurology of Language and the Left Hemisphere

If suprapharyngeal structures were not a major source of limitations on speech capability in the hominids, what can we say about the central nervous system structures needed to drive the peripheral apparatus? As we have already seen, lateral asymmetries and growth in the region of the major speech areas seem to have been present from an early date. There are several ways to address the neurological underpinnings of speech and language in living humans. The oldest approach, and which has provided the greatest amount of useful data, documents the breakdown of language-related functions as a consequence of damage or insult to discrete areas, for example, by tumor, penetrating injury, or infarction caused by a more or less localized disruption of blood supply.

Aphasia, acquired loss of the capacity to speak normally or at all, has been known as a left-hemisphere syndrome since the days of Paul Broca in the 1860s, who enunciated the famous dictum, "Nous parlons avec l'hemisphère gauche." ("We speak with the left hemisphere"). Whereas he reported the effects of frontal damage with a largely motor form of aphasia, the German Karl Wernicke shortly afterwards described a left-hemisphere language syndrome of a somewhat different nature, consequent on damage in more posterior sensory regions. The two regions, the former in the frontal operculum and adjacent to the face motor area, and the latter in the vicinity of the temporoparietal junction and adjacent to the primary auditory cortex, are now known to have no clear cytoarchitectonic, topographic, or physiological limits, but may constitute the irreduceable core for spoken language, subserving the cognitive operations of phonetics, syntax, and semantics (Mesulam 1990). See Figure 9.3.

Damage to Wernicke's area is associated with fluent well-articulated, melodically intact, paraphasic speech, with reduced use or misuse of content words (semantic paraphasia) or replacement by circumlocutions or neologisms. Thus the patient replaces nouns, verbs, or adjectives with related or nonsense words, which are nevertheless strung together according to apparently normal grammatical rules and inflectional patterns. Speech sounds superficially normal, but is devoid of content. Electrostimulation of Wernicke's area, a procedure sometimes necessary before surgery to determine safe or desirable limits, leads to interference in comprehension and naming. The region lies at the semantic–lexical pole of a language network and is an entry point for "chunking" auditory sequences into neural word representations that can trigger associations underlying meaning and thought. The lexicon, represented by a distributed multidimensional information network, is not a word bank so much perhaps as a nodal bottleneck for accessing a distributed grid of connectivity, containing information on sound, word, and meaning relationships (Mesulam 1990). Its output, conversely, "chunks" thoughts into words which match underlying meanings, the words themselves being largely realized in Broca's (motor speech) area in the frontal lobe.

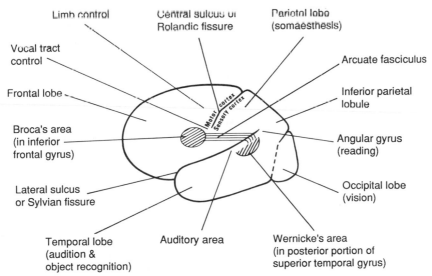

Figure 9.3 The major surface features and language-related areas of the left hemisphere.

Damage to Broca's area is associated with nonfluent speech (four- to five-word phrases as a maximum, usually) that is effortful, dysarthric, and telegraphic, and favors *content* words which are usually appropriate for the intended meaning. Thus it is in many ways the complement of Wernicke's aphasia, with aberrant word order and omission of grammatical "functor" words like the article, prepositions, and conjunctions. It is associated with agrammatical speech which is not merely a strategy to reduce the effort consequent on the dysarthria resulting from frontal injury. Note again that regions just *anterior* to the classical Broca's area are almost certainly also implicated, regions which we now know are also involved in *other* complex coordinated motor behaviors such as manual praxis. Again, this observation underlines the close neuroanatomical and evolutionary relationships between speech and praxis.

Stimulation of Broca's area leads to extreme speech arrest. According to Mesulam (1990), it lies at the syntactic–articulatory pole of the language network and provides a bottleneck for transforming neural word representations (originating from Wernicke's area, and transmitted at least partially via the arcuate fasciculus, the transcortical route running from Wernicke's to Broca's areas) into corresponding articulations. The latter are achieved via the syntactic sequencing of phonemes, morphemes and words themselves. Whereas Wernicke's area provides meaning-appropriate content words, Broca's area determines ordering and actual articulation. Neither (as traditionally described) are exclusively receptive–sensory and expressive–motor, respectively. Thus, for example, Broca's aphasia results in problems in comprehension especially involving grammatical functor words, whereas stimulation of Broca's area affects phoneme identification (Ojemann 1983), findings that are compatible with the motor theory of speech perception. (According to this account, the listener understands speech by reproducing, covertly or in precursor form, the sequences of articulatory gestures that would have been necessary to (re)produce that pattern of sounds.)

Positron emission tomography (PET scans, which permit the visualization of brain areas metabolically active during a particular task or behavior) or electro-recordings from Broca's and Wernicke's areas during speech tasks or listening, show that both are simultaneously rather than successively active; word selection occurs in parallel with the anticipatory programming of syntax and articulation. The articulatory envelope and grammatical structure provide the vehicle for delivering words. There is no sequential hierarchy, but rather the grammatical structure influences word choice, just as word choice influences syntax. The speech and language system of the human brain does not consist of discrete centers or modules (Fodor 1983) for comprehension, articulation and grammar, whatever might be the case with respect to other capacities. Rather, it has an architecture consisting of a parallel, distributed network wherein nodal foci of relative specialization work in concert with each other (Mesulam 1990).

The concept of parallel distributed processing is relatively new and underlies the differences between brains and (conventional) computers in how they process information. Traditional computers handle information in a serial manner, and although outstanding at computational tasks such as arithmetic, they are relatively poor at the complex pattern and speech recognition tasks at which we excel. Our superiority seems to stem from the ability of the brain to compute in a massively parallel manner, though recently there have been dramatic advances in constructing devices capable of parallel processing. These devices are of considerable interest to neurobiologists and cognitive scientists because they offer a model for understanding how we function. The architecture essentially consists of networks of highly interconnected, simple processing elements (Mitchell et al. 1991), which individually can only perform trivial mathematical operations, for example, adding the values of inputs or subtracting threshold values. The activity of an element is determined by the sum of the inputs; elements can influence each other via the interconnections which are positively or negatively weighted, and which can change according to certain contingencies, for example, simultaneous coactivation of two or more elements. Thus, as Farah (1990) observes, in such a system stimulus representation consists of a pattern of activation across the set of interconnected units. The extent to which activation in one unit induces activation in other, connected units depends on their connection strengths (weights). Units representing mutually consistent hypotheses will tend to activate (reinforce) each other via positive weights, and those representing inconsistent hypotheses will inhibit each other via negative weights. Consequently, any input to the system will change its initial pattern of activation levels, or "memory," so that it will settle down ("relax") to a new stable state representing the most consistent set of hypotheses about the input, combining information from both the current stimulus input and the previously stored experience (and see also McClelland, Rumelhart, and the PDP Research Group 1986).

There is also evidence for simultaneous processing of neural information in cortical networks of parallel columnar aggregates (Goldman-Rakic 1988). Moreover, Mesulam (1990) describes how multiple brain areas are involved in directed attention; the parietal, frontal, and cingulate cortices corresponding to sensory, motor, and motivational components. Damage to any one area leads to a partial attention deficit, and damage to more than one results in a far more profound disturbance. Indeed, recent evidence of plastic reorganization of brain function, even in adult animals (traditionally only extreme juveniles were thought capable of reassigning functional connectivities in the event of localized brain insult or trauma), is itself compatible with the concepts of parallel distributed processing. Thus, Merzenich (1987) found that a region in the parietal cortex topographically mapping somatosensory information from a finger can subsequently respond to stimulation of *adjacent* fingers if that finger is lost. Even more dramatically, Pons et al. (1991) found that in adult macaques years after section of sensory nerves from the forelimbs at the point of entry into the spinal cord, the parietal zone that had originally responded to the limb now responded to *facial* stimulation; the area that was now newly responsive to the face was over 1 cm beyond the original face-responsive zone.

Not only may the brain establish dynamic new connections in the event of altered sensory input due to trauma, but its modes of processing, when intact, may vary from time to time. Thus, contrary to traditional neurological accounts, recent PET scan studies indicate that the patterns of activation consequent on, for example, reading a word, are not linear and invariant, from visual to auditory association areas and Wernicke's area, and on to Broca's area; instead, depending on task demands, levels of reading expertise, processing set, and so on, activation may pass directly from visual to motor regions. Such a flexible information processing system is confirmed by a new subtractive PET technique (Petersen et al. 1990); the computer subtracts the PET image of active brain areas when a subject merely inspects a word from those active when the word is spoken, and the latter from those active when the task is to think up associations. It can thus determine and display those areas specifically recruited when the subject simply has to *say* something, irrespective of looking, or to think up associations independently of looking or talking.

There are of course also paleocortical (limbic) and subcortical centers involved in language. The cingulate cortex provides a substrate for the interaction between the more primitive vocalization systems as found, for example, in nonhuman primates, and the more complex speech control systems of the human neocortex (Deacon 1986a,b). The basal ganglia and thalamic nuclei on the left are respectively involved in motor planning, programming, and execution on the one hand, and attention, activation, and alerting on the other. Crosson, Novack, and Trenerry (1988) provide a comprehensive and convincing account of the integrated functioning of all these systems, none of which operates in isolation from the others. The arcuate fasciculus bidirectionally interconnects Broca's and Wernicke's areas directly, which also have reciprocal connections with the thalamic nuclei. The latter receive input from both the basal ganglia and Broca's and Wernicke's areas. The basal ganglia receive input from both of these cortical speech centers, but can only output to them via the thalamus (see Figure 9.4). Crosson et al. see the thalamus as activating two separate systems in Broca's area, a language formulator and a motor programmer. Thalamic pathways in conjunction with Wernicke's area perform *semantic* monitoring before articulatory release, whereas the arcuate fasciculus, in conjunction with Wernicke's area, performs *phonological* monitoring before release. The basal ganglia in conjunction with the thalamus hold up (for monitoring) and release (after monitoring) the formulated information. In this way, formulated language passes from Broca's to Wernicke's area via the thalamus for semantic monitoring. If correct, it is released by the basal ganglia (and thalamus) back to Broca's area for programming, and if not it is reformulated. After phonological formulation in Broca's area, the information is sent via the arcuate fasciculus to Wernicke's area for phonological monitoring. Again, if correct, it is released back by the basal ganglia (and thalamus) to Broca's area for articulation. All systems must simultaneously operate for smoothly flowing conversation; while one segment is being executed in speech, the next is being monitored for phonological accuracy, the next is being motor programmed, the next is being verified for semantic accuracy, and the next is being initially formulated. Nonhu-

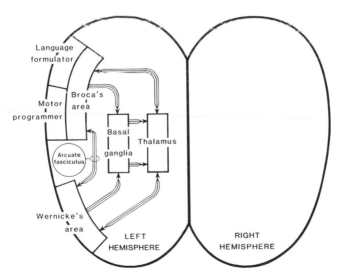

Figure 9.4 Schematic representation of cortical and subcortical language structures and their interconnections.

man primates presumably lack the full elaboration of this complex, recursive, and interactive system of multiple feedback loops, which permit rapid, smooth interplay between a conversational dyad.

Spoken language is traditionally regarded as a more or less exclusive property of left hemisphere function, perhaps, in part, reflecting—or maybe even bringing about—that hemisphere's superiority in analytic, sequential, time-dependent processes (Bradshaw and Nettleton 1981). However, extreme perinatal damage to the left hemisphere can result in the right taking over language functions (for review, see Bradshaw 1989). The earlier and the more complete the takeover, as reflected for example in left-handedness, usually the better is the prognosis for recovery of language function. Partial recovery of reading functions in adults after major left-hemisphere injury may, in certain cases, take the form of deep dyslexia, again thought by some to reflect right-hemisphere involvement (Bradshaw 1989). Deep dyslexia is characterized by semantic paralexias, where, for example, the word "rock" is read out aloud as the semantically related "stone," "house" as "home," "spectacles" as "telescope," and so on. It is as if the visual representation of the target word cannot gain direct access to the left-hemisphere articulatory system, but instead accesses semantic associates in the right hemisphere, one of which is then transfered across the commissures to the left hemisphere for articulation (see Figure 9.5). Not everyone agrees with this interpretation. Other evidence of right-hemisphere mechanisms in deep dyslexia comes from patients' occasionally preserved ability to read uncommon, irregular words, for example, "yacht," and "sword," which cannot be spoken simply by applying normal phonological rules of letter–sound correspondences. However, they are often unable to read out loud common *regular* words (e.g., "hat" and "bone").

Presumably, such patients recognize irregular words by word shape in the right hemisphere, and are not disturbed by their inability to employ left-hemisphere grapheme–phoneme correspondence rules. Thus they cannot pronounce legal nonwords ("grife" or "crobe'), or say if "sail" and "sale" rhyme, but are happy to claim that "sew" and "few" do. They have great difficulty with grammatical functor words like "to," "and," and "be" which do not refer to visualizable (in the right hemisphere?) objects, but can easily read out "bee" (with the extra "e") which now becomes an object.

Right-hemisphere damage has long been known to affect speech melody, intonation, accent, stress, and pauses—the suprasegmental and prosodic aspects of speech; recently, however, it has also been shown to be associated with some other subtle dysfunctions. Thus patients may miss the point in humor and jokes, and make inappropriately literal

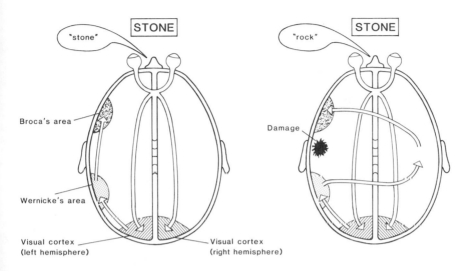

NORMAL BRAIN   **BRAIN OF DEEP DYSLEXIC**

Figure 9.5 When a normal reader sees the word "stone," the signal, originally projected to the occipital visual cortex, passes forward to Wernicke's area in the left hemisphere, probably via the angular gyrus, and on to Broca's frontal speech area, when "stone" may be spoken. In a deep dyslexic, a break between the two speech areas may require a bypass via connotative language areas in the right hemisphere, with the result that the left hemisphere eventually utters the semantic paralexic response "rock." From J. L. Bradshaw, *Hemispheric Specialization and Psychological Function*. Chichester: Wiley, 1989, with permission.

interpretations when asked to explain metaphors and aphorisms like "make hay while the sun shines" and "do not sail too close to the wind." They may have problems in detecting absurdities and in understanding figurative speech. Their difficulties in understanding the whole utterance leads to rigidity, literalness, insensitivity, loss of ability to recognize connotative significance, and problems with sarcasm, hyperbole, bathos (like "a single nuclear blast can spoil your holiday"), and context (see for review, Brownell 1988). Indeed, in some respects the right hemisphere contributes to quintessentially human behaviors, such as metaphor, humor, creativity, and fluency of ideas, pragmatic understanding and an awareness of social nuances (Cutting 1990). With a right-hemisphere injury, patients may be less able to think up alternative uses of an object (Diggs and Basili 1987), or even to judge the prices of everyday objects (Smith and Milner 1984). Minor frontal lobe damage of either side is associated with general disinhibition, impulsiveness, laughter and jocularity, and a lack of restraint which is often found by the observer to be quite amusing. Tangentiality (flight of ideas or the linking of normally dissociated ideas, itself the key to true humor and even creativity) may also occur. It is of course this frontal syndrome which may be disinhibited or released in schizophrenia (and see Joseph, 1990, for review). More severe frontal damage, however, is typically associated with apathy, rigidity, inattentiveness, and lack of emotionality, the akinetic-abulic syndrome.

Whereas the left hemisphere is closely involved in the denotative and expressive or productive aspects of speech and language, and the right contributes to connotative aspects of meaning, speech perception is perhaps only slightly less lateralized to the left hemisphere than speech production. Nevertheless, according to Annett's (e.g., Annett 1991) influential theory, in the absence of the factor which normally induces their concordance—the genetic "right-shift" factor which underlies our normal patterns of dextrality and speech lateralization—speech perception and speech production may be independently lateralized, as indeed seems to occur with left-handers. Moreover, Annett argues that the *absence* of the right-shift factor may be associated with risks to speech production, rather than to hand skill or praxis. Conversely, the factor's *presence* may be associated with risks to left-hand skill. It is certainly the case that while the nonpreferred (left) hand of dextrals is usually considerably inferior to that (the right) of sinistrals, the preferred hands of both groups are about on par. She claims that the balance of costs and benefits with respect to speech and praxis is in accordance with the hypothesis that evolutionary selection pressures for left-hemisphere specialization depended on advantages in speech, but that these advantages were bought at a cost to the left hand and the right hemisphere; they seem to be part of a human balanced polymorphism for cerebral asymmetry.

While spoken language is predominantly a left-hemisphere specialization in most people, at least at the level of actual articulatory control, we can ask whether this is also true of other forms of output. This includes signing, given the argument (critically reviewed in preceding section) that gesture may have preceded spoken language and is inextricably tied to it. We can address this issue by examining congenitally deaf individuals who must communicate by American Sign Language (ASL), and who subsequently, late in life, experience left- or right-hemisphere strokes or damage. Poizner, Klima, and Bellugi (1987) review the evidence from such cases, noting that ASL is a true formal language with grammar, but it is not based on any spoken language. ASL seems to be mediated by the left hemisphere like normal articulate speech, and is much more severely disrupted by left- than by right-hemisphere damage. Any reduction in the magnitude of these asymmetries may reflect the large spatial component of the signs, their relative meaningfulness, the patient's familiarity or otherwise with the system, and the fact that lack of phonological rehearsal in ASL may downplay certain left-hemisphere mechanisms.

Indeed, in the absence of normal speech, there seems to be an innate drive to communicate by alternative means such as signing. Thus Petitto and Marentette (1991) report an early stage of manual "babbling" in very young children born deaf and exposed from birth to sign language— exactly comparable to the articulatory babbling of a normal prespeech infant. Babbling may therefore be an amodal language capacity under maturational control, wherein phonetic and syllabic units, or their equivalents, are produced as a first step to building a mature language system. Indeed, the gestural systems *invented* by the deaf children of hearing parents (and *not* exposed to sign language) show that certain linguistic properties emerge even when the child is raised in a virtual language vacuum. As noted by Meier (1991), children approach the task of language acquisition with expectancies about how languages are organized, and with an innate capacity to acquire language, relatively independent of modality, according to a single maturational schedule with critical periods.

Similar conclusions can be drawn from the study of the parallel evolution of pidgin and creole languages (Diamond 1991a,b) in many places around the world where a "superior" invading culture, small in numbers, imposes itself on a large number of native inhabitants for purposes of colonization, trade, or military control. The vocabularies are generally derived from the invaders and syntax is minimal at the initial, pidgin stage. The vocabularies consist largely of referents—nouns, adjectives, and verbs—and are used by adults for basic communicatory purposes. However, children caught at this interface, often with parents from different cultures and communicating (across the interface) only by pidgin, *invent and impose grammars* which, although still simple and rarely inflected, convert the pidgin into a true (creole) language. Most creole languages have similar grammatical structures, supporting the idea of an innate syntactic mechanism, like sets of switches each with alternative positions which can be set to correspond to the local grammar of a child's particular culture. This idea, which has in one form or another been applied in other contexts to a variety of biological and genetic issues, is compatible with Chomsky's universal grammar, though there is no reason for him to deny prior biological precedent for the evolution of a syntactic mechanism. Therefore, children construct the developing creole, of which modern Indonesian is a spectacular example. Creolization, a natural experiment in language evolution that has independently occurred many times, may be viewed as a model of the evolution of language during phylogeny. Indeed, our propensity to (re)invent language in terms of imposing grammar and syntax on referents of the sort employed by vervet monkeys—and in pidgin—suggests that it was this faculty which was at least partly responsible for the changes that occurred between the origin of anatomically modern humans in Africa and the Upper Paleolithic revolution (Lieberman 1991).

It is unclear to what extent true (as opposed to pidgin, or even creole) grammar is present in most symbolic behaviors by apes (Greenfield and Savage-Rumbaugh 1990). Thus the relationship between symbols should be reliable, meaningful, and productive, generating a wide variety of new and spontaneous combinations if we are to securely assign this faculty to apes. Consequently syntax, in human language, permits a small number of items at one level (phonemes or words) to be legitimately recombined to generate a larger number of meaningful items at the next hierarchical level (i.e., words or phrases). This clustering helps overcome the spatial and temporal limitations of our short-term memories, and helps us to communicate at a faster rate and in a more economical and precise fashion than otherwise. It might also facilitate retrieval of information from long-term memory stores.

Do speech and manual praxis share common, or closely related or adjacent cerebral mechanisms? Apraxia (for review, see, e.g., De Renzi 1989) has traditionally been divided into ideational and ideomotor. The former refers to disturbances, following brain trauma, in *planning* voluntary purposeful movements, and the latter is held to encompass difficulties in *executing* these movements, although this distinction is unclear to some and may not always be useful. Generally, an apraxic patient has difficulties in performing purposeful movements deliberately (as opposed to automatically), and on request and out of context, in the absence of elementary sensory or motor deficits, or dementia, or problems in comprehension. Nevertheless, the patient can perform individual components, often in the wrong sequence. Thus when pretending to light a cigarette, he or she may strike the match and place it in the mouth. However, in a smoking context, the action sequence may be performed automatically without error. Thus a patient may make the sign of the cross on entering a church, but be unable to do so if requested in the clinic.

Left-hemisphere parietal damage may lead to bilateral apraxia, that is of both arms, just as left-hemisphere damage in these regions may affect language. Kimura (1982) notes how close the two regions are to each other, language and praxis, and describes the conditions of clinical association and dissociation of the two syndromes, aphasia and apraxia. She notes how closely our manual gestures follow the rhythmic structure of spoken speech. It is tempting to consider that the two functions, respectively subserving communication and tool use, have coevolved, or at least during evolution could have mutually facilitated each other. Thus the two functions have much in common, for example, the skilled serial motor activity of a syntactic nature which involves the precise timing and sequencing of actions to be inserted at appropriate points within a developing framework. Thus speech, when it appeared, could have "parasitized" the preadapted neural mechanisms for praxis in the left hemisphere (see Bradshaw and Nettleton 1981; Morgan and McManus 1988). Nevertheless, the recursive and embedded nature of speech syntax is quite different from the largely linear sequentiality of most kinds of manual praxis.

The Neurology of Art: A Bihemispheric Contribution

The neurology of left-hemisphere functions in general, and of language and praxis in particular, is reasonably well understood as befits the single most important faculty for integrating the individual within society—an observation which itself supports the view that social communication (Machiavellian intelligence) played a major role in the process of hominization. Much less well understood is the neurology of the right hemisphere—long regarded as "silent" and submissive to the dominant left—and of artistic and musical behaviors. That is *not* to say that all such behaviors belong to the "minor" right hemisphere; it is just that many aspects are not so firmly lateralized to the time-dependent, sequential, analytic functions of the left hemisphere which mediate language and praxis. Moreover, many people who experience brain trauma that affects artistic behaviors do not know what they have lost, particularly if they have never previously practiced such behaviors. As we shall, parietal damage may result in the artist ignoring one spatial region on the canvas, temporal damage is associated with problems in recognition, and frontal damage with deficits of praxis (i.e., performance). Left-hemisphere damage may lead to oversimplification of drawings, with loss or impoverishment of detail, and poor discrimination of elements, but with an intact overall perception of configuration and pattern of spatial relationships. Conversely, right-hemisphere injury results in haphazard, inappropriate combinations of correctly discriminated features, with rich internal detail but faulty overall configurations, proportions, and spatial interrelationships (Bradshaw 1989). See Figure 9.6. Patients with left-hemisphere injury may be able to sort paintings by style but not by subject matter,

Target stimulus   Left CVA   Right CVA

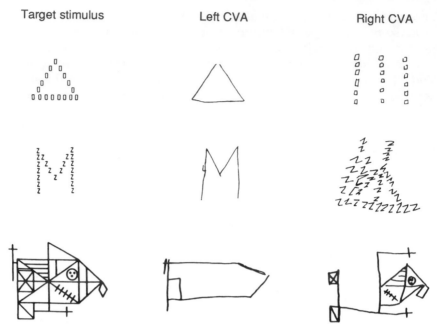

Figure 9.6 Drawings of target figures composed of subfigures, and the Rey-Osterreith figure, from memory by a right- and a left-brain damaged patient. CVA = cerebro-vascular accident (stroke). From L. C. Robertson and M. R. Lamb, Neuropsychological contributions to theories of part/whole organization. *Cognitive Neuropsychology*, 1991, 23:299–330, with permission.

whereas the reverse deficit may be found with right-side damage (Gardner 1982).

Further insights into the neurology of art come from the agnosias (see, e.g., Brown 1989; Farah 1990; Stiles-Davis, Kritchevsky, and Bellugi 1988). Agnosia may be defined as an acquired (through insult or disease) inability to recognize objects or their meanings and significance, or to correctly manipulate them, in the absence of apraxia (see previously) or anomia (word or name finding difficulties). Apperceptive agnosia is a disorder of object *perception;* and subsequent to damage to the parieto-occipital right hemisphere, it is associated with problems in perceiving anomalies in sketchily drawn or incomplete figures, in recognizing degraded or fragmented letters, overlapping figures, or unusual views of objects taken from noncanonical orientations or with uneven lighting. A patient may, despite claiming an inability to recognize an object, nevertheless when picking it up, correctly mold the hand to make the grasp, implying covert recognition or disconnection of conscious recognition from automatic processes.

With associative agnosia, there is a disorder of object meaning or *significance* in that patients may treat hitherto familiar objects as unfamiliar or novel; an affected physician described his stethoscope as "a long cord with a disk at the end," but could in fact immediately recognize it for what it was by touch. Objects may be perfectly drawn or copied, and still result in an inability to say what they are. Sometimes broad categorical statements may be made about unrecognized objects, for example, a cup as "some sort of container" and a tractor as "some sort of vehicle," again indicating partial or incomplete knowledge. Occipitotemporal regions of the left hemisphere are largely involved; the syndrome is demonstrable when patients can no longer match objects or pictures by *use*, *function*, or *category*, as opposed to by *appearance* as in apperceptive agnosia. There may therefore be three stages or elements in object recognition: visual sensory analysis which is largely bilateral; a fully organized structural description of the object or scene with major right-hemisphere components; and finally an assignation of meaning which is largely mediated by the left hemisphere. However just as we saw in language processing, all three processes may go on at least to some extent in parallel, with feedback between them. At this stage absolutely nothing is known about possible contributions from subcortical (thalamic and basal ganglia) centers, though limbic regions including the cingulate are likely to input affective or emotional contributions.

Prosopagnosia is a particular kind of agnosia where the patient displays a selective loss in the ability to recognize *familiar* faces, even though they are still seen as faces, and *unfamiliar* faces can still be discriminated from each other. We have already reviewed recent findings on this topic in another context, so will merely summarize relevant material. Patients suffering from prosopagnosia may be forced to rely on the presence of a mole, for example, in recognizing a spouse. Normal face recognition depends on two processes (Bradshaw 1989). First, there is the ability to recognize the face as a whole, as something more than just the sum of the component features, but via the pattern of interrelationships *between* and otherwise *independent* of individual features; this is an ability that may be largely mediated by the right hemisphere. Second, there is the ability to perform piecemeal, feature-by-feature analysis, a left-hemisphere process which is perhaps more important with unfamiliar material. Early clinical reports emphasized right-hemisphere involvement in the syndrome; later, both hemispheres were thought to be implicated, and recently clinicians are again stressing that posterior damage of the right hemisphere is necessary and probably sufficient to cause prosopagnosia. Thus the retained ability to recognize the spouse if he or she has an identifying feature like a mole implies intact, residual mechanisms in the left hemisphere to perform analytic feature detection.

Recently it has been shown that prosopagnosia may not in fact be restricted to faces, a finding which earlier supported the idea that there might be a specialist processor dedicated to face recognition in the right hemisphere, analogous to a language processor in the left. Thus face recognition is only slightly less complex than language processing given the huge number of faces, of all ages and with varying degrees of "paraphernalia" like spectacles, make-up, and beards, viewed under a wide range of conditions, which we can successfully recognize over a lifetime. Moreover, faces convey information not only on personal identity but also on the target's current emotional state. There is now an indication that prosopagnosics may also have difficulties with other forms of generic stimuli requiring individuation, where category members are ambiguous or confusingly similar or require arbitrary differentiation and labeling. Thus a farmer may no longer be able individually to identify his cows, a dog fancier may now confuse different breeds of dog, an auto enthusiast may have problems with different makes of car (for review, see Bradshaw 1989). As with agnosia and apraxia, automatic unconscious recognition or performance may be preserved, even with the loss of conscious, deliberate, and volitional aspects. Thus a prosopagnosic may display autonomic indications of recognition (e.g., changes in body sweat) when a familiar face appears, all the while claiming nonrecognition.

In this context, we note that in monkeys evidence is now mounting of cell populations in the superior temporal sulcus (TPO and PGa) and the inferior temporal (IT) cortex which are selectively responsive to faces and others which respond to *parts* of faces, for example, eyes, mouth, or hair, the direction of gaze, or the inter-eye and eye-to-mouth distances (Harries and Perrett 1991; Simone 1991). Monkeys respond to real faces, plastic models, photographs, and (to a lesser extent) to line drawings. Indeed, Kendrick and Baldwin (1987) report that cells in the temporal cortex of conscious sheep respond preferentially to the sight of faces—sheep, human, and dog—with responses that are influenced by factors relevant to social interaction (e.g., familiarity and threat). Thus face-specific cells may be found in a wide range of species. Today the prevailing theory favors the idea of distributed neural assemblies or neural networks, rather than gnostic units as single cells at the top of a pyramid beginning with line and edge detectors and ending with "grandmother cells" to recognize something as discrete and unique as one's own grandmother. However, these face-selective cells are reminiscent of an earlier approach. There seem to be two populations of such cells, one group responding to expression, and the other to identity, residing in the superior temporal sulcus and the inferior temporal gyrus, respectively.

Why do monkeys treat faces so differently from other classes of stimuli? Faces are of course important for recognizing individuals and for social communication; this will be discussed again in the context of the evolution of Machiavellian intelligence. We can compare these specialized cortical structures in the monkey with analogous speech-mediating structures in the human left hemisphere and also in the left-hemisphere structures mediating song in songbirds. All three occur in supramodal association areas on the left hemisphere. Thus Perrett et al. (1988) reports that macaque face cells are more prevalent in the left hemisphere; whereas this is opposite to our own visual pattern of right-hemisphere mediation, and to the behavioral studies by Vermeire and Hamilton with macaques reviewed in Chapter 6, the apparent contradiction may perhaps be resolved if we view such face processing by monkeys as a communicative rather than as a pattern-recognizing function.

A final right-hemisphere phenomenon relevant to considerations of pictorial or artistic representations is unilateral neglect, hemineglect, or hemi-inattention. In this case, the patient may display neglect or inattention toward events occurring in the side of space contralateral to the lesion; this is usually the left side, after right-hemisphere damage, right neglect after left-hemisphere damage being uncommon. The patient may brush up against a doorpost on the left, ignore food on the left side of the plate (until it is rotated), shave, groom, or dress only the right side, leave a large margin on the left, copy or produce only the right sides of drawings (or the right side of all objects in a scene) and so on (see Figure 9.7). As we saw with prosopagnosia, there may be a dissociation between conscious and unconscious aspects; if a patient is shown a picture of a house with flames emerging from the left side, the flames may not be noticed, although they are seen quite normally if they are shown on the right side of the house. However if the patient is asked which picture she or he prefers, one with *no* flames or one with flames on the left, the former is chosen, even though the latter is claimed to be a perfectly normal picture (see Figure 9.8). Thus there must have been a subconscious registration of the flames on the left. (For a further discussion of such a dissociation between conscious and unconscious processing, see Bradshaw 1989.)

Left neglect after right-hemisphere injury is commoner, longer lasting, and more severe than right neglect after left-hemisphere injury, leading to suggestions that the right hemisphere may mediate attention for

Figure 9.7 Paintings by a well-known French artist after left hemiplegia. Hemi-neglect may involve the whole of the left part of the drawing (upper left), or some of its elements, such as the basket (upper right), or the left side of the figures (lower left). It is also found in the left portions of some objects or figures: the flower bouquet (upper left), the peasant's shoulder (upper right), the female customer seated in a bar (lower left). The hand is poorly structured on the left side (lower right). From R. A. Vigouroux, B. Bonnefoi and R. Khalil, Réalizations picturales chez un artiste peintre présentant une héminégligence gauche. *Revue Neurologique*, 1990, 146:665–670, with permission.

events in extracorporeal space not only contralaterally (like the left hemisphere) but (unlike the left) ipsilaterally as well, so that if the right hemisphere is damaged the left can only handle contralateral right hemispace. Indeed, the same situation may even apply to rats; Vargo, Richard-Smith and Corwin, (1989) produced unilateral neglect by damage to the contralateral dorsomedial prefrontal cortex. Contralesional neglect, which subsequently abated, was reinstated with spiroperidol, a dopamine receptor agonist, in animals with left side-damage. However, in recovered animals with right-side damage, not only did lower doses of spiroperidol suffice to reinstate neglect, but it did so even for events occurring in space ipsilateral to the lesion. The parallel in asymmetries between the two species, human and rat, is remarkable. However another observation (Bisiach and Luzzatti 1978) suggests that neglect may also involve a distortion of representational space: left neglecting pa-

Figure 9.8 A patient suffering from left hemineglect and asked to choose between two houses, one of which appears to be on fire on the left side, apparently sees nothing odd about the latter house, but nevertheless typically prefers to choose the top one.

tients, when asked to imagine themselves standing at one end of the Piazza del Duomo in Milan, could only name buildings on the right, and when asked next to imagine themselves at the other end could now only name the row (again on the right) on the *opposite* side of the Piazza.

According to Mesulam (1990), at least three cortical brain areas are involved: the posterior parietal cortex of the right hemisphere, the pre-motor prefrontal cortex and the cingulate regions. The first of these may be responsible for mapping extracorporeal space in predominantly perceptual terms, and the second for mapping space in terms of possible orienting (with head and eyes) and exploration (with eyes and hands) responses. The cingulate gyrus may assign value and affective meaning to spatial events or coordinates. Subcortical (tectal) centers like the superior colliculi are also involved, at a more automatic and unconscious level, with orienting and alerting. However, as so often in the human brain, sensory and motor processes are inextricably intertwined, and the conscious–unconscious distinction is often made more easily at a cognitive level than at a physiological level.

At a cortical level, spatially guided behavior, the breakdown of which manifests in unilateral neglect, is, as we saw, mediated by prefrontal and parietal cortices. The former, responsible more for peripersonal space within which reaching and grasping may be directed, may in fact play an overall executive role, regulating posterior parietal representations of extrapersonal events. Goldman-Rakic (1988) describes for monkeys and humans massively parallel connectivities between corresponding structures within these two cortical areas. The whole system, especially in the human right hemisphere, may represent a parallel distributed network to subserve attentional processes (and see also Pandya and Yeterian 1990), comparable and analogous to fronto–parietal systems subserving language in our left hemispheres.

Visual Imagery

Visual imagery is one activity where a right hemisphere involvement might be predicted (and compare left neglect of imagined street scenes as just discussed). However, according to Farah (1989), posterior regions of the left hemisphere instead are involved; this is shown in the loss of image generation ability after posterior left-hemisphere injury, by electrophysiological (evoked potential) analysis, and by studies of changes in regional blood flow during visualization tasks. Moreover, when mental images are used as templates to facilitate discrimination of laterally presented visual stimuli, effects may be greater with presentations to the left hemisphere (via the right visual field). Of course, many computational processes are involved in imaging, perhaps differentially in the two hemispheres, for example, storage and maintenance of visual memories from which images can be generated, their actual generation, and inspection of such images and matching to current visual input. The work of Kosslyn (1987) offers a further resolution of this conflict. Left-hemisphere specialization for image generation may occur only when the locations of image parts are specified precisely and categorically in absolute terms; a right-hemisphere superiority may instead supervene when the loci of image parts are specified less precisely and with reference to an overall common frame. Kosslyn emphasizes the processing commonalities between veridical visual processing and its substitute visual imagery (seeing without the actual sensory input), as in mental rotation and image scanning; imagery, he says, serves to simulate or model reality, and to permit navigation.

The linguistic encoding of categorical relations may have served to attract similar, though nonlinguistic, processing operations. This would thus account for the left hemisphere's advantage in easily processing categorized (and named) shapes, line orientations (horizontal and vertical, whereas the right hemisphere is better with fine gradations beyond such categorical anchor-points) and judgments regarding contact. Conversely, the right hemisphere is better at judgments of relative proximity. If the left hemisphere develops a catalog of categorical representations, the right hemisphere, according to Kosslyn, encodes spatial relationships whether for determining the location of objects, or for focusing and directing attention (compare unilateral neglect).

The virtue of this approach is that just as we cannot claim that all aspects of language are mediated by the left hemisphere, or that all aspects of shape recognition are mediated by the right hemisphere, we also cannot ask whether all aspects of imagery are mediated by a single hemisphere; the phenomenon decomposes into numerous subabilities, each differentially lateralized. Indeed the exact nature of human cerebral asymmetry is far from fully understood. Suffice it to say that the left hemisphere is perhaps preeminent in making functional associations, in the formal operations of speech such as phonology and syntax, in sequential time-dependent processes where events are ordered and programmed in time, and in a discrete, detailed categorical approach. Conversely, the right is perhaps concerned more with physical relationships, with associative and connotative meaning, and with a configurational, relational, and integrative approach (Bradshaw and Nettleton 1981). Nevertheless, strategic aspects are also important: With closely spaced temporal events, the right hemisphere can use sensory quality to discriminate one sequence from another, and with widely spaced temporal events it can learn and retain order by spatial remapping. We have already encountered similar theories (e.g., by Bianki) in studies of the rat in Chapter 4.

Why do we take pleasure in aesthetics? Why do we perceive as beautiful a sunset, flower or portrait? Why do we *create* objects or patterns of aesthetic appeal? Eibl-Eibesfeldt (1988) notes that both activities employ perceptual systems that evolved to subserve pattern recognition and spatial orientation. Perception is an active, constructive process (as indeed is speech perception, as we saw), involving a search for order, clarity, and regularity, an active categorization and interpretation process. We are biased to see clear distinct forms (*gute Gestalten*) when we briefly observe incomplete geometric figures like circles, squares or triangles; we strive for symmetry and regularity, and automatically complete incomplete patterns—as shown by proofreaders' errors, and the patterns we see in clouds, the fire, and Rorschach inkblots. Finding such patterns and completing slight imperfections is rewarding, like successful problem solving—hence the joy of puzzles. Art may be especially satisfying when it requires such problem solving, or is an expression of our perceptual biases, or when it induces surprise, puzzlement, or awe. However our species may not be totally unique in this regard. Elephants spontaneously make drawing motions in the dust with their trunks and can be taught to wield a paint brush on paper (Diamond 1991b), as can chimpanzees (Eibl-Eibesfeldt 1988). The latters' artwork is judged by us as aesthetic, symmetrically, and harmoniously filling available space in a manner of controlled composition. We may even ascribe similar qualities to the bowers constructed by the male bower bird of Australia and New Guinea for the purpose of enticing females. This, of course, has a direct utilitarian role in the breeding cycle. Ancient rock art may also have had important social purposes of group cohesion and cultural tradition, as with modern religious art and the totems of northwestern coastal America.

Music and the Brain

Just as not all aspects of language, art, or imagery are lateralized to a single hemisphere, we shall see that music also consists of a number of discrete components all differentially lateralized to the left and right, and differentially localized in an anterior and posterior direction. However, as Henson (1985, p. 489) observes, "There is an ultimate mystery of musical experience which is not susceptible to neurological study." Amusia, the loss of the ability to create, perform, or assess music after localized brain trauma, has been one major source of evidence concerning the neurology of music. Yet unlike speech, which is acquired spontaneously if the individual is exposed to the correct experiential situation at an appropriate developmental stage, music seems to require more deliberate learning, at least in our western culture. (It is entirely possible that under different circumstances, e.g., using the Suzuki method, music might mimic language with respect to easy spontaneous acquisition, but as yet these circumstances have not been explored.) Thus many individuals may never *know* that their injury resulted in amusia, if they never previously engaged in musical activities. Its lower communicatory and social importance also means that the clinician sees fewer cases, and those may be predominantly of a particular kind, for example, of professional performers. Nevertheless, just like language, to which music has many curious and interesting parallels, music consists of many dissociable elements, for example, rhythm, pitch, timbre, intensity, tempo, and duration, each associated with their own specific amusia and brain area. Indeed we can broadly categorize the disorder into *receptive* amusia (posterior) with impairments in pitch perception, instrument recognition, chord analysis, melody recognition, and reading musical notation, and *expressive* amusia (anterior); the latter involves impaired production of new or known melodies, vocally or instrumentally produced, with or without inabilities to transcribe music (Bradshaw 1989; Henson 1985; Zatorre 1984). However, posterior temporal regions may be the more important.

At the normal level, music requires perception of both simultaneous (e.g., tonal, and possibly right-hemisphere mediated) and serial (e.g., rhythmic, and left hemisphere?) events. Music can also be experienced, with different laterality effects, by subjects with different levels of musical sophistication. As a generalization, increasing musical sophistication has been shown to draw more and more processing resources from the right to the left hemisphere (Gates and Bradshaw 1977; Henson 1985). Thus as observed by Peretz and Morais (1988), hemispheric dominance for musical processing depends on stimulus characteristics, the subject's evaluation of the task, perceived demand characteristics, expertise, and adopted strategies. Musical education may facilitate the extraction of local features (left hemisphere) at the expense of contour information (right hemisphere), the latter being more readily available to nonmusicians. There is indeed much clinical evidence that global aspects (pitch, melody, and timbre) may appeal to right-hemisphere processors, while temporal aspects (rhythm, timing, and duration) may appeal more to the left hemisphere.

This situation is reminiscent of speech and language–right hemisphere mediation of prosodic and connotative aspects, with the temporal, sequential, and syntactic aspects being the province of the left. For this reason, a global–local dichotomy for the lateralization of music has found favor (Gardner 1982; Peretz and Morais 1988). Thus a right temporal lobectomy damages performance much more than a left temporal lobectomy on the Seashore Test of Musical Talent (Zatorre 1984), with deficits in timbre, tonal memory, loudness and pitch discrimination, complex tone discrimination, and melody recognition. Conversely, left-hemisphere damage affects recognition of familiar tunes and lyrics. (Timbre is cued by changes in the upper partials of a complex tone, and in its onset and offset characteristics; thus timbre may be viewed as a simultaneous pattern of pitches—chord—whereas melody is a sequential pattern of pitches.)

Music is, of course, a fundamental component of the arts and dates back millennia in various rites and practices, quite apart from its possible esthetic value. (Flutes, as we saw, date back to the Aurignacian.) It is nonsemantic, noncommunicatory, apparently nonadaptive, so why is–was it undertaken? Singing may have been its earliest preinstrumental form, and may have been important for mating behaviors, mother–infant exchanges, and emotional expression (Peretz and Morais 1988). Otherwise, group music making often appears at any social or work grouping: sea chanties, weaving songs, religious and celebratory songs. In speech, prosody and intonation contours convey important paralinguistic, suprasegmental, and nondeclarative information about the emotional state of the speaker, and this right-hemisphere mediated function may have been an important element in the evolution of musical behaviors. Music may therefore be a "playing of the emotions"—like the visual and plastic arts, but in a modality closer to that of speech and linguistic communication (and see Gardner 1982; Sidtis 1984). However no single explanation is likely to capture all the factors which probably operate in the elaboration of musical behaviors in different societies and over time.

The Prehistory of Spoken Language

If we know little about the evolution of music, what do we know about the prehistory of languages currently spoken? Philologists can compare extant languages in an attempt at reconstructing their immediate progenitors and, perhaps, eventually their ultimate ancestors or even the original protolanguage, if indeed one ever existed. The approach developed from the study of the Indo-European superfamily of languages, which is by far the largest in terms of numbers both of languages (which include most European languages and many spoken in India and Persia) and of speakers—nearly half the world's population. The study of Indo-European languages commenced over 200 years ago when in 1786 Sir William Jones noted affinities, due, he presumed, to common descent, between Gothic, Greek, Latin, and Sanskrit (Lewin 1988b). Since then, the nature of affinities has been documented between descendants of Latin (the Romance group of languages) and of Sanskrit (languages of the Indian subcontinent). The patterns of affinities between Sanskrit and its cognates have also been studied; for example, Hittite and

Avestan, and other superfamilies have been established, cognate with Indo-European, for example, Uralic, Altaic, Dravidian, Kartvelian, and so on.

As observed by Gamrelidze and Ivanov (1990) and Lewin (1988b) there are many similarities between language change and genetic change over time, divergence between isolated populations being the key to both processes. When biologists reconstruct extinct organisms, they can use both extant and relevant fossil species. There are no equivalent fossils for the reconstructing linguist, except for writing which dates back at most to 6 kya in Mesopotamia. So the linguist or philologist must create dictionaries from commonalities observed in later tongues, although most such reconstructions are atopic, achronic, and aphonic, which means that we do not know where, when, or how the words were spoken. Moreover, just as molecular biologists identify molecules as groups that perform similar functions in widely divergent species to infer the characteristics of the primordial cell from which the species presumably descended, so too the linguist seeks correspondences in grammar, syntax, pronunciation, and especially vocabulary, among known languages. In this way an attempt is made to reconstruct their immediate forebears and ultimately the original tongue, working backwards according to the laws governing phonological change. Chance similarities and loan words are the major problems.

Thus, Grimm's law of *Lautverschiebung* (sound shift) postulates that sets of consonants in Indo-European languages displace one another in a predictable and regular fashion. Interestingly, the techniques of comparative linguistics can even help locate where and when an ancestral language might have been spoken. Thus the words of Indo-European (as derived by backwards extrapolation from existing languages in that superfamily) describe a landscape, agricultural society, and a climate that place it in eastern Anatolia (Turkey) 7 to 6 kya (although see also Renfrew 1987). Such words include those for certain trees, crops, animals, and landscapes. The agricultural revolution, starting in the nearby Mesopotamia, created the food surplus that permitted the Indo-Europeans to found city states from which 6 kya they commenced migrating westward and eastward over Eurasia to the British Isles, Scandinavia, Russia, Iran, India, and even Sinkiang (the ancient Tocharian language is Indo-European).

The techniques of reconstructive linguistics, as applied to Indo-European, even resulted in predictions that, in accordance with good scientific method, could be positively tested against scientific evidence. Thus it seemed certain that Indo-European languages were descended from a tongue that employed a class of laryngeal consonants that have failed to survive in any known tongue. Later, with the discovery and decipherment of Hittite, their presence was confirmed in a historical tongue.

Indo-European is currently the most thoroughly reconstructed proto-language. However, Nostratic at around 15 kya has been proposed as an antecedent to include Indo-European, Afro-Asiatic (Arabic and Berber), Kartvelian, Uralic (Finnish and Samoyed), Altaic (Turkish and Mongolian), and Dravidian (Southern Indian). Two Soviet scholars, working independently, published reconstructions of Nostratic, amounting to several hundred words based on common key items. These included personal pronouns, body parts, sun and moon, and so on. The corpus suggests that the "Nostrates" were hunter–gatherers, without agriculture, although with domesticated animals, including the dog. The oldest domestic dog bones, at 14 kya, support this interpretation. Later, it is suggested, Nostratic fragmented into Afro-Asiatic languages whose people (from the proposed vocabulary) built fortifications, cultivated land, hunted with bow and arrow, and marketed. Around 9 kya, proto-Semitic may have arisen and provided many loan words to the Indo-Europeans in Anatolia. The latter group are postulated by Shevoroshkin (1990) to have split up around 5 kya into the various branches of Celts, Baltic people, Germanic people, Slavs, Greeks, Italians, and Indo-Iranians.

On a level with Nostratic, other super-superfamilies have been proposed, for example, Dene-Caucasian (or Sino-Caucasian) of 14 kya composed of Sino-Tibetan (Chinese), Yeniseian (Siberian), and Caucasian (Basque, Sumerian, and Etruscan). American languages likewise have been grouped into Amerind (whose speakers crossed to Alaska around 12 kya) and Na-dene (who populated northwestern Canada and Alaska around 9 kya). See Shevoroshkin (1990). Other scholars link all of these with proto-Australian, Austro-Asiatic, Congo-Saharan (central and north Africa), Austronesian (Indonesian, Phillipines, and Pacific Islands), and Thai. These are seen to have descended from one original tongue spoken around 35 kya (although this *excludes* Khoisan of the Bushmen and Hottentots of south Africa), and of which 150 to 200 words are said to have been reconstructed. This "vanishing point" of around 35 kya (and see Miller 1981, and Ruhlen 1987) is of course the time of the west European Upper Paleolithic cultural explosion.

We mentioned the similarities between reconstructive linguistics and molecular biology. Cavalli-Sforza et al. (1988, 1990) combined the two approaches. They ordered 42 "aboriginal" human populations by summary of selected "non-DNA" (i.e., enzyme) allele frequencies (120) using Nei's genetic distance technique. They then superimposed linguistic phyla upon the phenogram, concluding that the two patterns of evolution, molecular and linguistic, effectively followed the same history in terms of sequences of fissions. Khoisan can even be accommodated now, in terms of the first split of around 100 kya (and see also Cann et al. 1987). Barbujani and Sokal (1990) similarly found that zones of abrupt genetic change (in terms of gene frequencies) and linguistic boundaries coincide in Europe, indicating that the genetic structure of a population reflects its language affiliations and vice versa. Linguistic differences therefore seem to have acted as reproductive barriers or isolating mechanisms (the mechanics of racism?), causing spatially close populations to diverge in gene frequencies. Of course, unlike genes, which are constrained to the reproductive cycle via the genome, words can be acquired (borrowed) or modified during an individual's lifetime, and can in fact spread throughout a human population within a single generation, especially slang words or words labeling a new technology. Moreover, while languages, like subspecies, can hybridize, unlike the biological situation, closely related languages are no more likely to do so unless they happen (as indeed may often be the case) to be mutually adjacent. Thus we should perhaps observe some caution in seeking close parallels between molecular biology and language evolution.

The Evolution of Intelligence and Encephalization

Tool use, art, and language, as discussed in earlier sections, and complex social behaviors incorporating ritual, myth, religion, and affiliative activity where the individual subjugates his or her individual aims to collective socio-political goals, all characterize *Homo sapiens sapiens*. Precursors of at least some of these human behaviors may be found in other primates, a group which has evolved separately from the other mammals since the late Cretaceous. However, it is unlikely that mammalian brains have evolved by advancing along a single continuum from simple to complex; there is no *a priori* reason to think that the primate solution is necessarily the most advanced one. (It is certainly a mistake, as proposed by Eccles, 1991, to suggest that evolution is somewhere subject to a higher guiding force.) Indeed, as Gould (1991a) persuasively argues, were time's videotape of the whole history of life to be rewound to a fresh start, and played over again from that new beginning, all the stories would have a different outcome. Human existence and consciousness, even vertebrate life in general, would be unlikely to reappear on the second take; evolution is no inevitable ladder of directed progress or *scala naturae* tending for ever "upwards," a working out of cosmic emergent complexity. We might of course be tempted to argue that the nature of the constraints which can be imposed upon a dynamically interacting and evolving ecology are such as to tend to reduce the range of possibilities and lead to common solutions to similar evolutionary pressures. Thus we witness the homoplasies (parallel evolution) between so many structures in the cephalopod and vertebrate.

On a grand scale, evolution behaves in accordance with the predictions of chaos theory as exemplified in the short term in meteorological phenomena—a very complex interacting system the states of which cannot be predicted except for a very brief period ahead. It may not be an exaggeration to note that the flap of a butterfly's wing may be almost enough to cause changes to a weather system in a matter of days, because of the highly multiplicative effects of even the smallest events which feed back positively in a cascade fashion. On the other hand, were we to speculate on what a differently evolved, technological creature might look like, we perhaps would have to stay reasonably close to our current plan; it would probably have to be land- rather than water-based, of a certain size, and possessed of good powers both of communication and manipulation. Again, we see strange similarities between our limbs and quite differently evolved structures—a prehensile, manipulable and controllable elephant's trunk, or a cephalopod's "arm." Divers experience the uncanny feeling of intelligent sentience when a large squid reaches out its arm tentatively to explore the diver's face, watching the while with an almost human eye, and then fearfully withdrawing its limb with a rapid jerk. Squids, of course, despite a separation from vertebrate (chordate) stock since the Cambrian seas of more than half a billion years ago, are highly intelligent, not far behind a rat, and they are only molluscs (Young 1964).

Earlier attempts at comparing the "intelligence" of different species and producing a "*scala intellectūs*", perhaps from amoeba to humans, with pigeons, rats, and cephalopods jockeying for relative positions, failed because of the lack of comparable tests or tasks for species with radically different habits and habitats. A test must be ecologically valid and address behaviors within an animal's normal repertoire—and almost by definition each species has its own repertoire, as well as highly individual sensory, cognitive, and motor capacities. Appropriately tested, a pigeon can outperform a human, and not just in innate homing abilities but in pattern-recognition tasks and problems requiring training. Even attempts to compare performance at a more general level have generally foundered, for example, in speed of reversal learning. In this paradigm different species are tested in their own habitat with suitable stimuli and responses within their repertoire. They are assessed with respect to how quickly they can learn to reverse their responses to previously learned discriminations. Whatever we in fact mean by intelligence, the term is inappropriate at the level of comparative (i.e., animal) psychology, and many other capacities including innovativeness, insight, and self-awareness also need consideration (Griffin 1984; Radner and Radner 1989; Walker 1985). (At a purely practical level, the ability to invent novel solutions to new problems may be a useful index of intelligent behavior; as we saw, Hauser, 1988, reviewed the conditions leading to a wild vervet monkey using dry *Acacia* pods as a tool to extend nutritive sap, and transmitting this solution to other troop members.)

There is in fact no single definition of intelligence, and many are concerned that a hypothetical abstract or even a set of performance scores have been "reified" into an entity which really has no independent existence. This, and other considerations, notably those relating to different human populations (and possibly gender), and the contribution of environmental factors, will not be considered here, nor will the history of the various scales, their standardization, reliability, validity, or even utility in everyday life. For many years the observation that performance on a wide range of tasks and tests tended to correlate, seemed to confirm that a common general ability was being tapped, one to do with perceiving and manipulating relationships, and relating to abstract reasoning in a wide variety of situations, rather than to the recall of factual information. Later, a distinction was made between "fluid" problem-solving ability in dealing with unfamiliar situations, and the application of "crystalized" intelligence in employing existing skills and knowledge. In contrast, Sternberg's (1985) triarchic theory proposes three "intelligences": one subserving standard problem-solving performance (analytic); one subserving creativity, lateral thinking, and the strategic development of novel solutions (synthetic); and finally one involving the application of procedures and learning how to acquire new skills (a practical component).

Gardner (1983), however, proposed a multiplicity of unrelated forms of intelligence, including language abilities, musical abilities, logical and mathematical reasoning, spatial reasoning, body movement skills, and social sensitivity. Support for this approach comes from Fodor's (1983) modularity of mind hypothesis (see above) where the mind is seen not as a single general-purpose mechanism, but as a cluster of highly task-specific modules, linked one to another, but quasi-autonomous in operation. Support also comes from Gardner's own neurological research, where, as we saw, localization of function is indicated from the discrete effects of lesions in generating, for example, amusia, alexia, aphasia, acalculia, agnosia, apraxia, and so on. Further support comes from the "idiot-savant" syndrome where a person with a generally very low functioning level possesses an island of outstanding ability in perhaps computation, calendrical calculation, or drawing (Treffert 1989). Indeed such individuals challenge the very idea that one should expect a relatively high correlation between different abilities (Howe 1989). On the other hand, "idiots savants" are often also extremely withdrawn or autistic, with restricted communicative ability and obsessive personalities. It may be that an extraordinary obsessive interest in numbers, calendars, racing lists, and so on, as indeed evinced by such individuals, naturally leads to apparently supernormal abilities or achievements, even in otherwise handicapped individuals.

Perhaps the most radical departure from traditional approaches is the idea that intelligence is either directly assessed by—or is even identical to—speed of functioning or processing in the central nervous system, as indexed by decision time tasks (see e.g., Vernon 1987). Such a task may involve monitoring a bank of eight lights arranged in a circle, *any* three of which might come on simultaneously, the required response being to depress the adjacent button of the light *furthest* from the other two. If from this choice reaction time is subtracted simple reaction time, time taken to press the button when only one light ever comes on, and it is always the same light, that is, sensorimotor response time, the difference—decision time—seems to correlate highly with real-world performance. Thus high intelligence might involve keeping many things in mind simultaneously and processing them interactively before any of them fade. Such an approach certainly seems to tap mental agility. However, unlike the approach of Gardner, this ignores the possibility of specific abilities which may have their counterpart in *nonhuman* "intelligence," with each species having its own particular ecological adaptations, or a given lineage adapting to changing environmental pressures during evolution. At the hominid level such abilities as speed of skill learning, efficiency of information retrieval from long-term memory, decision making, practical problem solving, spatial abilities, concept formation, rule learning, tool use, communication, and so on, might be expected to loom large with, in many cases, dedicated neurological areas for subserving them, for example, the prefrontal, parietal, and temporal cortices, and the cerebellum. Moreover (and see Hodos 1986), intellectual differences between species (or taxa in a lineage) are likely to involve qualitatively as well as quantitatively different ecological adaptations. Indeed, certain *quantitative* differences in appropriate *combinations* of elements in a mosaic of abilities can give rise to real or apparent *qualitative* intellectual differences at a more global level.

According to the principle of allometry (see e.g., Harvey 1986), brain size (BR) increases with body size (BO), but not in a one-to-one ratio. In fact it increases more slowly along any given taxonomic continuum, for example, the mammalian spectrum from mouse to elephant. A power function best describes the relationship, with a straight line when logarithms are used on both axes (brain and body size):

$$BR = a \, BO^b$$
$$\log BR = \log a + b \log BO$$

where a is the allometric coefficient, and b is the exponent. An exponent of around 2/3 is typically found, meaning that brains increase in mass roughly 2/3 as fast as bodies. Encephalization is the residual capacity that remains after basic allometric requirements ("a large brain to drive a large body") are satisfied, and it represents the additional information that can be handled beyond that used in routine control of body functions. Thus it is the phenotypic "expression" of intelligence (Jerison 1986).

While gestation length, metabolic rate, precocial–altricial state at birth, and longevity all correlate with (and may determine) relative brain size, mammals are allometrically "brainier" than reptiles, and reptiles brainier than amphibia. Primates and cetaceans stand out above the other mammals (Martin 1982), and H. sapiens sapiens has about 3.6 times the brain size "expected" for a "normal" Old World simian primate of our body size (Falk 1980). The adult brain is around 2.3 times bigger than that of the neonate in the case of the ape, whereas the ratio is 3.5 in the case of humans (Lewin 1989a). While humans and apes weigh about the same and possess similar gestation periods, the human neonate is about twice as big (as is its brain) as the newborn ape. However, we get our large adult brains through prolonged growth rather than via an unusual growth rate. As we have seen, Homo habilis was the first to evidence increased encephalization. While we cannot be sure to what extent an upright bipedal posture contributed (together with pelvic adaptations and obstetric consequences), we can be sure that tool use was not a major factor, at least in the early stages. Thus, as we saw, chimpanzees are capable of making and using tools of an Oldowan level, and encephalization proceeded at a far faster pace than lithic technology.

Hodos (1986) asks whether increased encephalization necessarily increases information processing functions (attention, learning, memory, reasoning, and concept formation) or only perhaps other aspects such as motivation, instinct, and similar such instinctual, stereotyped, and fixed action patterns. He raises two other questions, which we shall address in detail. First, can we can relate global intelligence to global brain measures at an intraspecific level (i.e., crudely, are large-headed people generally somewhat brighter than their fellows who take a smaller hat size? Second, can gross studies of the brain predict the nature of internal complexity, connectivities, and differential growth or reduction in size of specific areas or centers?

It would seem that a large brain is not a sine qua non for articulate speech. Jensen-Jazbutis (1970, cited in Jerison 1982) notes that microcephalic dwarves with a brain of around 300 cm^3 (i.e., in the chimpanzee range), though severely retarded, possess some speech and reasonable sensorimotor abilities. Dorman (1991) reviews the whole question of microcephaly and intelligence. The condition is defined in terms of a head circumference of more than 2 (or 3, according to some authorities) standard deviations (SDs) below the age mean. It is a symptom rather than a diagnostic entity and may be due to many causes, but generally it appears to be the result of a defect in brain growth, as the latter determines skull growth rather than vice versa. Whether the etiology is genetic or secondary to some other insult, the most likely period of disruption of brain growth likely to result in microcephaly is during the first half of the second trimester of pregnancy, although it can occur beyond the neonatal period. While some studies report that most microcephalics are mentally retarded, others, taking the criterion of head size as 2 or more SDs below the mean for age, find that 2% of school-age children qualify as microcephalics, and yet are indistinguishable from their normocephalic peers on tests of intelligence. While some diminution in intelligence is usually reported, certain subgroups (in particular autosomal dominant and recessive microcephalics) may, according to several studies, have *extremely* small head circumferences and normal IQs. Thus *intraspecifically*, an excessively small head size does not itself greatly affect intelligence; there is no straightforward relationship between diminished head size and lowered intelligence; although see recent controversial claims to the contrary by Willerman, reviewed by Holden (1991), and any effect on IQ in microcephalics may depend on the extent and type of the underlying pathology present in microcephalics.

Why, then, do humans generally have such large heads, with consequent obstetric consequences? Either an enlarged brain is accompanied by reorganization of internal components and *this* factor determines intellectual capacity–a possibility which we shall next address–or a normal intelligence in a microcephalic merely reflects the enormous spare capacity in the human brain; architectural reorganization of the brain and increase in encephalization during evolution may have continued to maintain this level of spare capacity. That we *do* have an enormous and often untapped level of spare capacity is evident from the findings of infantile and later hemispherectomy (e.g., Goodman and Whitaker 1985);

when total decortication (e.g., to alleviate otherwise intractable epilepsy or other effects of widespread unilateral brain damage) of one hemisphere is undertaken at an early age (such that plasticity still permits remaining structures to assume the affected function), there may be surprisingly little effect on overall intellectual capacity.

Deacon (1986b) argues that the internal mosaic reorganization of the brain resulting from selection for specialized modular functions may be a more fundamental determinant of intellectual capacity (intelligence, in humans), with an *overall* brain enlargement as a secondary consequence of a selective reorganization and enlargement of just a few structures. He also notes that enlargement of a particular brain structure may not necessarily indicate selection for the functions of that structure; rather, due to embryological coupling, the selective enlargement of one structure may produce a range of disproportionate changes elsewhere, thus obscuring the true nature of any evolutionary changes. Kien (1991) takes these ideas further via an augmentative feedback loop, whereby an increase in encephalization results in the potential for new abilities. These then select for further encephalization, with technology and language (and thereby ritual, myth, culture, and religion, permitting information transmission between individuals and across generations) being consequences of, rather than the initial driving forces for, both encephalization and reorganization. Thus, after a certain optimal size is attained, further increases may be counterproductive, leading to inefficient processing. However after reorganization to permit more efficient data management, further increases in both encephalization and levels of performance again become possible. Such a model could predict a step increase via a series of plateaus, rather than a steady growth. While, to our knowledge, this effect has not been specifically reported, our reading of the literature suggests that it might in fact have occurred, with changes manifesting between rather than within hominid taxa.

An autocatalytic feedback process of the sort just envisaged, in which neural reorganization in response to evolutionary pressure to process more information faster maintains the need for further changes and improvements, may of itself generate symbolic and referential systems like language. Language, as we saw, is not just a means of communication; it is a tool for thinking and problem solving, for inferential learning and logic. Indeed, the best argument against *H. erectus* possessing language is the poor technological showing of that taxon in the fossil record, bearing in mind the perishability of many ancient artifacts. The pygmy people of central Africa leave little evidence of their culture. Language may be a necessary prerequisite for technical progress, but not a sufficient one, and so cannot of itself be the main driving force on toward the world of technology. Agriculture and pastoral behaviors could not have appeared in so short a time without language and its powers of representation—although as Bickerton (1990) observes ants (and termites) over millions of years have evolved the capacity to cultivate fungal gardens, and to "herd" aphids for their sweet secretions. Territoriality of course is the price of agriculture—you must guard what you are growing or have stored. It converted a social mammal (whose loose-knit bands of hunter–gatherers probably rarely exceeded around 50 in number) into a "social insect" (Bickerton 1990) with a strict hierarchical chain of command, high technology, and a war machine. We are not of course any more *intelligent* now than were our Aurignacian ancestors, as evidenced by the fact that two generations of education may be sufficient to convert uprooted tribespeople into technocrats.

Blakemore (1991) notes that more than half of the cerebral cortex in Old World monkeys, and probably about the same in ourselves, is occupied by a number of separate visual areas, each representing all or part of the visual field. Some of these appear committed to the interpretation of high-level perceptual aspects of the visual scene, and others to lower level information such as orientation of edges, spatial frequencies or dimensions of patterns, retinal disparity of binocular targets, motion velocity, contrast, and color. While humans have a particularly high percentage of neocortex relative to the rest of the brain, the prefrontal association cortex, which we saw subserves planning, intention, and response modeling, is particularly strongly represented (Deacon 1986b). So, too, though perhaps to a somewhat lesser extent, are parieto-temporal association cortices. It is interesting to note that it is just such recently evolved secondary and tertiary association cortices that are particularly vulnerable to that uniquely human and devastating degenerative disorder, Alzheimer's disease (Damasio, Van Hoesen, and Hyman 1990), which has been said to result in the gradual "vanishing" of the mind. (Only humans show the neurofibrillary tangles and amyloid-containing plaques so characteristic of the disease.) That said, according to Armstrong (1989) the degree of cortical folding and the surface topography of the brains of chimpanzees and humans are very similar posteriorly, with little evidence of major reorganization, and with evolutionary changes largely restricted to anterior, prefrontal regions. Thus the inferior frontal regions are involved in the sequencing of orofacial movements, and the inferior prefrontal areas are involved in the syntactic and articulatory aspects of speech itself.

Deacon (1986a,b) notes that his studies of patterns of connectivity in the brains of nonhuman primates in regions homologous to our own language circuits, suggest that the evolution of human language abilities did not result from major connectivity changes. Rather, it came from a quantitative increase in the relevant structures, that is, an increase in the relative number of cells and projections comprising the neural pathways involved in interrelating auditory, facial, laryngeal, and lingual functions. These functions converge in the prefrontal region, especially in the vicinity of Broca's area, which is particularly enlarged in humans. Indeed (see Falk 1986, and Tobias 1987) Broca's area is discernible on the endocasts of *H. habilis*, as is also Wernicke's area in the inferior parietal lobule, suggesting the possibility of some form of speech. Finally, massive increases in the size of the human cerebellum (together with frontal and prefrontal regions) may underlie our fine levels of eye-hand coordination; when we reach toward a target (or wield a weapon) the eyes briefly lead the hand, and there is a sophisticated intermeshing of control of head and eye turn with hand and arm movements (Goodale 1990a,b). However certain insectivores are quite skilled in such behaviors without the human level of neural machinery.

Encephalization is associated not just with body size; in fact, both correlate with length of gestation, helplessness at birth, a long juvenile phase, delayed maturity and reproduction, fewer (and more spaced) offspring, obstetric complications, lower mortality rates as an infant, greater longevity, and a lower metabolic rate (Foley and Dunbar 1989; Smith 1990; and see also Harvey and Nee 1991). In humans, the prenatal phase of rapid brain growth continues somewhat longer after birth (12 months) than before (9 months). During this period of altriciality, demands are made on the mother not only in terms of resources but also with respect to socialization. An enlarged brain, moreover, needs a longer life to realize its potential, and this has consequences in terms of energy and the ecology. Indeed retarded aging may itself be an important factor in human evolution, permitting the survival of repositories of knowledge and information, before the advent of writing and other artificial storage systems.

Because growth rate, especially before birth, helps determine ultimate brain size, and growth rate is largely determined by ecological pressures (growth rate tending to increase in a patchy unpredictable environment, Foley and Dunbar 1989), a patchy, changing environment of the sort we described earlier may have influenced growth rate, brain size, and intellectual capacity. Be this as it may, the factor or factors which brought about the progressive increase in hominid encephalization and intellectual capacities would have been at least in part similar to the ones that we reviewed (Chapter 7, and earlier in this chapter) in connection with the origin of bipedalism, tool use, and language, even though all four essentially human phenomena arose at very different moments in the evolutionary trajectory of the hominids. Again, we would emphasize that although a lengthy list of individual factors have been proposed as sources of evolutionary pressure, current biomedical thinking in many domains increasingly has to appeal to several mutually interacting or facilitating components or contributors to a given phenomenon.

Since many of the single-factor accounts of increasing encephalization and "intellectualization" in hominids have already been discussed earlier in the contexts of an upright bipedal posture and language, they will be mentioned only briefly here. They include, for example, hunting (which also involves planning, strategy, and often group coordination and cooperation), throwing and aiming (with its reliance upon frontal regions of the right hemisphere to mediate the spatial aspects of arm movement; see, e.g., Leonard and Milner 1991), butchering, tool use and technology (Wynn 1988), extractive foraging (Parker and Gibson 1979), and feeding strategies involving complex bimanual manipulation (Wundram 1986). According to this last proposal, cortical asymmetry was boosted, given its very long evolutionary history, as a result of feeding strategies which required both hands to be employed simultaneously in different modes. In this way, a wide variety of foods could be exploited—shoots, flowers, bark, berries, nuts, fruit, roots, insects, other terrestrial and aquatic vertebrates and invertebrates, eggs, and so on, present in the changing and patchy ecology of those glacial times, which included lakes, streams, forest, savannah, and grassland. Two hands working together would be needed to gather and process foods, remove seeds from pods, pick berries from thorny bushes, select fresh leaves from beneath dry ones, shell eggs, pluck birds, skin game, scale fish, clean roots, and crack nuts. Only primates *extensively* use both hands for feeding, and humans have the most *varied* diet of practically all animals. Again frontal areas, particularly the regions in the left hemisphere involved in praxis, would be called on and would be likely to enlarge (compare the frontal regions of the right hemisphere concerned in connection with aiming and throwing).

In a wider context, there is the general question of feeding niches and possible changes in feeding strategies. While as we saw all early hominids (along with the great apes) were essentially frugivores, dental and mandibular analyses suggest (Andrews and Martin 1991) that the manufacture and use of tools by the genus *Homo*, in particular the use of tools to *make* tools, made possible a dietary expansion into areas unattainable by our ape ancestors, with a reduction in tooth size, robustness, and body strength. Further shifts seem to have occurred at the time of *H. erectus*, along with significant brain expansion, reduced sexual dimorphism in body size, more systematic tool making, fire, and increased migration.

The other side of this feeding-driven approach to the growth of brain structures and cognitive capacity invokes selective pressures to remember the *locations* and availability of widely dispersed and often ephemeral food sources (Milton 1988; Steele 1989). These pressures would address posterior parietal regions of the right hemisphere. The need to *communicate* the location of such sources, conversely, to other individuals would lead to the development of periSylvian structures in the left hemisphere. Increased intellectual capacities would then feed back on obtaining a high-quality diet and its preparation (including cooking to detoxify and soften). Thus the dentition is spared (worn teeth impose longevity limits in many species), and the consequences are observed in the unique human gut, which is adapted for a nutrient-dense and rapidly digested diet (Milton 1988).

Another food-related hypothesis involves food sharing and a mixed economy or division of labor between male hunters and female gatherers. This demands cooperation and communicating (and "social intelligence"), a hypothesis which extends (Lovejoy, 1981) to females remaining in a nurturing and educational capacity at a home base with young, to be provisioned by males. However, as we, saw sexual dimorphism in size was considerable at least up to *H. erectus*, a phenomenon generally found in polygynous species, whereas monogamous pair bonding would perhaps facilitate the previously discussed scenario.

It may be possible to determine the exact age of teeth by counting the striae of Retzius within the enamel (Bromage and Dean 1985; although see also Mann, Lampl, and Monge 1990), such analyses, linked to observed skeletal growth patterns, indicate that prolonged infant care and significant postnatal brain growth may have begun with *H. erectus*, while prior hominids followed the shorter apelike pattern. At this point, we should perhaps remind the reader of Marshack's view of Upper Paleolithic intelligence as vision-centered, imaging and conceptual; however, with Mellars (1989a,b) we should be wary of claiming that there was a shift in basic cognitive *abilities* at the transition from archaic to fully modern human. As he notes, it is fallacious to equate the *expression* of culture with its potential. Otherwise the absence of nuclear physics, and space and computer technology would wrongly relegate the inhabitants of the nineteenth century (or of the Pacific islands) to an intellectually inferior level.

While we again emphasize the importance of a multifactorial interactive model incorporating a range of elements of the sort just discussed, one particular factor should perhaps be given special prominence, the theory of Machiavellian intelligence (Whiten and Byrne 1988). The interests of individuals within groups are rarely identical, unless the individuals are clones or near clones like social insects, or eusocial mammals like the naked mole rat, *Heterocephalus glaber* (Eisenberg 1981, p. 109). (These mammals, distantly related to the porcupine, live like social insects in tunnels, with a single large queen giving birth to up to 25 young at a time with siblings caring for the offspring.) Group membership may lead to greater access to resources and reduced susceptibility to predation, at least at the *individual* level, and more efficient hunting. However, group membership involves the constant need to reassess our standing in a changing social network of peers, which itself changes as a consequence of our own actions. Social skills are needed to enable us to form alliances and to switch tactics should it be advantageous. We therefore spend increasing periods of time gathering information on social relationships and attempting to deduce the intentions of others and their attentional states. We may attribute several levels of intentionality to others; for example, "X thinks that Y knows that X wants Y to do Z." Thus low-ranking monkeys will apparently deliberately mislead other high-ranking individuals to gain sexual access to females under their control (Cheney and Seyfarth 1990), a high-ranking chimpanzee will cultivate the company of a knowledgeable low-ranking one to improve its feeding rate via deception (Menzel 1988), or a female noticing a good food item will leave the troop and pretend to groom so as not to have to share it with others (Savage-Rumbaugh and McDonald 1988). Intense social competition will result in manipulation and deception—and perhaps account for our own fascination with gossip and biography. We can therefore contrast social intelligence (and compare Gardner 1983) with that adapted for the physical world. The latter includes tool behavior and technical or object intelligence which involves general problem solving, discovery, exploration, and innovation, and in fact the latter may derive from the former. Thus chimpanzees are preeminent in imitative learning, not just *copying* others, but focusing on how others solve problems and then improving thereon.

Indeed, proponents of the Machiavellian intelligence hypothesis claim that tool, technical, and lithic technology would never have demanded such complex behaviors and intelligence as are manifested in balancing competitive and cooperative options, reciprocity, deceit, tactics, plot and counterplot, and the gamesmanship associated with dominance, alliances, and access to food, mates, and shelter (Whiten and Byrne 1988). All of these may permit the weak to beat the strong. In a world where one's adversaries are intellectual adversaries and conspecifics, social or Machiavellian intelligence, which developed initially to cope with interpersonal relationships at a local level, has perhaps extended now to find expression in the initial "institutional creations of the savage mind," such as the in structures of kinship, totemism, myth, and religion (Humphrey 1976), issues to which we shall return shortly.

Further support for the Machiavellian hypothesis comes, as we have already seen, from the special sensitivity that human infants show almost from birth to social stimuli. They respond to the sounds of speech in quite a different way than to other forms of auditory stimulation, they prefer to look at pictures of faces than at other equally complex patterns, and they readily imitate expressions. Indeed, the physical attractiveness of faces seems not to be solely defined by culture and to be gradually learned; even infants inspect for a longer period photographs of faces which adults judge to be attractive, possessing delicate and regular features (Langlois et al. 1987).

The hypothesis of Machiavellian intelligence originated from the observation that primates differ from other taxa more in terms of their social environment than their subsistence economy; we might nevertheless ask why gorillas and orangutans, with their considerable intellectual capacities and leisurely, undemanding lives surrounded by easily accessible food with little to do but eat, sleep, and play, nevertheless in fact play comparatively little and are relatively nonsocial (Milton 1988). Conversely, African meerkats are highly social, but not noticeably intelligent. Indeed as Milton observes, our own modern social environment is no more complex than that of chimpanzees and baboons, all of which suggests, as we have already observed, that other factors must also be taken into account. However, as we shall see, only the great apes, apart from ourselves, may possess conscious self-awareness in terms of recognizing themselves in a mirror and not seeing the image as simply that of another animal. Consciousness is a necessary tool for seeing oneself and hence others as individuals, to be able to predict their behavior by putting oneself in their place. Thus Premack (1988) argues that chimpanzees can attribute states of mind (wants, knowledge, beliefs, and intentions) to another animal and thereby predict and manipulate its behavior, if necessary by deliberate deception. This deliberate behavior contrasts with the innately released wounded-bird behavior of a plover leading a potential predator away from its nest. (Ristau, 1991, however, notes that such feigning of injury may depend on the intruder actually *looking towards* the nest, as if the bird imputes intentionality to the intruder.) Premack describes a demonstration involving a chimpanzee; the chimpanzee needed the assistance of a human who alone possessed the key to open a box of bananas on the other side of a field. The chimpanzee would pull the human with the key toward the box, and if the human was blindfolded and unable to negotiate the journey, the chimpanzee would try to remove the blindfold to facilitate matters, ignoring other cloths over, for example, the person's nose or mouth. Gomez (1991) similarly describes a hand-reared juvenile gorilla who, to reach up to a high latch, would either roughly drag over a box or gently lead an experimenter by the hand while looking into his eyes and then to the box, and back again, as if to establish both attention and intentionality. Thus the gorilla seemed to understand that an agent's behavior may depend on what it is looking at and attending to. Whiten (1991) likewise reviews a study where a chimpanzee was apparently able to differentiate between guessing and knowing in an observer. Food was hidden from the animal in such a way that a human "knower" could, in view of the chimpanzee, observe the hiding, while a "guesser" left the room at the critical moment. The chimpanzee subsequently preferred the knower over the now-returned guesser in using pointing by these individuals as a guide to its own selection of a food location.

Observations of the sort just described, and the idea that chimpanzees can attribute states of mind to other animals, and in effect have a "theory of mind," could only occur after the demise of behaviorism. Chimpanzees of course are good behaviorists; they are skilled at interpreting and manipulating another's behavior, although humans perhaps are preeminent at deliberate prevarication, that is, at telling lies. It is still unclear to what extent chimpanzees regularly try to make a conspecific believe that they want to do one thing, when they really intend to do something quite different. Autistic children, however, seem to lack some or all of these social abilities (for reviews, see Frith 1989; Leslie 1990, 1991). They are notably deficient at understanding the behavior of others, at being able to predict their actions, and to operate within human society, capacities that normally develop by the age of 3 or 4 years. The capacity to pretend, play games of pretence, understand pretence in others, know when others are wrong, and impute beliefs, knowledge, expectancies or intentions in others all seem deficient in autistic children, as demonstrated by games played between child and psychologist. They have little difficulty in inferring *physical* causality (as judged, e.g., by tests with comic book scenarios), something at which in fact chimpanzees are notably poor, but they lack the "theory of mind" and *social* competence of our closely related cousin, the chimpanzee. Of course, it may be no accident that autistic children lack both communicative capacities and the ability to "mind read" others. Pinker and Bloom (1990) note that our lifestyle depends on extended cooperation for food, nurturance, safety, and so on, which benefit all participants, though at the expense of vulnerability to freeloading by cheats. This situation puts pressure on the ability to persuade others by *argument* and to change someone else's behavior, rather than simply to "communicate." Social competence can, however, exist without language (it is present in a wide range of species) and almost certainly preceded the evolution and development of language. Indeed, language impairment need not necessarily accompany autism; when inherited, it may be associated with anomalous asymmetries in the periSylvian region, and such anomalies may also be found in otherwise asymptomatic parents and siblings (Plante 1991).

Self-Awareness and Self-Recognition in Mirrors

The ability to impute desires and capacities to others, to put oneself in their place, and if necessary deliberately to deceive them, may be a product of Machiavellian intelligence and may be a capacity of only the most advanced primates, although so far sufficient testing of other species has not been done. The ability to recognize oneself as an individual apart from other conspecifics, and to look at oneself as part of a socially interacting group, is therefore a closely related faculty. Indeed the capacity to look at oneself, to introspect one's own operations and behaviors, and to perceive the past and future as distinct from the present, permits the development of hypotheses which can be tested internally before committing oneself irrevocably to a certain course of action. Gallup (1983) noted that when chimpanzees first look in a mirror, they treat the reflection as if it is another chimpanzee, but after several days they start to exhibit more self-directed activity, grooming themselves in it and using it to inspect hitherto and otherwise inaccessible or unseen parts of the body. If they are anesthetized and their foreheads or ears are marked with a nonirritating dye, on recovery and seeing their reflection, chimpanzees and orangutans (but not gorillas) show interest in the marks and touch them. Gallup concludes that self-directed behavior guided by the mirror demonstrates self-awareness in these two species of great apes and ourselves (when 12 or more months old).

Others (see e.g., Epstein, Lanza, and Skinner 1981) however have commented that just because you can make self-directed behavior it does not *necessarily* mean that you are *self-aware*—you *could* be touching those parts of your body which correspond to those of the image of the "other" creature in the mirror. Ultimately, of course, self-awareness in anyone but yourself can never be proven, as philosophers have long argued. Similarly babies smiling in response to their mother's smile, or protruding their tongues in apparent imitation of a similar action by their mothers (as discussed above) may well be simply exhibiting hard-wired reflex behavior adaptive to mother–child bonding. However, it should be noted that elephants can spontaneously use mirrors to find otherwise hidden food, but do not apparently self-recognize; they respond to their image as if it were that of another elephant (Povinelli 1989). Such a difference in apparent self-recognition between the two species, chimpanzee and elephant, despite their similar use of mirrors to gain access to otherwise hidden regions, is at least suggestive that true self-recognition may underlie the chimpanzees' responses to mirrors. Indeed, intuitively, it is very hard to conclude otherwise when we take into account recent reports (Rumbaugh and Savage-Rumbaugh 1990) involving the use by the chimpanzee, Austin, of closed-circuit TV to explore otherwise inaccessible parts of his anatomy. He even used a flashlight to enhance the image of his throat.

Consciousness of the self—self-awareness—is perhaps a higher level of awareness than consciousness of events and objects, exclusive of the self, in the surrounding world. Both can only be inferred, and again one can never be sure that any other creatures, human or animal, apart from oneself, possess these faculties. Indeed the behavioristic tradition (Skinner 1974) assumes that consciousness, if it ever exists, is little more than an epiphenomenon; it need not be invoked to guide or explain behavior, no matter how complex, which can be run off as an automatic sequence of responses to external or environmental triggers. Consciousness, however, is one way that the success or failure of ongoing responses can be assessed or monitored. Indeed, the literature on human skills (see, e.g., Schmidt 1988) emphasizes that conscious monitoring of responses is a feature of nonautomatic performance of unfamiliar or difficult tasks. Under these circumstances, only one task—the attended one—can be undertaken at a time. Later, with familiarity and the ability to run off a ballistic series of unmonitored responses to match the changing environmental demands, such automaticity can be achieved without conscious awareness (indeed introspection can spoil an experienced golfer's swing); moreover another task can be undertaken simultaneously without interfering with the original task. In fact, conscious awareness may supervene at the point of the selective-attention bottleneck where response monitoring occurs (compare LaBerge and Samuels 1974). Conscious awareness need not require the substrate of language, and indeed the nonverbal right hemispheres of split-brain patients, if appropriately interrogated by nonverbal means, exhibit a full range of human emotions, desires, fears, and ambitions (Bradshaw 1989). In fact, if *H. erectus* was indeed bereft of articulatory language, the nonverbal right hemisphere of a modern commissurotomy patient could be regarded as a realistic model of *erectus* thought. Conversely, the productive and innovative thought of the artist or scientist need not depend on language; the apocryphal introspections of original thinkers are replete with accounts of preverbal images providing solutions to artistic or scientific problems or initiatives. Of course words and language may follow, to clothe the newly emerging concepts in transmissible or archival form.

In conclusion, we may follow Rumbaugh and Savage-Rumbaugh (1990) noting that human conscious awareness involves an understanding of the existence of behavioral alternatives, the concomitant knowledge of having chosen one rather than another of these alternatives, and an ongoing introspective monitoring of progress toward that alternative.

The idea that consciousness supervenes at the level of nonautomatic monitoring is not new. Tulving and Thomson (1973) divided long-term memory into semantic (rule-governed, as with language, mathematics, and chess and similar games) and episodic (autobiographical and arbitrary events). Only the latter are open to introspection and conscious awareness. Subsequently, Baddeley (1986) reworked the old distinction of long- versus short-term memory across the boundary of which episodic memory had uneasily sat; he viewed working (or "scratch pad") memory as a kind of short-term memory into which information from either the present (via the senses) or the past (via long-term memory) could be fed for moment-to-moment processing. Under certain circumstances such material could reach conscious awareness, again particularly so if not being performed automatically, as with highly overpracticed skills.

Baars (1988) takes these ideas further, drawing an analogy between consciousness and a blackboard information system that acts as a central information exchange. His theoretical framework contains a global workspace, specialized unconscious processors, and contexts (Kellog 1990). Consciousness involves a central information exchange, the global workspace, into which specialist processors send their input. A thought or image achieves conscious awareness when the processor generating it gains access to the global workspace. A specialist processor corresponds to one of Fodor's independent processing modules (Fodor 1983)—a specialist unit automatically and unconsciously subserving a particular and discrete function or cognitive ability. In modern computer parlance it corresponds to an expert system, and in the context of our earlier discussion of prefrontal strategic intelligence and posterior specialized intelligences, it obviously corresponds to one of the latter. (It should be noted that we have earlier indicated that Fodor's modularity-of-the-mind hypothesis is already giving way, both in cognitive and neural science, to alternate formulations which instead view both the mind and the brain in terms of massively-parallel interactive networks, rather than discrete modules, see, e.g., McClelland and Rumelhart 1986). These unconscious specialized processors of Baars fail us when faced with novel, degraded, or ambiguous input, for which no single processor can provide a solution. The global workspace then allows many processors to cooperate. Baars' third element, contexts, are in effect clusters of unconscious processors that have previously cooperated so often that it has become habitual. Without our awareness an entire hierarchy of contexts governs our information processing. It goes without saying that conscious awareness is only going to supervene when processing is no longer automatic and decisions have to be made. If nonhuman primates possess consciousness, it is presumably in such out-of-the-ordinary circumstances that it manifests, for example, during problem solving or social interaction, rather than during routine living. Again we might see consciousness as an emerging capacity when a creature and its interaction with its environment, including its conspecifics, exceeds a certain level of complexity, when automaticity is no longer sufficient for adequate functioning. There is however no need to impute a special role for language, or for language-related left-hemisphere processors, in its operation.

Alexander (1989) lists five clusters of capacities of the human psyche which differ quantitatively or qualitatively from our closest living (and presumably ancestral) relatives. These include, first, conscious self-awareness together with foresight, intent, will, planning, purpose, scenario-modeling, thought, reflection, imagination, representation, and the abilities to deceive (as also possessed by the apes, as just discussed) and to self-deceive. The second cognitive group includes learning, logic, reasoning, intelligence and problem solving. (Increasingly, nonhuman species are accepted as having to varying extents these capacities; see, e.g., Griffin 1984; Walker 1985). Linguistic ability comprises the third faculty, followed by the human emotions of grief, embarrassment, despair, guilt, uncertainty, and so on, in contrast presumably with the more fundamental "animal" emotions of fear, anger, and pleasure. Interestingly, there is strong evidence for a predominantly right-hemisphere mediation of emotions in humans (Bradshaw 1989), chickens (Chapter 2), and rats (Chapter 3). Finally, there are the human personality traits of stubbornness, pliancy, timidity, arrogance, persistence, audacity, and so on. Dog and horse owners might be tempted to impute distinct personalities to their charges, but they are unlikely to reach such levels of subtlety as those ascribed to our own species, but of course much work remains to be done in this area. We would like to list certain other quintessentially human activities which, apart from tool use and language (which belong in a somewhat different domain and have already been fully explored), can be distributed in part along a rough continuum:

Superstition–Individual and unsystematized irrational beliefs used to explain, predict, or alter events or other people.

Religion–Organized superstitions operating at a group level.

Ritual–The systematized practice of superstitious and religious beliefs.

Morality–Codified beliefs concerning one's relationships with others, and including (on the positive pole) ethics, idealism, and self-sacrifice.

Politics–Systematized group beliefs to organize society.

War–A way of enforcing the preceding.

Game-playing–A practical way of sublimating the preceding.

Humor–An intellectual way of sublimating the preceding, which may be truly unique to our species, and which serves to reduce stress.

The arts–A subjective way of representing the universe including ourselves.

Science–A logical (rather than a superstitious or a subjective) way of explaining and predicting the universe.

Technology–A way of applying the preceding to alter, manipulate, or control the individual or the environment.

Pigeons, if given food at regular intervals regardless of what they are doing, will come to behave as if a causal relationship exists between their arbitrary acts and food presentation; they will continue to exhibit such arbitrary behaviors in a "superstitious" fashion, perhaps providing a model for our own ritual behaviors. We have direct archaeological evidence for ritual and the representational arts as far back as *II. erectus*. Religion may be the common factor to both. Indeed, the skeptic might argue that the near universality of spiritual or religious beliefs does not necessarily prove the existence of a god—or of the supernatural. Instead, these phenomena may be manifestations of the workings of our cognitive systems, which in all forms of pattern recognition are designed to impose order and meaning upon initially chaotic input. Proofreaders' errors (not seeing typographical errors, but only what is expected), and figures seen in Rorschach inkblots, in clouds, and in the fire are instances of an intrinsic effort after meaning, and the spiritual sense may be no different. Religiosity, excessive practice or observance of religious beliefs, may be associated with temporal lobe dysfunction, including epilepsy and schizophrenia, and therefore may even represent a functionally lateralized syndrome (see, e.g., Gruzelier and Flor Henry 1979). Thus abnormal activation (through epilepsy, toxins, or hallucinogenic drugs) of the hippocampus within the temporal lobes, especially in the right hemisphere (Hodoba 1986) can lead to "visions" and religiosity (Schiff et al. 1982). It is known in its milder (and presumably nonpathological) manifestations to be a heritable trait, from studies of common behavior patterns in identical twins reared apart (Bouchard et al. 1990). Religiosity is perhaps the greatest single source of interpersonal divisiveness, including racism—a form which it frequently takes. It would be ironic if we were shortly to face Armageddon because of a combination of religiosity, politics, and technology.

Where does all this leave us with respect to the origins and evolution of language? We have already seen that Neanderthals were probably capable, in terms of peripheral anatomy, of a range of speech sounds at least approaching that of modern humans, and that H. habilis possessed adaptations in the central nervous system which at least presaged the speech-related areas of H. sapiens sapiens. As Noble and Davidson (1991) remark, however, structures precede function in evolution, and the presence of a larger brain, parieto-temporal asymmetries, and a hyoid does not prove that these structures were already in use for the functions that they now serve, even though they may well have provided the necessary preadaptive mechanisms. Indeed, Noble and Davidson (1991; and see also Foley 1991b) argue that language only appeared in the last 40,000 years, 60,000 years or more after the emergence of anatomically modern people. They claim that the principle characteristic of human language is its symbolic nature, whereby arbitrary (i.e., defined by mutually agreed convention) symbols designate other objects or actions. Animal, including primate, communication, even as acquired in the laboratory, lacks the symbolic potential of our therefore unique language systems. They argue that even though *all* communication systems involve gestures or signs which convey meanings, only with language do its users *discover* that this is the case, thereby possessing a higher-order awareness, a meta-awareness, of the meaning or symbolic nature of language. By contrast, they claim that unlike our context-free, mutually agreed on symbol systems, the elements of, for example, primate communication are hard-wired. They go on to argue that only in the European Upper Paleolithic of around 40 kya do we find iconic signs or representations in art, reflecting this meta-awareness, and a necessary precursor to language. They offer the scenario of the evolution of aimed throwing leading to the capacity to indicate or point, followed by imitative gesture (e.g., of an animal's outline), which may have become "frozen" on the plastic surface of a mud floor or cave wall. These iconic signs, they say, permitted the appearance of language.

While we agree that before 40 kya lithic and other technologies remained remarkably stable for thousands or even millions of years, we do not agree that this necessarily indicates that before then, humans (or hominids) were effectively without language, any more than that before recent historic times we were without consciousness (Jaynes 1977). Indeed, we find the scenario of Noble and Davidson speculative at best, and unlikely at several nodal points in its argument. We are unconvinced that the symbolic nature of human language depends on its users having "discovered" the arbitrary and conventionalized nature of the sign–meaning relationships, and that this contrasts with the hard-wired nature of nonhuman communication. The existence of local dialects in bird song and cetacean signaling, and the fact that passerines must be exposed at critical periods to the song of their conspecifics (Nottebohm 1979), provides evidence to the contrary. The argument by Noble and Davidson that speech arose from the decoupling of vocalizations that happened to accompany gestural and iconic activity we find particularly unconvincing, and we have already rejected the idea that gesture probably played a major intermediary role in language evolution. We agree that human language, unlike all other communicatory systems, is "productive" and open-ended and capable of producing an infinite number of utterances from a finite number of elements. We also agree that in this sense it resembles tool manufacture and use, and that tool behaviors (and gesture) probably accompanied, in a mutually facilitatory way, language evolution, along with the evolution of society. However we do not see why the attainment of a "critical mass" in art, technology, and society may not have led to the cultural explosion of the European Upper Paleolithic (without invoking the sudden appearance of language).

Indeed, we see a long evolutionary continuity in language-like behaviors before then, building on preexisting primate communication systems. As Foley (1991b) notes, a sudden appearance of language around 40,000 years ago would be "strange," given the recent evidence of complex communicatory systems in social mammals and birds, which are highly sensitive both to other individuals' behavior and its context. Thus sophisticated animal communication is strongly embedded in social behavior and organization; primates are highly sociable, we above all, so it is very unlikely that early hominids did not have some form of language, even though all existing languages may in fact for one reason or another be traceable back to 40 kya, or even 100 kya. Thus, just as we are all probably descended from an African form of modern *H. sapiens sapiens* (although see contrary multiregional views discussed earlier) which replaced existing forms of *Homo*, so too were preexisting languages probably replaced by those of new invading peoples.

In conclusion, we see language as having a long evolutionary tradition continuous with primate communicatory systems. It has "bootstrapped" to its present level of sophistication via autocatalytic multifactorial feedback involving the simultaneous development, along their own fronts, of tool use and technology, social interchange and society, consciousness and understanding, and intellect. It is a mistake to seek single origins either in causality or in time for any of these essentially human attributes. When they achieved a certain level of attainment and society, a new "critical mass" was reached which manifested itself in the step change of the European Upper Paleolithic. Since then, we have seen other revolutions, of agriculture, industry, science and technology, and communications, culminating in the present "age of information." All, like language and tool use, had precedents, and periods of development and stasis, and in their present manifestations, though continuous with those of earlier hominids and even hominoids, are orders of magnitude more complex (but in evolutionary terms not necessarily better adapted) than those of our forebears.

Summary and Conclusions

It is indeed inconceivable that the makers of the exquisite stone tools of the Upper Paleolithic, and the artists and sculptors of that same period who left us equally tantalizing and evocative images of their times, were insensitive to esthetics and incapable of sophisticated communication. However, the paucity or absence of similar relics from prior eras should not be taken as negative evidence for the existence of language. Ultimately, all evidence, for or against something as ephemeral and evanescent as language, which can leave no traces or fossils, is bound to be circumstantial, including even the arguments from paleoneurology; the early presence of a structure known today to subserve a certain function cannot prove that that structure was so used before now. However, we believe that just as you do not *need* language to construct and use tools and to create great art—modern case histories in behavioral neurology and psychiatry testify to the impressive capacities of aphasic adults and autistic children—so too language, at a complex modern level, can occur in the absence of a complex modern culture. What might a paleoanthropologist from another planet conclude about the !Kung people of the Kalahari, whose sandy environment until recently left little tangible evidence of their passage other than transient campgrounds little different from ancient Olduvai?

Art, tool use, language, intellect, and self-awareness have all undoubtably interacted to their mutual evolutionary benefit. So too have many other factors. Indeed, in this book we have emphasized multifactorial autocatalytic feedback models, be it for the evolution of an upright bipedal posture and locomotion, praxis, tool use, or intellect. It is a mistake to seek any single line of causality, and equally mistaken to exclude the significant influence of *other extraneous* factors like "mind reading" and "Machiavellian intelligence" in seeking to account for any one of our now no longer uniquely human capacities, such as intellect and conscious self-awareness.

We must reject Chomsky's view of the absolute uniqueness of human speech, of a unique language organ without evolutionary precedent. There are cytoarchitectonically homologous regions in other primates to our Broca's and Wernicke's speech areas, and all the patterns of connectivity seem to be in place. Nor can we maintain the old distinction, at either level, that human speech is essentially cortical and voluntary, whereas the phonation of other primates is limbic and automatic. Indeed, it is apparent that monkeys are capable of considerable information transfer, and comparable analyses of chimpanzee calls has only just begun. We must stress the evolutionary continuity of human speech with primate call systems. However, there is a drive to communicate in humans which is subject in its development to maturational milestones, and which may be frustrated under conditions of *total* deprivation during childhood. Indeed deaf–mute children spontaneously develop their own signing systems which possess their own syntax; autistic children without communication may seek an expressive outlet via, for example, art; and the evolution of pidgin to creole languages around the world, with a corresponding growth in syntactic complexity, usually via children caught between mutually near-unintelligible societies, emphasizes the special role of human language.

Language, of course, as Lieberman (1991) observes, allows us to transcend the constraints of biological evolution. We can transmit information across generations from the past, and jointly plan future courses of action. Chimpanzees, both wild and captive, employ gestures in various contexts and often with new meanings, and seem to be on the verge of developing a system of representational communication (Kendon 1991). They share with us the ability to conceptualize, a sine qua non for the emergence of true language, but may not have crossed a linguistic Rubicon because they have not *needed* to; thus, while our social lives involve complementary behavior and mutual sharing, chimpanzees tend to lead largely parallel social lives.

Fire was used as far back as the time of *H. erectus*, and played an important role in providing heat (for exploring colder climates), in protection against predators, and in preparing food. Equally as important, it would have extended the daily time available for, and favored the development of, cultural activities and language. Nevertheless only *H. sapiens sapiens* showed unambiguous evidence of ritual associated with burial, and a rich cultural and artistic life, and the latter perhaps only in the last 40,000 years or so. We must therefore probably seek a multifactorial explanation for the evolution of language, while noting that it undoubtedly had both behavioral and neural precedents in earlier taxa. Recent findings that the bonobo, *Pan paniscus*, can naturalistically acquire considerable powers to comprehend our spoken speech emphasize this.

Many of the preadaptations (posture, encephalization, social change, juvenile altriciality, neoteny and delayed maturity, and increasing longevity) were also important in the evolution of intellect and culture. The basic neurological structure of earlier primates was probably retained, although with some reorganization and considerable local growth. In particular, we evidence massive evolution of prefrontal structures associated with an overall executive role for strategic planning, coordinating more automatic (and often posterior) problem-solving systems associated with language, object and face recognition, imagery, space perception, and so on. These "posterior intelligences" then come under the guiding and decision-making control of frontal structures.

Therefore the growth of intelligence has probably involved autocatalytic feedback processes in response to environmental pressures to process ever more information ever faster, thereby inevitably leading to the evolution (via art?) of symbolic and referential systems like language. We can therefore explain our own pleasure in aesthetics as a successful problem-solving experience, all perception being part of an effort to impose meaning on sensory input. Even music, which shares numerous features with language, at least at the paralinguistic, suprasegmental, prosodic, and emotional levels, may be seen in the context of communication and social bonding, perhaps via ritual.

The evolution of intellect is necessarily bound up with the concept of intelligence. The latter is bedeviled with problems of definition and reification, and probably should be considered in terms of a number of more or less independent (modular? posterior?) intelligences, perhaps under loose (prefrontal?) executive control. Thus, we probably should not seek to compare intelligence between species, at least without reference to ecological contexts. Nevertheless speed of decision making and the ability to hold many things in mind simultaneously, may be useful indices of a general information processing capacity at the human level. While more encephalized species, bearing in mind allometric relationships between brain and body sizes, seem generally to be capable of more complex behavior, at the level of the individual and the species, brain size seems to bear little relation to ability. Thus microcephalics are not necessarily greatly disadvantaged, emphasizing the enormous spare capacity of the human brain.

A factor which has recently received considerable emphasis with respect to the evolution of primate intellect is the growth of so-called Machiavellian intelligence. Among primates, one's conspecifics may be one's own worst enemies, and the need to deceive and to plot, to build alliances, and to guess the thoughts or intentions of others—to develop a "theory of mind"—may have driven the evolution of intellect and led to consciousness of the self as a separate individual within society. Such a faculty may be subject to defective maturation in autistic children, who despite normal problem-solving capacities in more mechanical contexts seem unable to attribute beliefs to others.

What then is consciousness? Is it an exclusive prerogative of humans, and our only remaining unique distinction, or like language and tool use does it spread into other species? The commonly accepted evidence for consciousness includes self-recognition, long-term planning (e.g., chimpanzees preparing tools and carrying them some distance to the site where they will be used), and symbolic play (Jolly 1991). Thus juvenile, home-reared chimpanzees may apparently pull imaginary toys the imaginary strings of which require disentangling from obstacles. However, as Jolly observes, if we more readily concede consciousness to chimpanzees than to other species, it may not necessarily be because their minds are so like ours; instead, it may be because their problems and preoccupations seem so similar.

We have already emphasized the evolutionary growth of frontal and prefrontal structures which seem to mediate executive decisions on problem-solving strategies. It has long been known that whereas easy, automatic, and practiced responses may take place with little subjective awareness, when behavior has to be carefully monitored because of difficulty or unfamiliarity, we become acutely aware of what we are doing. The idea of a short-term working or scratch-pad memory, or of a working space into which we feed the output of specialized, automatic, and unconscious processors, especially when they are unable to proceed on their own, is itself closely associated with conscious awareness. That does not mean that consciousness is necessarily associated with language, or that specialist dedicated processors are necessarily modular in nature; the brain is increasingly being seen as a massively parallel system of interconnected networks. Indeed, in cognitive science, the neurosciences, and artificial intelligence and computer architecture, parallel networking seems to be the way that we shall in the future model the brain and construct computational systems.

It is easy to overemphasize the role of cerebral asymmetry and behavioral lateralties in human evolution, despite a popular fascination with left-handedness, double consciousness in the commissurotomized brain, and the near universal association between left-hemisphere injury and language disorder. Cerebral asymmetry is just one dichotomy among, for example, cortical–subcortical, anterior–posterior, and so on. However, it is a neat and clear division, leaving two *nearly* mirror-image structures, and the *functions* so divided *do* seem to address some so-called uniquely human characteristics like consciousness, language versus artistic behavior, and manual praxis, including tool use. On the one hand, we have tried to show that these characteristics may not be nearly so uniquely human after all; on the other hand, we have *also* tried to show that, contrary to common misconceptions, neither are we nearly so unique, as previously thought, in being overtly or covertly asymmetrical. Asymmetry may be a fundamental characteristic of living (and maybe even nonliving) matter, and behavioral asymmetries are common in all genera, being evident even half a billion years ago.

A general conclusion from a survey such as this seems to be that, at least in vertebrates where visual processes may have played an important role in the evolution of lateral asymmetries, the left hemisphere is specialized for sequential, feature-analytic behaviors which may underlie communication. The right hemisphere, conversely, seems to address spatial and emotional aspects. (Corballis, 1991, and in his earlier writings, comments that such right-hemisphere functions constitute a less coherent set than those of the left, which may therefore be prior; we would disagree, and would argue on evolutionary grounds that right-hemisphere spatial behaviors associated with the acquisition of food and a mate seem to constitute a good evolutionary springboard for all subsequent lateralized behaviors.) Either way, we would conclude that human beings are at most quantitatively, not qualitatively, different from other vertebrate genera in nearly all behaviors, except just possibly art and esthetics; nor, indeed, are they unique in their lateral asymmetries.

We have not crossed a Rubicon, linguistic or otherwise; we are merely rather further out on a number of evolutionary limbs—dangerously so, perhaps, if our subcortical (limbic) structures governing our drives and emotions have failed to keep pace with cortical developments.

Further Reading

Bickerton, D. Language and Species. Chicago: University of Chicago Press, 1990.

Byrne, R., and Whiten, A. (eds.). Machiavellian Intelligence: Social Expertise and the Evolution of Intellect in Monkeys, Apes and Humans. Oxford: Oxford University Press, 1988.

Cavalli-Sforza, L. L.; Piazza, A.; Menozzi, P.; and Mountain, J. Reconstruction of human evolution: Bringing together genetic, archaeological, and linguistic data. *Proceedings of the National Academy of Science,* 1988, 85:6002–6006.

Cheney, D. L., and Seyfarth, R. M. How Monkeys See the World: Inside the Mind of Another Species. Chicago: Chicago University Press, 1990.

Corballis, M. C. The Lopsided Ape. Oxford: Oxford University Press, 1991.

Diamond, J. The Rise and Fall of the Third Chimpanzee. London: Radius, 1991.

Jerison, H. J., and Jerison, I. (eds.). Intelligence and Evolutionary Biology. New York: Springer, 1986.

Lieberman, P. Uniquely Human: The Evolution of Speech, Thought and Selfless Behavior. Cambridge, Mass.: Harvard University Press, 1991.

Lindly, J. M., and Clark, C. A. Symbolism and modern human origins. *Current Anthropology*, 1990, 31:233–261.

Noble, W., and Davidson, I. The evolutionary emergence of modern human behavior: Language and its archaeology. *Man*, 1991, 26:223–253.

Parker, S. T., and Gibson, K. R. Language and Intelligence in Monkeys and Apes: Comparative and Developmental Perspectives. Cambridge: Cambridge University Press, 1990.

Pinker, S., and Bloom, P. Natural language and natural selection. *Behavioral and Brain Sciences*, 1990, 13:707–784.

Radner, D., and Radner, M. Animal Consciousness. Buffalo, N. Y.: Prometheus Books, 1989.

Rumbaugh, D. M., and Savage-Rumbaugh, E. S. Chimpanzees, competence for language and numbers. In W. C. Stebbins and M. A. Berkley (eds.), Comparative Perception, Vol II: Complex Signals. New York: Wiley, 1990, pp. 409–439.

Savage-Rumbaugh, S.; Romski, M. A.; Hopkins, W. D.; and Sevcik, R. A. Symbol acquisition and use by *Pan troglodytes, Pan paniscus, Homo sapiens*. In P. G. Heltne and L. A. Marquardt (eds.), Understanding Chimpanzees. Cambridge, Mass.: Harvard University Press, 1989, pp. 266–295.

Snowdon, C. T. Language capacities of nonhuman animals. *Yearbook of Physical Anthropology*, 1990, 33:215–243.

References

Able, K. P. and Bingman, V. P. The development of orientation and navigation behavior in birds. *The Quarterly Review of Biology*, 1987, 62:1–29.

Ackerman, S. European prehistory gets even older. *Science*, 1989, 246:28–30.

Adelstein, A. and Crowne, D. P. Visuospatial asymmetries and interocular transfer in the split-brain rat. *Behavioral Neuroscience*, 1991, 105:459–469.

Adret, P. and Rogers, L. J. Sex difference in the visual projections of young chicks: A quantitative study of the thalamofugal pathway. *Brain Research*, 1989, 478:59–73.

Afzelius, B. A. and Eliasson, K. Flagellar mutants in man: On the heterogeneity of the immotile cilia syndrome. *Journal of Ultrastructure Research*, 1979, 69:43–52.

Aitken, P. G. Cortical control of conditioned and spontaneous vocal behavior in rhesus monkeys. *Brain and Language*, 1981, 13:171–184.

Alba, F.; Ramirez, M.; Iribar, C.; Cantalejo, E.; and Osorio, C. Asymmetrical distribution of aminopeptidase activity in the cortex of rat brain. *Brain Research*, 1986, 368:158–160.

Albright, J. L.; Yungblut, D. H.; Arave, C. W.; and Wilson, J. C. Bovine laterality: Resting behavior. *Proceedings of the Indiana Academy of Science*, 1975, 85:407.

Alesci, R.; LaNoce, A.; Bagnoli, P.; and Ghelarducci, B. Dark-rearing modifies serotonin and 5-hydroxyindolacetic acid contents in specific brain regions of rabbit brain. *Behavioural Brain Research*, 1988, 31:115–120.

Alexander, R. D. Evolution of the human psyche. In P. Mellars and C. Stringer (eds.), The Human Revolution: Behavioral and Biological Perspectives on the Origins of Modern Humans. Princeton: Princeton University Press, 1989, pp. 455–513.

Allen, L. S.; Richey, M. F.; Chai, Y. M.; and Gorski, R. A. Sex differences in the corpus callosum of the living human being. *The Journal of Neuroscience*, 1991, 11:933–942.

Alonso, J.; Castellano, A.; and Rodriguez, M. Behavioral lateralization in rats: Prenatal stress effects on sex differences. *Brain Research*, 1991, 539:45–50.

Alter, I. A cerebral origin for dextrality. *Neuropsychologia*, 1989, 27:563–573.

Anderson, D. K.; Rhees, R. W.; and Fleming, D. E. Effects of stress on differentiation of the sexually dimorphic nucleus of the preoptic area (SND-POA) of the rat brain. *Brain Research*, 1985, 332:113–118.

Anderson-Gerfaud, P. Aspects of behaviour in the Middle Palaeolithic: Functional analysis of stone tools from south west France. In P. Mellars (ed.), The Emergence of Modern Humans: An Archaeological Perspective. Edinburgh: Edinburgh University Press, 1990, pp. 389–413.

Andrew, R. J. Precocious adult behaviour in the young chick. *Animal Behaviour*, 1966, 14:485–580.

Andrew, R. J. Lateralization of emotional and cognitive function in higher vertebrates, with special reference to the domestic chick. In J-P. Ewert, R. R. Capranica and D. Ingle (eds.), Advances in Vertebrate Neuroethology. New York: Plenum Press, 1983, pp. 477–509.

Andrew, R. J. The development of visual lateralization in the domestic chick. *Behavioural Brain Research*, 1988, 29:201–209.

Andrew, R. J. The nature of behavioural lateralization. In R. J. Andrew, (ed.), Neural and Behavioural Plasticity: The Use of the Domestic Chick as a Model. Oxford: Oxford University Press, 1991, pp. 536–554.

Andrew, R. J. and Brennan, A. The lateralization of fear behaviour in the male domestic chick: A developmental study, Animal Behaviour, 1983, 31:1166–1177.

Andrew, R. J. and Brennan, A. Sex differences in lateralization in the domestic chick. Neuropsychologia, 1984, 22:503–509.

Andrew, R. J. and Brennan, A. Sharply timed and lateralized events at time of establishment of long term memory. Physiology and Behavior, 1985, 34:547–556.

Andrew, R. J.; Mench, J.; and Rainey, C. Right-left asymmetry of response to visual stimuli in the domestic chick. In D. J. Ingle, M. A. Goodale and R. J. W. Mansfield, (eds.), Analysis of Visual Behavior. Cambridge: MIT Press, 1982, pp. 197–209.

Andrews, P. and Martin, L. Hominoid dietary evolution. Philosophical Transactions of the Royal Society of London B, 1991, 334:199–209.

Annett, J.; Annett, M.; Hudson, P. T. W.; and Turner, A. The control of movement in the preferred and nonpreferred hands. Quarterly Journal of Experimental Psychology, 1979, 31:641–652.

Annett, M. Laterality of childhood hemiplegia and the growth of speech and intelligence. Cortex, 1973, 9:4–33.

Annett, M. Left, Right, Hand and Brain: The Right Shift Theory. Hillsdale, N.J.: Erlbaum, 1985.

Annett, M. Speech lateralization and phonological skill. Cortex, 1991, 27:583–593.

Annett, M. and Annett, J. Handedness for eating in gorillas. Cortex, 1991, 27:269–275.

Antrobus, J. Cortical hemispheric asymmetry and sleep mentation. Psychological Review, 1987, 94:359–368.

Ardrey, R. The Hunting Hypothesis. New York: Atheneum, 1976.

Arensburg, B.; Schepartz, L. A.; Tillier, A. M.; Vandermeersch, B.; and Rak, Y. A reappraisal of the anatomical basis for speech in Middle Palaeolithic hominids. American Journal of Physical Anthropology, 1990, 83:137–146.

Arensburg, B.; Tillier, A. M.; Vandermeersch, B.; Duday, H.; Schepartz, L. A.; and Rak, Y. A middle palaeolithic human hyoid bone. Nature, 1989, 338:758–760.

Armstrong, E. A comparative review of the primate motor system. Journal of Motor Behavior, 1989, 21:493–517.

Arnold, A.; Nottebohm, F.; and Pfaff, D. W. Hormone concentrating cells in vocal control and other areas of the brain of the zebra finch (Poephila guttata). Journal of Comparative Neurology, 1976, 165:487–512.

Aruguete, M. S.; Ely, E. A.; and King, J. E. Laterality in spontaneous motor activity of chimpanzees and squirrel monkeys. American Journal of Primatology, in press.

Avise, J. C. Matriarchal liberation. Nature, 1991, 352:192.

Avise, J. C.; Neigel, J. E.; and Arnold, J. Demographic influences on mitochondrial DNA lineage survivorship in animal populations. Journal of Molecular Evolution, 1984, 20:99–105.

Baars, B. J. A Cognitive Theory of Consciousness. Cambridge: Cambridge University Press, 1988.

Babcock, L. H. and Robison, R. A. Preferences of Palaeozoic predators. Nature, 1989, 337:695–696.

Baddeley, A. Working Memory. Oxford: Clarendon Press, 1986.

Bahn, P. G. Early teething troubles. Nature, 1989, 337:693 (a).

Bahn, P. G. Getting into the groove. Nature, 1989, 339:429–430 (b).

Bahn, P. G. and Vertut, J. Images of the Ice Age. Leicester: Windward, 1988.

Bakalkin, G. Ya.; Tsibezov, V. V.; Sjutkin, E. A.; Veselova, S. P.; Novikov, I. D.; and Krivos-heev, O. G. Lateralization of LH-RH in rat hypothalamus. *Brain Research*, 1984, 296:361–364.

Baker, R. Human Navigation and Magnetoreception. Manchester: Manchester University Press, 1989.

Ball, N. J.; Amlaner Jr., C. J.; Shaffery, J. P.; and Opp, M. R. Asynchronous eye-closure and unihemispheric quiet sleep of birds. In W. P. Koella, H. Schulz, F. Obála, and P. Visser (eds.), Sleep '86. Gustav Fischer Verlag: Stuttgart, New York, 1988, pp. 127–128.

Baltin, S. Zur Biologie und Ethologie des Talegalla-Huhns (*Alectura lathami* Gray) unter besond-erer Berücksichtigung des Verhaltens während der Brutperiode. *Zeitschrift für Tierpsycholo-gie*, 1969, 6:524–572.

Barber, A. J. and Rose, S. P. R. Amnesia induced by 2-deoxygalactose in the day-old chick: Lateralization of effects in two different one-trial learning tasks. *Behavioral and Neural Biology*, 1991, 56:77–88.

Barber, J. T.; Ellgaard, E. G.; Thien, L. B.; and Stack, A. E. The use of tools for food transportation by the imported fire ant, *Solenopsis invicta*. *Animal Behaviour*, 1989, 38:550–552.

Barbujani, G. and Sokal, R. R. Zones of sharp genetic change in Europe are also linguistic boundaries. *Proceedings of the National Academy of Sciences*, 1990, 87:1816–1819.

Bard, K. A.; Hopkins, W. D.; and Fort, C. L. Lateral bias in infant chimpanzees (*Pan troglodytes*). *Journal of Comparative Psychology*, 1990, 104:309–321.

Barfield, R. J. Activation of copulatory behavior by androgen implanted into the preoptic area of the male fowl. *Hormones and Behavior*, 1969, 1:37–52.

Barfield, R. J.; Ronay, G.; and Pfaff, D. W. Autoradiographic localization of androgen concen-trating cells in the brain of the male domestic fowl. *Neuroendocrinology*, 1978, 26:297–311.

Barnéoud, P.; Le Moal, M.; and Neveu, P. J. Asymmetrical effects of cortical ablation on brain monoamines in mice. *International Journal of Neuroscience*, 1991, 56:283–294.

Barnéoud, P.; Neveu, P. J.; Vitiello, S.; and Le Moal, M. Functional heterogeneity of the right and left cerebral neocortex in the modulation of the immune system. *Physiology and Behavior*, 1987, 41:525–530.

Baylis, G. C.; Rolls, E. T.; and Leonard, C. M. Selectivity between faces in the responses of a population of neurons in the cortex in the superior temporal sulcus of the monkey. *Brain Research*, 1985, 342:91–102.

Bear, D.; Schiff, D.; Saver, J.; Greenberg, M.; and Freeman, R. Quantitative analysis of cerebral asymmetry. *Archives of Neurology*, 1986, 43:598–603.

Beck, B. B. Animal Tool Behavior: The Use and Manufacture of Tools by Animals. New York: Garland STPM Press, 1980.

Beck, C. H. M. and Barton, R. L. Deviation and laterality of hand preference in monkeys. *Cortex*, 1972, 8:339–363.

Becker, J. B.; Robinson, T. E.; and Lorenz, K. A. Sex differences and estrous cycle variations in amphetamine-elicited rotational behavior. *European Journal of Pharmacology*, 1982, 80:65–72.

Bednarik, R. G. On the Pleistocene settlement of South America. *Antiquity*, 1989, 63:101–111 (a).

Bednarik, R. G. More to Palaeolithic females than meets the eye. *Rock Art Research*, 1989, 6:105–115 (b).

Bednarik, R. G. The Galgenberg figurine from Krems, Austria. *Rock Art Research*, 1989, 6:118–125 (c).

Bednarik, R. G. An Acheulian haematite pebble with striations. *Rock Art Research*, 1990, 7:75.

Beecher, M. D.; Petersen, M. R.; Zoloth, S. R.; Moody, D. B.; and Stebbins, W. C. Perception of conspecific vocalisations by Japanese macaques: Evidence for selective attention and neural lateralization. *Brain, Behaviour and Evolution*, 1979, 16:443–460.

Bellugi, U.; Bihrle, A.; Jernigan, T.; Trauner, D.; and Doherty, S. Neuropsychological, neurological and neuroanatomical profile of Williams syndrome. *American Journal of Medical Genetics Supplement*, 1990, 6:115–125.

Benowitz, L. Conditions for bilateral transfer of monocular learning in chicks. *Brain Research*, 1974, 65:203–213.

Benson, D. F. Disorders of visual gnosis. In J. W. Brown (ed.), Neuropsychology of Visual Perception. Hillsdale, N. J.: Erlbaum, 1989, pp. 59–78.

Benson, D. F. and Geschwind, N. The alexias. In P. J. Vinken and G. W. Bruyn (eds.), Handbook of Clinical Neurology, Vol. 4. Amsterdam: North Holland, 1969, pp. 112–140.

Benson, D. F. and Geschwind, N. Aphasia and related disorders: A clinical approach. In M.-M. Mesulam (ed.), *Principles of Behavioral Neurology*. Philadelphia: F. A. Davis, 1985, pp. 193–238.

Bentivoglio, M.; Kuypers, H. G. J. M.; Catsman-Berrevoets, C. E.; and Dann, O. Fluorescent retrograde labeling in the rat by means of substances binding specifically to adenine-thymine rich DNA, *Neuroscience Letters*, 1979, 12:235–240.

Berbel, P. and Innocenti, G. M. The development of the corpus callosum: A light- and electron-microscopic study. *The Journal of Comparative Neurology*, 1988, 276:132–156.

Bermúdez de Castro, J. M. B.; Bromage, T. G.; and Jalvo, Y. F. Buccal striations on fossil human anterior teeth: Evidence of handedness in the Middle and early Upper Pleistocene. *Journal of Human Evolution*, 1988, 17:403–412.

Berrebi, A. S.; Fitch, R. H.; Ralphe, D. L.; Denenberg, J. P.; Friedrich, V. L. Jr.; and Denenberg, V. H. Corpus callosum: Region-specific effects of sex, early experience, and age. *Brain Research*, 1988, 438:216–224.

Bessette, B. B. and Hodos, W. Intensity, colour, and pattern discrimination deficits after lesions of the core and belt regions of the ectostriatum. *Visual Neuroscience*, 1989, 2:27–34.

Betancur, C.; Neveu, P. J.; and Le Moal, M. Strain and sex differences in the degree of paw preference in mice. *Behavioural Brain Research*, 1991, 45:97–101.

Beynon, P. and Rasa, O. A. E. Do dwarf mongooses have a language? Warning vocalisations transmit complex information. *South African Journal of Science*, 1989, 85:447–450.

Bianki, V. L. Lateralization of functions in the animal brain. *International Journal of Neuroscience*, 1981, 15:37–47.

Bianki, V. L. Lateralization of concrete and abstract characteristics analysis in the animal brain. *International Journal of Neuroscience*, 1982, 17:233–241.

Bianki, V. L. Hemisphere lateralization of inductive and deductive processes. *International Journal of Neuroscience*, 1983, 20:59–74 (a).

Bianki, V. L. Hemisphere specialization of the animal brain for information processing principles. *International Journal of Neuroscience*, 1983, 20:75–90 (b).

Bianki, V. L. Simultaneous and sequential processing of information by different hemispheres in animals. *International Journal of Neuroscience*, 1983, 22:1–6 (c).

Bianki, V. L. The Right and Left Hemispheres: Cerebral Lateralization of Function. *Monographs in Neuroscience*, Vol. 3. New York: Gordon and Breach, 1988.

Bianki, V. L. and Filippova, E. B. Comparative topography of functional interhemispheric asymmetry in the visual cortex during stimulation of different intensity. *Neuroscience and Behavioural Physiology*, 1977, 8:184–192.

Bianki, V. L. and Filippova, E. B. Hemisphere lateralization of the extrapolation reflex. *International Journal of Neuroscience*, 1985, 25:195–205.

Bianki, V. L. and Shramm, V. A. New evidence on the callosal system. *International Journal of Neuroscience*, 1985, 25:175–193.

Bickel, D. Sex with a twist in the tail. *New Scientist*, 1990, August 25, 24–27.

Bickerton, D. Language and Species. Chicago: University of Chicago Press, 1990.

Bihrle, A. M., Brownell, H. H. and Gardner, H. Humor and the right hemisphere: A narrative perspective. In H. A. Whitaker (ed.), Contemporary Reviews in Neuropsychology. New York: Springer, 1988, pp.109–126.

Binford, L. Human ancestors: Changing views of their behavior. *Journal of Anthropological Archaeology*, 1985, 4:292–327.

Birky, C. W. Relaxed cellular controls and organelle heredity. *Science*, 1983, 222:468–475.

Bishop, P. O.; Jeremy, D.; and Lance, J. W. The optic nerve: Properties of a central tract. *Journal of Physiology*, 1953, 121:415–432.

Bisiach, E. and Luzzatti, C. Unilateral neglect of representational space. *Cortex*, 1978, 14:129–133.

Blakemore, C. Computational principles of the visual cortex. *The Psychologist*, 1991, 14:73.

Boaz, N. T. Status of *Australopithecus afarensis*. *Yearbook of Physical Anthropology*, 1988, 31:85–113.

Boehm, C. Methods for isolating chimpanzee vocal communication. In P. G. Heltne and L. A. Marquardt (eds.), Understanding Chimpanzees. Cambridge, Mass.: Harvard University Press, 1989, pp. 2–21.

Boesch, C. Teaching among wild chimpanzees. Animal Behaviour, 1991, 41:530–532.

Boesch, C. and Boesch, H. Mental map in wild chimpanzees: An analysis of hammer transports for nut cracking. *Primates*, 1984, 25:160–170.

Bogen, J. E. and Bogen, G. M. Wernicke's region—Where is it? *Annals of the New York Academy of Sciences*, 1976, 280:834–843.

Borod, J. C. The neuropsychology of emotion: Evidence from normal, neurological, and psychiatric populations. In E. Perecman (ed.), Integrating Theory and Practice in Clinical Neuropsychology. Hillsdale, N.J.: Erlbaum, 1989, pp. 175–215.

Bottjer, S. W.; Schoonmaker, J. N.; and Arnold, A. P. Auditory and hormonal stimulation interact to produce neural growth in adult canaries. *Journal of Neurobiology*, 1986, 17:605–612.

Bouchard, T. J.; Lykken, D. T.; McGue, M.; Segal, N. L.; and Tellegen, A. Sources of human psychological differences: The Minnesota study of twins reared apart. *Science*, 1990, 250:223–228.

Box, H. O. Observations on spontaneous hand use in the common marmoset (*Callithrix jacchus*). *Primates*, 1977, 18:395–400.

Boxer, M. I. and Stanford, D. Projections to the posterior visual hyperstriatal region in the chick: An HRP study. *Experimental Brain Research*, 1985, 57:494–498.

Bracha, H. S.; Seitz, D. J.; Otemaa, J.; and Glick, S. D. Rotational movement (circling) in normal humans: Sex differences and relationship to hand, foot, eye, and ear preferences. *Brain Research*, 1987, 411:321–335.

Bradley, P.; Horn, G.; and Bateson, P. Imprinting: An electron microscopic study of the chick hyperstriatum ventrale. *Experimental Brain Research*, 1981, 41:115–120.

Bradshaw, J. L. The evolution of human lateral asymmetries: New evidence and second thoughts. *Journal of Human Evolution*, 1988, 17:615–637.

Bradshaw, J. L. Hemispheric Specialization and Psychological Function. Chichester: Wiley, 1989.

Bradshaw, J. L. and Bradshaw, J. A. Rotational and turning tendencies in humans: An analog of lateral biases in rats? *International Journal of Neuroscience*, 1988, 39:229–232.

Bradshaw, J. L.; Bradshaw, J. A.; and Nettleton, N. C. Abduction, adduction, and hand differences in simple and serial movements. *Neuropsychologia*, 1990, 28:917–931.

Bradshaw, J. L.; Burden, V.; and Nettleton, N. C. Dichotic and dichaptic techniques. *Neuropsychologia*, 1986, 24:79–90.

Bradshaw, J. L. and Nettleton, N. C. The nature of hemispheric specialization in Man. *Behavioral and Brain Sciences*, 1981, 4:51–91.

Bradshaw, J. L. and Nettleton, N. C. Monaural asymmetries. In K. Hugdahl (ed.), Handbook of Dichotic Listening: Theory, Methods and Research. New York: Wiley, 1988, pp.45–69.

Bradshaw, J. L.; Phillips, J. G.; Dennis, C.; Mattingley, J. B.; Andrews, D.; Chiu, E.; Pierson, J. M.; and Bradshaw, J. A. Initiation and execution of movement sequences in those suffering from and at risk of developing Huntington's disease. *Journal of Clinical and Experimental Neuropsychology*, 1992, 14:179–192.

Bradshaw, J. L and Sherlock, D. Bugs and faces in the two visual fields: The analytic/ holistic processing dichotomy and task sequencing. *Cortex*, 1982, 18:211–226.

Braitenberg, V. and Kemali, M. Exceptions to bilateral symmetry in the epithalamus of lower vertebrates. *Journal of Comparative Neurology*, 1970, 138:137–146.

Brass, C. A. and Glick, S. D. Sex differences in drug-induced rotation in two strains of rats. *Brain Research*, 1981, 223:229–234.

Bräuer, G. The evolution of modern humans: A comparison of the African and non-African evidence. In P. Mellars and C. Stringer (eds.), The Human Revolution: Behavioral and Biological Perspectives on the Origins of Modern Humans. Princeton: Princeton University Press, 1989, pp. 123–154.

Bräuer, G. and Rimbach, K. W. Late archaic and modern *Homo sapiens* from Europe, Africa, and Southwest Asia: Craniometric comparisons and phylogenetic implications. *Journal of Human Evolution*, 1990, 19:789–807.

Bresard, B. and Bresson, F. Reaching or manipulation: Left or right? *Behavioral and Brain Sciences*, 1987, 10:265–266.

Brewer, S. M. and McGrew, W. C. Chimpanzee use of a tool set to get honey. *Folia Primatologica*, 1990, 54:100–104.

Bromage, T. G. and Dean, M. C. Reavaluation of the age at death of immature fossil hominids. *Nature*, 1985, 317:525–527.

Brown, J. V.; Schumacker, U.; Rohlmann, A.; Ettlinger, G.; Schmidt, R. C.; and Skreczeck, W. Aimed movements to visual targets in hemiplegic and normal children: Is the "good" hand of children with infantile hemiplegia also "normal"? *Neuropsychologia*, 1989, 27:283–302.

Brown, J. W. (ed). Neuropsychology of Visual Perception. New York: Erlbaum, 1989.

Brown, T. and Molnar, S. Interproximal grooving and task activity in Australia. *American Journal of Physical Anthropology*, 1990, 81:545–553.

Brownell, H. H. Appreciation of metaphoric and connotative word meaning by brain damaged patients. In C. Chiarello (ed.), Right Hemisphere Contributions to Lexical Semantics. New York: Springer, 1988, pp. 18–31.

Bryden, M. P. Laterality: Functional Asymmetry in the Intact Brain. New York: Academic Press, 1982.

Bullock, S. P. and Rogers, L. J. Glutamate-induced asymmetry in the sexual and aggressive behavior of young chickens. *Physiology and Biochemistry Behavior*, 1986, 24:549–554.

Burchuladze, R., Potter, J. and Rose, S. P. R. Memory formation in the chick depends on membrane-bound protein kinase C. *Brain Research*, 1990, 535:131–138.

Butler, C. R. and Francis, A. C. Specialization of the left hemisphere in baboons: Evidence from directional preferences. *Neuropsychologia*, 1973, 11:351–354.

Byrne, R. W. and Byrne, J. M. Hand preferences in the skilled gathering tasks of mountain gorillas (*G. gorilla berengei*). *Cortex*, 1991, 27:521–546.

Cain, D. P. and Wada, J. A. An anatomical asymmetry in the baboon brain. *Brain, Behaviour and Evolution*, 1979, 16:222–226.

Calvin, W. H. The Throwing Madonna: From Nervous Cells to Hominid Brains. New York: McGraw-Hill, 1983.

Calvin, W. H. On the evolutionary expectations of symmetry in tool making. *Behavioral and Brain Sciences*, 1987, 10:267–268.

Camhi, J. M. Neuroethology: Nerve Cells and the Behavior of Animals. Sutherland, M.A: Sinauer Associates, 1984, pp. 211–255.

Cann, R. L. DNA and human origins. *Annual Review of Anthropology*, 1988, 17:127–143.

Cann, R. L., Stoneking, M. and Wilson A. C. Mitochondrial DNA and human evolution. *Nature*, 1987, 325:31–36.

Cannon, C. E. Descriptions of foraging behaviour of Eastern and Pale-headed Rosellas. *Bird Behaviour*, 1983, 4:63–70.

Carlson, J. N. and Glick, S. D. Cerebral lateralization as a source of interindividual differences in behavior. *Experimentia*, 1989, 45:788–798.

Carlson, J. N. and Glick, S. D. Brain laterality as a determinant of susceptibility to depression in an animal model. *Brain Research*, 1991, 550:324–328.

Carlson, J. N.; Glick, S. D.; and Hinds, P. A. Changes in d-amphetamine elicited rotational behavior in rats exposed to uncontrollable footshock stress. *Pharmacology, Biochemistry and Behavior*, 1987, 26:17–21.

Carson, R. G.; Chua, R.; Elliott, D.; and Goodman, D. The contribution of vision to asymmetries in manual aiming. *Neuropsychologia*, 1990, 28:1215–1220.

Cassells, B.; Collins, R. L.; and Wahlsten, D. Path analysis of sex difference, forebrain commissure area and brain size in relation to degree of laterality in selectively bred mice. *Brain Research*, 1990, 529:50–56.

Catania, C. C. What good is 5% of a language competence? *Behavioral and Brain Sciences*, 1990, 13:729–731.

Cavalli-Sforza, L. L.; Piazza, A.; Menozzi, P.; and Mountain, J. Reconstruction of human evolution: Bringing together genetic, archaeological, and linguistic data. *Proceedings of the National Academy of Science*, 1988, 85:6002–6006.

Cavalli-Sforza, L. L.; Piazza, A.; Menozzi, P.; and Mountain, J. Reply to Bateman et al., Speaking of forked tongues. *Current Anthropology*, 1990, 31:1–24.

Chapple, W. D. Role of asymmetry in the functioning of invertebrate nervous systems. In S. Harnad, R. W. Doty, L. Goldstein, J. Jaynes and G. Krauthamer (eds.), Lateralization in the Nervous System. New York: Academic Press, 1977 (a).

Chapple, W. D. Central and peripheral origins of asymmetry in the abdominal motor system of the hermit crab. *Annals of the New York Academy of Sciences*, 1977, 299:43–58 (b).

Charles, D.; Dickson, D.; Joyce, C.; Lewin, R.; and Pain, S. Chimp learns stone age use of tools. *New Scientist*, 1991, February 23, 11.

Chase, P. G. and Dibble, H. L. Middle Paleolithic symbolism: A review of current evidence and interpretations. *Journal of Anthropological Archaeology*, 1987, 6:263–296.

Cheney, D. L. and Seyfarth, R. M. Social and nonsocial knowledge in vervet monkeys. *Philosophical Transactions of the Royal Society of London B*, 1985, 308:187–201.

Cheney, D. L. and Seyfarth, R. M. How Monkeys See The World: Inside The Mind of Another Species. Chicago: Chicago University Press, 1990.

Chengappa, K. N.; Cochran, J.; Rabin, B. S.; and Ganguli, R. Handedness and autoantibodies. *The Lancet*, 1991, 338:694.

Chevalier-Skolnikoff, S. Spontaneous tool use and sensorimotor intelligence in *Cebus* compared with monkeys and apes. *Behavioral and Brain Sciences*, 1989, 12:561–627.

Chevalier-Skolnikoff, S. Tool use by wild *Cebus* monkeys at Santa Rosa National Park, Costa Rica. *Primates*, 1990, 3:375–383.

Cheverud, J. M.; Falk, D.; Hildebolt, C.; Moore, A. J.; Helmkamp, R. C.; and Vannier, J. Heritability and association of cortical petalias in rhesus macaques (*Macaca mulatta*). *Brain, Behaviour and Evolution*, 1990, 35:368–372.

Chi, J. G.; Dooling, E. C.; and Gillies, F. H. Left–right asymmetries of the temporal speech areas of the human fetus. *Archives of Neurology*, 1977, 34:346–348.

Chiarello, C. A house divided: Cognitive functioning with callosal agenesis. *Brain and Language*, 1980, 11:128–158.

Chiarello, C.; Senehi, J.; and Nuding, S. Semantic priming with abstract and concrete words: Differential asymmetry may be postlexical. *Brain and Cognition*, 1987, 31:43–60.

Chomsky, N. Rules and representations. *The Behavioral and Brain Sciences*, 1980, 3:1–61.

Chui, H. C. and Damasio, A. R. Human cerebral asymmetry evaluated by computed tomography. *Journal of Neurology, Neurosurgery and Psychiatry*, 1980, 43:873–878.

Cipolla-Neto, J.; Horn, G.; and McCabe, B. J. Hemispheric asymmetry and imprinting: The effect of sequential lesions of the hyperstriatum ventrale. *Experimental Brain Research*, 1982, 48:22–27.

Clarke, R. J. The Ndutu cranium and the origin of *Homo sapiens*. *Journal of Human Evolution*, 1990, 19:699–736.

Coates, M. I. and Clack, J. A. Polydactyly in the earliest known tetrapod limbs. *Nature*, 1990, 347:66–69.

Cobb, S. A comparison of the size of an auditory nucleus (n. mesencephalicus lateralis, pars dorsalis) with the size of the optic lobe in twenty-seven species of birds. *Journal of Comparative Neurology*, 1964, 122: 271–280.

Code, C. Language, Aphasia and the Right Hemisphere. Chichester: Wiley, 1987.

Cohen, L. G.; Meer, J.; Tarkka, I.; Bierner, S.; Leiderman, D. B.; Dubinsky, P. M.; Sanes, J. M.; Jabbari, B.; Branscum, B.; and Hallett, M. Congenital mirror movements: Abnormal organization of motor pathways in two patients. *Brain*, 1991, 114:381–403.

Cole, J. Paw preference in cats related to hand preference in animals and man. *Journal of Comparative Physiology and Psychology*, 1955, 48:137–140.

Collins, R. L. On the inheritance of handedness: II. Selection for sinistrality in mice. *Journal of Heredity*, 1969, 60:117–119.

Collins, R. L. When left-handed mice live in right-handed worlds. *Science*, 1975, 18:181–184.

Collins, R. L. Toward an admissible genetic model for the inheritance of the degree and direction of asymmetry. In S. Harnard, R. W. Doty, L. Goldstein, J. Jaynes, J. and G. Krauthamer (eds.), Lateralization in the Nervous System. New York: Academic Press, 1977, pp. 137–150.

Collins, R. L. On the inheritance of direction and degree of asymmetry. In Glick, S.D. (ed.), Cerebral Lateralization in Nonhuman Species. New York: Academic Press, 1985, pp. 41–71.

Collins, R. L. Observational learning of a left-right behavioral asymmetry in mice (*Mus musculus*). *Journal of Comparative Psychology*, 1988, 102:222–224.

Collins, R. L. Reimpressed selective breeding for lateralization of handedness in mice. *Brain Research*, 1991, 564:194–202.

Coltheart, M. The right hemisphere and disorders of reading. In A. W. Young (ed.), Functions of the Right Cerebral Hemisphere. New York: Academic Press, 1983, pp. 172–201.

Coltheart, M. Right hemisphere reading revisited. *Behavioral and Brain Sciences*, 1985, 8:363–365.

Conroy, G. C.; Vannier, M. W.; and Tobias, P. V. Endocranial features of *Australopithecus afarensis* revealed by 2- and 3-D computed tomography. *Science*, 1990, 247:838–840.

Contreras, C. M., Dorantes, M. E., Mexicano, G. and Guzmán-Flores, C. Lateralization of spike and wave complexes produced by hallucinogenic compounds in the cat. *Experimental Neurology*, 1986, 92:467–478.

Corballis, M. C. Laterality and myth. *American Psychologist*, 1980, 35:284–295.

Corballis, M. C. The origins and evolution of human laterality. In R. N. Malatesha and L. C. Hartlage (eds.), Neuropsychology and Cognition, Vol. 1. The Hague: Martinus Nijhoff, 1982, pp. 1–35.

Corballis, M. C. The Lopsided Ape. New York: Oxford University Press, 1991.

Corballis, M. C. and Beale, I. L. On telling left from right. Scientific American, 1971, 224:96–104.

Corballis, M. C. and Beale, I. L. The Ambivalent Mind. Chicago: Nelson Hall, 1983.

Corballis, M. C. and Morgan, M. J. On the biological basis of human laterality 1: Evidence for a maturational left–right gradient. Behavioral and Brain Sciences, 1978, 2:261–269.

Corbin, D. O. C.; Williams, A. C.; and White, A. C. Tardive dystonia: Which way do schizophrenias twist? The Lancet, 1987, January 31, 268–269.

Coren, S. and Porac, C. Fifty centuries of right-handedness: The historical record. Science, 1977, 198:631–632.

Cornford, J. M. Specialized resharpening techniques and evidence of handedness. In P. Calloway and J. Cornford (eds.), La Cotte de St. Brelade 1961–1978 Excavation. Norwich: Geo Books, 1986, pp. 337–351.

Cosgrove, R. Thirty thousand years of human colonization in Tasmania: New Pleistocene dates. Science, 1989, 243:1706–1708.

Costello, M. and Fragaszy, D. Prehension in Cebus and Saimiri: Grip type and hand preference. American Journal of Primatology, 1988, 15:235–245.

Cowan, W. M.; Adamson, L.; and Powell, T. P. S. An experimental study of the avian visual system, Journal of Anatomy (London), 1961, 95:545–563.

Crawley, J. C. W.; Crow, T. J.; Johnstone, F. C.; Oldland, S. R. D.; Owen, F.; Owens, D. G. C.; Poulter, M.; Smith, T.; Veall, N.; and Zanelli, G. Dopamine D2 receptors in schizophrenia studied in vivo. The Lancet, 1986, July 26, 224–225.

Crosson, B.; Novack T. A.; and Trenerry, M. R. Subcortical language mechanisms: Window on a new frontier. In H. A. Whitaker (ed.), Phonological Processes and Brain Mechanisms. New York: Springer, 1988, pp. 24–58.

Cuénod, M. Commissural pathways in interhemispheric transfer of visual information in the pigeon. In F. O. Schmitt, F. F. Worden (eds.), Neuroscience. Cambridge: MIT Press, 1974, pp. 21–30.

Cutting, J. The Right Cerebral Hemisphere and Psychiatric Disorders. Oxford: Oxford University Press, 1990.

Cynader, M.; Leporé, F.; and Guillemont, J. Inter-hemispheric competition during postnatal development. Nature, 1981, 290:139–140.

Damasio, A. R. and Damasio, H. The anatomical substrate of prosopagnosia. In R. Bruyer (ed.), The Neuropsychology of Face Perception and Facial Expression. Hillsdale, N. J.: Erlbaum, 1986, pp. 31–38.

Damasio, A. R.; Damasio, H.; and Tranel, D. Prosopagnosia: Anatomic and physiological aspects. In H. D. Ellis, M. A. Jeeves, F. Newcombe and A. Young (eds.), Aspects of Face Processing. The Hague: Martinus Nijhoff, 1986, pp. 268–277.

Damasio, A. R.; Van Hoesen, G. W.; and Hyman, B. T. Reflection on the selectivity of neuropathological changes in Alzheimer's disease. In M. F. Schwartz (ed.), Modular Deficits in Alzheimer-type Dementia. Cambridge, MA: Bradford/MIT, 1990, pp. 83–100.

Dantzer, R.; Tazi, A.; and Bluthé, R-M. Cerebral lateralization of olfactory-mediated affective processes in rats. Behavioural Brain Research, 1990, 40:53–60.

Dart, R. A. The predatory implemental technique of Australopithecus. American Journal of Physical Anthropology, 1949, 7:1–39.

Davey, J. E. and Horn, G. The development of hemispheric asymmetries in neuronal activity in the domestic chick after visual experience. Behavioural Brain Research, 1991, 45:81–86.

Davidson, E. H.; Hough-Evans, B. R.; and Britten, R. J. Molecular biology of the sea urchin embryo. Science, 1982, 217:17–26.

Davidson, I. Bilzingsleben and early marking. Rock Art Research, 1990, 7:52–56.

Davidson, I. The archaeology of language origins: A review. *Antiquity*, 1991, 65:39–48.

Davidson, I. and Noble, W. The archaeology of perception: Traces of depiction and language. *Current Anthropology*, 1989, 30:125–155.

Davies, M. N. O. and Green, P. R. Footedness in pigeons, or simply sleight of foot? *Animal Behaviour*, 1991, 42:311–312.

Davis, T. A. Reversible and irreversible lateralities in some animals. *Behavioral and Brain Sciences*, 1978, 2:291–293.

Deacon, T. W. Human brain evolution: I. Evolution of language circuits. In H. J. Jerison and I. Jerison (eds.), Intelligence and Evolutionary Biology. New York: Springer, 1986, pp. 363–381 (a).

Deacon, T. W. Human brain evolution: II. Embryology and brain allometry. In H. J. Jerison and I. Jerison (eds.), Intelligence and Evolutionary Biology. New York: Springer, 1986, pp. 381–415 (b).

Deacon, T. W. The neural circuitry underlying primate calls and human language. *Human Evolution*, 1989, 4:367–401.

de LaCoste, M. C.; Horvath, D. S.; and Woodward, D. J. Prosencephalic asymmetries in Lemuridae. *Brain, Behaviour and Evolution*, 1988, 31:296–311.

Delson, E. Oldest Eurasian stone tools. *Nature*, 1989, 340:96.

Denenberg, V. H. Critical periods, stimulus input, and emotional reactivity: A theory of infantile stimulation. *Psychological Review*, 1964, 71:335–351.

Denenberg, V. H. Hemispheric laterality in animals and the effects of early experience. *The Behavioral and Brain Sciences*, 1981, 4:1–49.

Denenberg, V. H. Behavioral asymmetry. In N. Geschwind and A. M. Galaburda (eds.), Cerebral Dominance: The Biological Foundations. Cambridge: Harvard University Press, 1984, pp. 114–133.

Denenberg, V. H.; Berrebi A. S.; and Fitch, R. H. A factor analysis of the rat's corpus callosum. *Brain Research*, 1989, 497:271–279.

Denenberg, V. H.; Cowell, P. E.; Fitch, R. H.; Kertesz, A.; and Kenner, G. H. Corpus callosum: Multiple parameter measurements in rodents and humans. *Physiology and Behavior*, 1991, 49:433–437.

Denenberg, V. H.; Fitch, R. H.; Schrott, L. M.; Cowell, P. E.; and Waters, N. S. Corpus callosum: Interactive effects of infantile handling and testosterone in the rat. *Behavioral Neuroscience*, 1991, 105, 562–566.

Denenberg, V. H.; Gall, J. S.; Berrebi, A.; and Yutzey, D. A. Callosal mediation of cortical inhibition in the lateralized rat brain. *Brain Research*, 1986, 397:327–332.

Denenberg, V. H.; Garbanati, J. A.; Sherman, G.; Yutzey, D. A.; and Kaplan, R. Infantile stimulation induces brain lateralization in rats. *Science*, 1978, 201:1150–1152.

Denenberg, V. H.; Hofmann, M.; Garbanati, J. A.; Sherman, G. F.; Rosen, G. D.; and Yutzey, D. A. Handling in infancy, taste aversion, and brain laterality in rats. *Brain Research*, 1980, 200:123–133.

Denenberg, V. H.; Hofmann, M. J.; Rosen, G. D.; and Yutzey, D. A. Cerebral asymmetry and behavioral laterality: Some psychobiological considerations. In N. A. Fox and R. J. Davidson (eds.), The Psychobiology of Affective Development. Hillsdale, N.J.: Erlbaum, 1984, pp. 77–117.

Denenberg, V. H.; Rosen, G. D.; Hofmann, M.; Gall, J.; Stockler, J.; and Yutzey, D. A. Neonatal postural asymmetry and sex differences in the rat. *Developmental Brain Research*, 1982, 2:417–419.

Denenberg, V. H. and Yutzey, D. A. Hemispheric laterality, behavioral asymmetry, and the effects of early experience in rats. In S. D. Glick (ed.), Cerebral Lateralization in Nonhuman Species. New York: Academic Press, 1985, pp. 109–133.

Denenberg, V. H.; Zeidner, L.; Rosen, G. D.; Hofmann, M.; Garbanati, J.A.; Sherman, G. F.; and Yutzey, D. A. Stimulation in infancy facilitates interhemispheric communication in the rabbit. *Developmental Brain Research,* 1981, 1:165–169.

De Renzi, E. Apraxia. In F. Boller and J. Grafman (eds.), Handbook of Neuropsychology, Vol. 2. New York: Elsevier, 1989, pp. 245–263.

D'Errico, F. Palaeolithic lunar calendars: A case of wishful thinking. *Current Anthropology,* 1989, 30:117–118.

Dettwyler, K. A. Can paleopathology provide evidence for compassion? *American Journal of Physical Anthropology,* 1991, 84:375–384.

Deuel, R. K. and Schaffer, S. P. Patterns of hand preference in monkeys. *Behavioral and Brain Sciences,* 1987, 10:270–271.

de Waal, F. B. M. Introduction: The Fourth Ape. In P. G. Heltne and L. A. Marquardt (eds.), Understanding Chimpanzees. Cambridge, Mass.: Harvard University Press, 1989, pp. 152–153.

Dewson, J. H. III. Preliminary evidence of hemispheric asymmetry of auditory function in monkeys. In S. Harnad, R. W. Doty, L. Goldstein, J. Jaynes and G. Krauthamer (eds.), Lateralization in the Nervous System. New York: Academic Press, 1977, pp. 63–71.

Dewson, J. H. III; Pribram, K. H.; and Lynch, J. C. Effects of ablations of temporal cortex upon speech sound discrimination in the monkey. *Experimental Neurology,* 1969, 24:579–591.

Diamond, J. Reinventions of human language. *Natural History,* 1991, 5(May), 22–28 (a).

Diamond, J. The Rise and Fall of the Third Chimpanzee. London: Radius, 1991 (b).

Diamond, M. C. Age, sex, and environmental influences. In N. Geschwind, and A. M. Galaburda (eds.), Cerebral Dominance. Cambridge, Mass.: Harvard University Press, 1984, pp.134–146.

Diamond, M. C. Rat forebrain morphology: Right–left: Male–female: Young–old: Enriched–impoverished. In S. Glick (ed.), Cerebral Lateralization in Nonhuman Species. New York: Academic Press, 1985, pp. 73–87.

Diamond, M. C., Dowling, G. A. and Johnson, R. E. Morphologic cerebral cortical asymmetry in male and female rats. *Experimental Neurology,* 1981, 71:261–268.

Diamond, M. C.; Johnson, R. E.; Young, D.; and Singh S. S. Age-related morphologic differences in the rat cerebral cortex and hippocampus: Male–female; Right–left. *Experimental Neurology,* 1983, 81:1–13.

Diamond, M. C.; Murphy, Jr., G. M.; Akiyama, K.; and Johnson, R. E. Morphologic hippocampal asymmetry in male and female rats. *Experimental Neurology,* 1982, 76:553–565.

Diamond, R. and Carey, S. Why faces are and are not special: An effect of expertise. *Journal of Experimental Psychology: General,* 1986, 115:107–117.

Diaz Palarea, M. D.; Gonzalez, M. C.; and Rodriguez, M. Behavioral lateralization in the T-maze and monoaminergic brain asymmetries. *Physiology and Behavior,* 1987, 40:785–789.

Dibble, H. L. The implications of stone tool types for the presence of language during the Lower and Middle Palaeolithic. In P. Mellars and C. Stringer (eds.), The Human Revolution: Behavioral and Biological Perspectives on the Origins of Modern Humans. Princeton: Princeton University Press, 1989, pp. 415–432.

Diggs, C. C. and Basili, A. G. Verbal expression of right cerebrovascular accident patients: Convergent and divergent language. *Brain and Language,* 1987, 30:130–146.

Dimond, S. J. and Harries, R. Face touching in monkeys, apes and man: Evolutionary origins and cerebral asymmetry. *Neuropsychologia,* 1984, 22:227–233.

Dingwall, W. O. The evolution of human communicative behavior. In F. J. Newmeyer (ed.),

Linguistics: The Cambridge Survey, Vol. 111 Language: Psychological and Biological Aspects. New York: Cambridge University Press, 1988, pp. 274–313.

Dodson, D. L.; Seltzer, C.; and Ward, J. P. Left whole-body turning biases in quadrupedal and bipedal prosimians. In preparation.

Dorman, C. Microcephaly and intelligence. *Developmental Medicine and Child Neurology*, 1991, 33:267–272.

Dorozynski, A. Looking for the father of us all. *Science*, 1991, 251:378–380.

Dorozynski, A. and Anderson, A. Collagen: A new probe into prehistoric diet. *Science*, 1991, 254:520–521.

Douglas-Hamilton, I. and Douglas-Hamilton, O. Among The Elephants. London: Collins, 1975.

Drew, K. L.; Lyon, R. A.; Titeler, M.; and Glick, S. D. Asymmetry in D-2 binding in female rat striata. *Brain Research*, 1986, 363:192–195.

Dubbledam, J. L. The avian and mammalian forebrain: Correspondences and differences. In R. J. Andrew (ed.), Neural and Behavioural Plasticity: The Use of the Domestic Chick as a Model. Oxford: Oxford University Press, 1991, pp. 65–91.

Duchin, L. E. The evolution of articulate speech: Comparative anatomy of the oral cavity in *Pan* and *Homo*. *Journal of Human Evolution*, 1990, 19:687–697.

Ducker, G.; Luscher, C.; and Schultz, P. Problemlöseverhalten von Stieglitzen (*Carduelis carduelis*) bei "manipulativen" Aufgaben. *Zoologisches Beitr.*, 1986, 23:377–412.

Eakin, R. M. and Brandenberger, J. L. Unique eye of probable evolutionary significance. *Science*, 1981, 211:1189–1190.

Eccles, J. C. Brain, speech, and consciousness. *Naturwissenschaften*, 1973, 60:167–176.

Eccles, J. C. Evolution of the Brain: Creation of the Self. London: Routledge, 1991.

Ehret, G. Left hemisphere advantage in the mouse brain for recognizing ultrasonic communication calls. *Nature*, 1987, 325:249–251.

Ehrlichman, H.; Zoccolotti, P.; and Owen, D. Perinatal factors in hand and eye preference: Data from the Collaborative Perinatal Project. *International Journal of Neuroscience*, 1982, 17:17–22.

Eibl-Eibesfeldt, I. The biological foundations of aesthetics. In I. Rentschler, B. Herzberger and D. Epstein (eds.), Beauty and the Brain: Biological Aspects of Aesthetics. Basel: Birkhauser-Verlag, 1988, pp. 29–67.

Eidelberg, D. and Galaburda, A. M. Inferior parietal lobule: Divergent architectonic asymmetries in the human brain. *Archives of Neurology*, 1984, 41:843–852.

Eisenberg, J. F. The Mammalian Radiations: An Analysis of Trends in Evolution, Adaptation and Behavior. Chicago: University of Chicago Press, 1981.

Ekhardt, R. B. The solution for teething troubles. *Nature*, 1990, 345:578.

Elliott, D. Human handedness reconsidered. *Behavioral and Brain Sciences*, 1991, 14:341–342.

Engbretson, G. A.; Reiner, A.; and Brecha, N. Habenular asymmetry and the central connections of the parietal eye of the lizard. *Journal of Comparative Neurology*, 1981, 198:155–165.

Engel, M. H.; Macko, S. A.; and Silfer, J. A. Carbon isotope composition of individual amino acids in the Murchison meteorite. *Nature*, 1990, 348:47–49.

Epstein, R.; Lanza, R. P.; and Skinner, B. F. "Self-awareness" in the pigeon. *Science*, 1981, 212:695–696.

Ettlinger, G. Hand preference, ability, and hemispheric specialization: In how far are these factors related in the monkey? *Cortex*, 1988, 24:389–398.

Evans, C. S. and Marler, P. On the meaning of alarm calls: Functional reference in an avcan vocal system. *Animal Behaviour*, 1992, in press.

Fagot, J. Ontogeny of object manipulation and manual lateralization in the Guinea baboon: Preliminary observations. In J. P. Ward (ed.), Current Behavioral Evidence of Primate Asymmetry. New York: Springer, 1992, pp. 235–250.

Fagot, J.; Drea, C. M.; and Wallen, K. Evidence for an asymmetrical hand usage in rhesus monkeys in tactually and visually regulated tasks. *Journal of Comparative Psychology*, 1991, 105:260–268.

Fagot, J. and Vauclair, J. Handedness and bimanual coordination in the lowland Gorilla. *Brain, Behaviour and Evolution*, 1988, 32:89–95 (a).

Fagot, J. and Vauclair, J. Handedness and manual specialization in the baboon. *Neuropsychologia*, 1988, 26:795–804 (b).

Fagot, J. and Vauclair, J. Manual laterality in nonhuman primates: A distribution between handedness and manual specialization. *Psychological Bulletin*, 1991, 109:76–89.

Falk, D. Hominid brain evolution: The approach from paleoneurology. *Yearbook of Physical Anthropology*, 1980, 23:93–107.

Falk, D. Endocranial casts and their significance for primate brain evolution. In D. R. Swindler (ed.), Comparative Primate Biology, Vol. 1: Systematics, Evolution and Biology. New York: Liss, 1986, pp. 477–490.

Falk. D. Hominid paleoneurology. *Annual Review of Anthropology*, 1987, 16:13–30.

Falk, D. Primate tool use: But what about their brains? *Behavioral and Brain Sciences*, 1989, 12:595–596.

Falk, D. Brain evolution in *Homo*: The radiator theory. *Behavioral and Brain Sciences*, 1990, 13:333–381.

Falk, D.; Cheverud, J.; Vannier, M. W.; and Conroy, G. C. Advanced computer graphics technology reveals cortical asymmetry in endocasts of rhesus monkeys. *Folia Primatologica*, 1986, 46:98–103

Falk, D.; Hildebolt, C.; Cheverud, J.; Vannier, M.; Helmkamp, R. C.; and Konigsberg, L. Cortical asymmetries in frontal lobes of rhesus monkeys (*Macaca mulatta*). *Brain Research*, 1990, 512:40–45.

Falk, D.; Pyne, L.; Helmkamp, R. C.; and DeRousseau, C. J. Directional asymmetry in the forelimb of *Macaca mulatta*. *American Journal of Physical Anthropology*, 1988, 77:1–6.

Fantz, R. L. and Fagan, J. F. Visual attention to size and number of pattern details by preterm infants during the first 6 months. *Child Development*, 1975, 46:3–18.

Farah, M. J. The neuropsychology of mental imagery. In F. Boller and J. Grafman (eds.), Handbook of Neuropsychology, Vol. 2. New York: Elsevier, 1989, pp. 395–414.

Farah, M. J. Visual Agnosia. Cambridge, MA: MIT Press, 1990.

Fersen, L. von and Güntürkün, O. Visual memory lateralization in pigeons. *Neuropsychologia*, 1990, 28:1–7.

Fischer, R. B.; Meunier, G. F.; and White, P. J. Evidence of laterality in the lowland gorilla. *Perceptual and Motor Skills*, 1982, 54:1093–1094.

Fitch, R. H.; Berrebi, A. S.; Cowell, P. E.; Schrott, L. M.; and Denenberg V. H. Corpus callosum: Effects of neonatal hormones on sexual dimorphism in the rat. *Brain Research*, 1990, 515:111–116.

Fitch, R. H.; Cowell, P. E.; Schrott, L. M.; and Denenberg, V. H. Corpus callosum: Ovarian hormones and feminization. *Brain Research*, 1991, 542:313–317.

Fitch, R. H.; Cowell, P. E.; Schrott, L. M.; and Denenberg, V. H. Corpus callosum: Demasculinization via perinatal anti-androgen. *International Journal of Developmental Neuroscience*, 1991, 1:35–38.

Fitts, P. M. The information capacity of the human motor system in controlling the amplitude of movement. *Journal of Experimental Psychology*, 1954, 47:381–391.

Fleming, D. E.; Anderson, R. H.; Rhees, R. W.; Kinghorn E.; and Bakaitis, J. Effects of prenatal stress on sexually dimorphic asymmetries in the cerebral cortex of the male rat. *Brain Research Bulletin*, 1986, 16:395–398.

Fletcher, P. On *Homo loquens:* review of P. Lieberman, Uniquely Human: The Evolution of Speech, Thought and Selfless Behavior. Harvard University Press, 1991. *Nature*, 1991, 350:535.

Flowers, K. Handedness and controlled movement. *British Journal of Psychology*, 1975, 66:39–52.

Fodor, J. A. The Modularity of Mind: An Essay on Faculty Psychology. Cambridge, MA: MIT Press, 1983.

Foley, R. A. The ecological conditions of speciation: A comparative approach to the origins of anatomically modern humans. In P. Mellars and C. Stringer (eds.), The Human Revolution: Behavioral and Biological Perspectives on the Origins of Modern Humans. Princeton: Princeton University Press, 1989, pp. 298–318.

Foley, R. A. Africa: Evolutionary hot spot for hominids. *Australian Natural History*, 1991, 23:596 (a).

Foley, R. A. Language origins: The silence of the past. *Nature*, 1991, 353:114–115 (b).

Foley, R. A. and Dunbar, R. Beyond the bones of contention. *New Scientist*, 1989, October 14, 21–25.

Foley, R. A. and Lee, P. C. Finite social space, evolutionary pathways, and reconstructing hominid behavior. *Science*, 1989, 243:901–906.

Forsythe, C.; Milliken, G. W.; Stafford, D. K.; and Ward, J. P. Posturally related variations in the hand preferences of the ruffed lemur (*Varecia variegata variegata*). *Journal of Comparative Psychology*, 1988, 102:248–250.

Forsythe, C. and Ward, J. P. Black lemur (*Lemur macaco*) hand preference in food reaching. *Primates*, 1988, 29:369–374.

Forward, E.; Warren, J. M.; and Hara, K. The effects of unilateral lesions in sensorimotor cortex on manipulation by cats. *Journal of Comparative and Physiological Psychology*, 1962, 55:1130–1135.

Fowler, C. A. and Rosenblum, L. D. Duplex perception: A comparison of monosyllables and slamming doors. *Journal of Experimental Psychology: Human Perception and Performance*, 1990, 16:742–754.

Fragaszy, D. M. and Adams-Curtis, L. E. What next for handedness research? *Behavioral and Brain Sciences*, 1988, 11:722–723.

Fragaszy, D. M. and Mitchell, S. R. Hand preference and performance on unimanual and bimanual tasks in capuchin monkeys (*Cebus apella*). *Journal of Comparative Psychology*, 1990, 104:275–282.

Fraser, F. C. Association of neural tube defects and parental nonright-handedness. *American Journal of Human Genetics*, 1983, 35:89A.

Freeman, B. M. and Vince, M. A. Development of the Avian Embryo. London: Chapman and Hall, 1974.

Freeman, G. and Lundelius, J. J. The developmental genetics of dextrality and sinistrality in the gastropod *Lymnaea peregra*. *Wilhelm Roux's Archives of Developmental Biology*, 1982, 191:69–83.

Fride, E.; Collins, R. L.; Skolnick, P.; and Arora, P. K. Immune function in lines of mice selected for high or low degrees of behavioral asymmetry. *Brain, Behavior, and Immunity*, 1990, 4:129–138.

Friedman, H. and Davis, M. "Left-handedness" in parrots. *Auk*, 1938, 80:478–480.

Frith, U. Autism: Explaining the Enigma. Oxford: Blackwell, 1989.

Fromkin, V.; Krashen, S.; Curtiss, S.; Rigler, D.; and Rigler, M. The development of language in Genie: A case of language acquisition beyond the critical period. *Brain and Language*, 1974, 1:81–107.

Frost, G. T. Tool behavior and the origins of laterality. *Journal of Human Evolution*, 1980, 9:447–459.

Frumkin, L. R. and Grimm, P. Is there a pharmacological asymmetry in the human brain? An hypothesis for the differential hemispheric action of barbiturates. *International Journal of Neuroscience*, 1981, 13:187–197.

Fukuda, M.; Yamanouchi, K.; Nakano, Y.; Furuya, H.; and Arai, Y. Hypothalamic laterality in regulating gonadotrophic function: Unilateral hypothalamic lesion and ovarian compensatory hypertrophy. *Neuroscience Letters*, 1984, 365–370.

Gabel, S. The right hemisphere in imagery, hypnosis, REM sleep and dreaming. *Journal of Nervous and Mental Diseases*, 1988, 176:323–329.

Galaburda, A. M. Anatomical asymmetries. In N. Geschwind and A. M. Galaburda (eds.), *Cerebral Dominance: The Biological Foundations*. Cambridge, Mass.: Harvard University Press, 1984, pp. 11–25.

Galaburda, A. M.; Aboitiz, F.; Rosen, G. D.; and Sherman, G. F. Histological asymmetry in the primary visual cortex of the rat: Implications for mechanisms of cerebral asymmetry. *Cortex*, 1986, 22:151–160.

Galaburda, A. M.; Corsigha, J.; Rosen, G. D.; and Sherman, G. F. Planum temporale asymmetry, reappraisal since Geschwind and Levitsky. *Neuropsychologia*, 1987, 25:853–868.

Galaburda, A. M.; LeMay, M.; Kemper, T. L.; and Geschwind, N. Right–left asymmetries in the brain. *Science*, 1978, 199:852–856.

Galaburda, A. M. and Pandya, D. N. Role of architectonics and connections in the study of primate brain evolution. In E. Armstrong and D. Falk (eds.), *Primate Brain Evolution: Methods and Concepts*. New York: Plenum, 1982, pp. 203–216.

Galdikas, B. M. F. Orang-utan tool use at Tanjung Putting Reserve, Central Indonesian Borneo (Kalimantan Tengah). *Journal of Human Evolution*, 1982, 10:19–33.

Gallesten, C. R. Animal navigation: The representation of space, time and number. *Annual Review of Psychology*, 1989, 40:155–189.

Galloway, J. A cause for reflection. *Nature*, 1987, 330:204–205.

Galloway, J. A handle on handedness. *Nature*, 1990, 346:223–224 (a).

Galloway, J. Putting a twist in the tale. *Nature*, 1990, 343:513–514 (b).

Gallup, G. G. Toward a comparative psychology of mind. In R. L. Mellgren (ed.), *Animal Cognition and Behavior*. Amsterdam: North Holland, 1983, pp. 473–510.

Gamkrelidze, T. V. and Ivanov, V. V. The early history of Indo-European languages. *Scientific American*, 1990, March, 82–90.

Ganslosser, U. Beobachtungen uber Doria-Baumkänguruhs (*Dendrolagus dorianus*) und Grauen Baumkänguruhs (*Dendrolagus inustus*) in Zoologischen Gärten. *Zoologischer Anzeiger*, 1977, 198:393–412.

Garay, A. S. Origin and role of optical isometry in life. *Nature*, 1968, 219:338–340.

Garbanati, J. A.; Sherman, G. F.; Rosen, G. D.; Hofmann, M.; Yutzey, D. A.; and Denenberg, V. H. Handling in infancy, brain laterality, and muricide in rats. *Behavioural Brain Research*, 1983, 71:351–359.

Gardner, H. Artistry following damage to the human brain. In A. W. Ellis (ed.), *Normality and Pathology in Cognitive Functions*. New York: Academic, 1982, pp. 299–323.

Gardner, H. *Frames of Mind: The Theory of Multiple Intelligence*. New York: Basic Books, 1983.

Gardner, H.; Brownell, H. H.; Wapner, W.; and Michelow, D. Missing the point: The role of the right hemisphere in the processing of complex linguistic materials. In E. Perecman (ed.), Cognitive Processing in the Right Hemisphere. New York: Academic Press, 1983, pp.169–191.

Gargett, R. H. Grave shortcomings: The evidence for Neanderthal burial. Current Anthropology, 1989, 30:157–177.

Garn, S. M.; Mayor, G. H.; and Shaw, H. A. Paradoxical bilateral asymmetry in bone size and bone mass in the hand. American Journal of Physical Anthropology, 1976, 45:209–210.

Gaston, K.E. Interocular transfer of pattern discrimination learning in chicks. Brain Research, 1984, 310:213–221.

Gaston, K.E. and Gaston, M.G. Unilateral memory after binocular discrimination training: Left hemisphere dominance in the chick? Brain Research, 1984, 303:190–193.

Gates, A. and Bradshaw, J. L. The role of the cerebral hemispheres in music. Brain and Language, 1977, 4:403–431.

Gazzaniga, M. S. Cerebral dominance viewed as a decision system. In S. J. Dimond and J.G. Beaumont (eds.), Hemisphere Function in the Human Brain. London: Elek Science, 1974, pp. 367–382.

Gazzaniga, M. S. and Le Doux, J. E. The Integrated Mind. New York: Plenum Press, 1978.

Gazzaniga, M. S. and Smylie, C. S. What does language do for a right hemisphere? In M. S. Gazzaniga (ed.), Handbook of Cognitive Neuroscience. New York: Plenum Press, 1984, pp.199–209.

Gee, H. A backbone for the vertebrates. Nature, 1989, 340:596–597.

Geschwind, N. and Behan P. Left-handedness: Association with immune disease, migraine, and developmental learning disorders. Proceedings of the National Academy of Science, U.S.A., 1982, 79:5097–5100.

Geschwind, N. and Galaburda, A. M. Cerebral lateralization: Biological mechanisms, associations, and pathology I-III: A hypothesis and program for research. Archives of Neurology, 1985, 42:428–459, 521–552, 634–654.

Geschwind, N. and Galaburda, A. M. Cerebral Lateralization: Biological Mechanisms, Associations and Pathology. Cambridge, Mass.: Bradford/MIT, 1987.

Geschwind, N. and Levitsky, W. Human brain: Left–right asymmetry in temporal speech region. Science, 1968, 161:186–187.

Gibbons, A. Our chimp cousins get that much closer. Science, 1990, 250:376.

Gibson, K. R. New perspectives on instincts and intelligence: Brain size and the emergence of hierarchical mental construction of skills. In S. T. Parker and K. R. Gibson (eds.), "Language" and Intelligence in Monkeys and Apes: Comparative and Developmental Perspectives. Cambridge: Cambridge University Press, 1990, pp. 96–128.

Gilbert, D. B.; Paterson, T. A.; and Rose, S. P. R. Dissociation of brain sites necessary for registration and storage of memory for a one-trial passive avoidance task in the chick. Behavioral Neuroscience, 1991, 105:553–561.

Ginùbili de Martinez, M. S.; Rodriguez de Turco, E. B.; and Barrantes, F. J. Endogenous asymmetry of rat brain lipids and dominance of the right cerebral hemisphere in free fatty acid response to electronconvulsive shock. Brain Research, 1985, 449:315–321.

Glick, S. D. Heritable differences in turning behavior of rats. Life Sciences, 1985, 36:499–503.

Glick, S. D. Rotational behavior in children and adults. In J. P. Ward (ed.), Current Behavioral Evidence of Primate Asymmetries. New York: Springer, 1992, pp. 169–182.

Glick, S. D. and Carlson, J. N. Regional changes in brain dopamine and serotonin metabolism induced by conditioned circling in rats: Effects of water deprivation, learning and individual differences in asymmetry. Brain Research, 1989, 504:231–237.

Glick, S. D. and Cox, R. D. Nocturnal rotation in normal rats: Correlation with amphet-amine-induced rotation and effects of nigrostriatal lesions. *Brain Research,* 1978, 150:149–161.

Glick, S. D.; Hinds, P. A.; and Shapiro, R. M. Cocaine-induced rotation; Sex-dependent differences between left- and right-sided rats. *Science,* 1983, 221:775–777.

Glick, S. D.; Jerussi, T. P.; and Zimmerberg, B. Behavioral and neuropharmacological correlates of nigro-striatal asymmetry in rats. In S. Harnard, R. W. Doty, L. Goldstein, J. Jaynes and G. Krauthamer (eds.), Lateralization in the Nervous System. New York: Academic Press, 1977, pp. 213–249.

Glick, S. D. and Ross, D. A. Right-sided population bias and lateralization of activity in normal rats. *Brain Research,* 1981, 205:222–225.

Glick, S. D.; Ross, D. A.; and Hough, L. B. Lateral asymmetry of neurotransmitters in human brain. *Brain Research,* 1982, 234:53–63.

Glick, S. D. and Shapiro, R. M. Functional and neurochemical asymmetries. In N. Gesch-wind and A. M. Galaburda (eds.), Cerebral Dominance: The Biological Foundations. Cambridge, Mass.: Harvard University Press, 1984, pp. 147–166.

Glick, S. D. and Shapiro, R. M. Functional and neurochemical mechanisms of cerebral lateralization in rats. In S. D. Glick (ed.), Cerebral Lateralization in Nonhuman Species. New York: Academic Press, 1985, pp. 158–184.

Glick, S. D.; Shapiro, R. M.; Drew, K. L.; Hinds, P. A.; and Carlson, J. N. Differences in spontaneous and amphetamine-induced rotational behavior, and in sensitization to amphetamine, among Sprague-Dawley derived rats from different sources. *Physiology and Behavior,* 1986, 38:67–70.

Goldberg, E. and Costa, L. Hemisphere differences in the acquisition of descriptive sys-tems. *Brain and Language,* 1981, 14:144–173.

Goldman, S. and Nottebohm, F. Neuronal production, migration, and differentiation in a vocal control nucleus of the adult female canary brain. *Proceedings of the National Academy of Sciences of the U.S.A.,* 1983, 80:2390–2394.

Goldman-Rakic, P. S. Topography of cognition. Parallel distributed networks in primate association cortex. *Annual Review of Neuroscience,* 1988, 11:137–156.

Goldstein, L.; Stolzfus, N. W.; and Gardocki, J. F. Changes in interhemispheric amplitude relationships in the EEG during sleep. *Physiology and Behavior,* 1972, 8:811–815.

Gomez, J. C. Visual behavior as a window for reading the mind of others in primates. In A. Whiten (ed.), Natural Theories of Mind: Evolution, Development and Simulation of Everyday Mindreading. Oxford: Blackwell, 1991, pp. 195–207.

Goodale, M. A. Vision and Action: The Control of Grasping. Norwood, N. J.: Ablex, 1990 (a).

Goodale, M. A. Brain asymmetries in the control of reaching. In M. A. Goodale (ed.), Vision and Action: The Control of Grasping. Norwood, N. J.: Ablex Publishing Corpora-tion, 1990, pp. 14–32 (b).

Goodale, M. A. and Graves, J. A. Interocular transfer in the pigeon: Retinal locus as a factor. In D. J. Ingle, M. A. Goodale and R. J. W. Mansfield, (eds.), Analysis of Visual Behavior. Cambridge, M.A.: MIT Press, 1982, pp. 211–240.

Goodall, J. Gombe: Highlights and current research. In P. G. Heltne and L. A. Marquardt (eds.), Understanding Chimpanzees. Cambridge, Mass.: Harvard University Press, 1989, pp. 2–21.

Goodall, J. van Lawick. The Behaviour of Free Living Chimpanzees in the Gombe Stream Reserve (Tanzania). *Animal Behavior Monographs,* 1968, 1:161–311.

Goodman, R. A. and Whitaker, H. A. Hemispherectomy: A review (1928–1981) with special reference to the linguistic abilities and disabilities of the residual right hemisphere. In C. T. Best (ed.), Hemispheric Function and Collaboration in the Child. New York: Academic Press, 1985, pp. 121–155.

Gould, S. J. Wonderful Life: The Burgess Shale and the Nature of History. London: Penguin, 1991 (a).

Gould, S. J. Abolish the recent. *Natural History,* 1991, May(5), 16–21 (b).

Gould, S. J. Eight (or fewer) little piggies. *Natural History,* 1991, January, 22–29 (c).

Gould, S. J. and Eldredge, N. Punctuated equilibria: Tempo and mode of evolution reconsidered. *Paleobiology,* 1977, 3:115–151.

Govind, C. K. Asymmetry in lobster claws. *American Scientist,* 1989, 77:468–474.

Grant, R. J.; Colenbrander, V. F.; and Albright, J. L. Effect of particle size of forage and rumen cannulation upon chewing activity and laterality in dairy cows. *Journal of Dairy Science,* 1990, 73:3158–3164.

Greenfield, P. M. and Savage-Rumbaugh, E. S. Grammatical combination in *Pan paniscus*: Processes of learning and invention in the evolution and development of language. In S. T. Parker and K. R. Gibson (eds.), "Language" and Intelligence in Monkeys and Apes: Comparative and Developmental Processes. Cambridge: Cambridge University Press, 1990, pp. 540–578.

Griffin, D. R. The Question of Animal Awareness. New York: Rockefeller University Press, 1981.

Griffin, D. R. Animal Thinking. Cambridge, Mass.: Harvard University Press, 1984.

Groves, C. P. and Humphrey, N. K. Asymmetry in gorilla skulls: Evidence of lateralized brain function? *Nature,* 1973, 244:53–54.

Gruber, D.; Waanders, R.; Collins, R. L.; Wolfer, D. P.; and Lipp, H-P. Weak or missing paw lateralization in a mouse strain (I/LnJ) with congenital absence of the corpus callosum. *Behavioural Brain Research,* 1991, 46:9–16.

Gruzelier, J. and Flor-Henry, P. Hemispheric Asymmetries of Function in Psychopathology. Amsterdam: Elsevier, 1979.

Guiard, Y.; Diaz, G.; and Beaubaton, D. Left hand advantage in right handers for spatial constant error: Preliminary evidence in a unimanual ballistic aimed movement. *Neuropsychologia,* 1983, 21:111–115.

Guidon, N. and Delibrias, G. Carbon-14 dates point to Man in the Americas 32,000 years ago. *Nature,* 1986, 321:769–771.

Güntürkün, O. Lateralization of visually controlled behavior in pigeons. *Physiology and Behavior,* 1985, 34:575–577.

Güntürkün, O. Embryonale Orientierung als Ontogenetischer Auslöser für Visuelle Lateralization. In D. Frey (ed.), Bericht über den 37. Kongreβ der Deutschen Gesellschaft für Psychologie in Kiel 1990, Band 1. Verlag für Psychologie, 1990, pp. 51–52.

Güntürkün, O. and Böhringer, P.G. Lateralization reversal after intertectal commissurotomy in the pigeon. *Brain Research,* 1987, 408:1–5.

Güntürkün, O.; Emmerton, J.; and Delius, J.D. Neural asymmetries and visual behavior in birds. In H.C. Lüttgau and R. Necker (eds.), Biological Signal Processing, FRG, Deutsche Forschungsgemeinschaft. Weinheim, Germany: VCH Verlagsgesellschaft, 1989, pp. 122–145.

Güntürkün, O. and Hoferichter, H-H. Neglect after section of a left telencephalotectal tract in pigeons. *Behavioural Brain Research,* 1985, 18:1–9.

Güntürkün, O. and Kesch, S. Visual lateralization during feeding in pigeons. *Behavioral Neuroscience,* 1987, 101:433–435.

Güntürkün, O.; Kesch, S.; and Delius, J. D. Absence of footedness in domestic pigeons. *Animal Behaviour,* 1988, 36:602–604.

Güntürkün, O. and Kischel, K-F. Is visual lateralization in pigeons sex-dependent? *Behavioural Brain Research*, 1992, 47:83–87.

Gurusinghe, C. J. and Ehrlich, D. Sex-dependent structural asymmetry of the medial habenular nucleus of the chicken brain. *Cellular and Tissue Research*, 1985, 240:149–152 (a).

Gurusinghe, C. J. and Ehrlich, D. Age, sex, and hormonal effects on structural asymmetry of the medial habenular nucleus of the chicken brain. *Neuroscience Letters*, 1985, 19:S67 (b).

Haaland, K. Y. and Harrington, D. The role of the hemispheres in closed loop movements. *Brain and Cognition*, 1989, 9:158–180.

Habgood, P. J. The origin of anatomically modern humans in Australasia. In P. Mellars and C. Stringer (eds.), The Human Revolution: Behavioral and Biological Perspectives on the Origins of Modern Humans. Princeton: Princeton University Press, 1989, pp. 245–273.

Haight, J. R. and Neylon, L. Morphological variation in the brain of the marsupial brushtailed possum, *Trichosurus vulpecula*. *Brain, Behaviour and Evolution*, 1978, 15:415–445.

Hale, W. R. Superior posterior. *New Scientist*, 1991, 1771, June 1, 3.

Halpern, D. F. Sex Differences in Cognitive Abilities. Hillsdale, N. J.: Erlbaum, 1986.

Hamilton, C. R. and Vermeire, B. A. Hemispheric differences in split-brain monkeys learning sequential comparisons. *Neuropsychologia*, 1982, 20:691–698.

Hamilton, C. R. and Vermeire, B. A. Discrimination of monkey faces by split-brain monkeys. *Behavioural Brain Research*, 1983, 9:263–275.

Hamilton, C. R. and Vermeire, B. A. Complementary hemispheric superiorities in monkeys. *Society for Neuroscience Abstracts*, 1985, 11:869.

Hamilton, C. R. and Vermeire, B. A. Complementary hemispheric specialization in monkeys. *Science*, 1988, 242:1691–1694 (a).

Hamilton, C. R. and Vermeire, B. A. Cognition, not handedness, is lateralized in monkeys. *Behavioral and Brain Sciences*, 1988, 11:723–725 (b).

Hamilton, C. R. and Vermeire, B. A. Functional lateralization in monkeys. In F. L. Kitterle (ed.), Cerebral Laterality. Theory and Research: The Toledo Symposium. Hillsdale, N. J.: Erlbaum, 1991, pp. 19–34.

Harding, C. F.; Walters, M. J.; and Parsons, B. Androgen receptor levels in hypothalamic and vocal-control nuclei in the male zebra finch. *Brain Research*, 1984, 306:333–339.

Hardyck, C. and Petrinovich, L. F. Left handedness. *Psychological Bulletin*, 1977, 84:385–404.

Harlap, S. Gender of infants conceived on different days of the menstrual cycle. *New England Journal of Medicine*, 1979, 300:1445–1448.

Harré, R. and Reynolds, V. The Meaning of Primate Signals. Cambridge: Cambridge University Press, 1984.

Harries, M. H. and Perrett, D. I. Visual processing of faces in temporal cortex: Physiological evidence for a modular organization and possible anatomical correlates. *Journal of Cognitive Neuroscience*, 1991, 3:9–24.

Harris, L. J. Laterality of function in the infant: Historical and contemporary trends in theory and research. In G. Young, S. J. Segalowitz, C. M. Corter and S. E. Trehub (eds.), Manual Specialization and the Developing Brain. New York: Academic, 1983, pp. 177–248.

Harris, L. J. Footedness in parrots: Three centuries of research, theory, and mere surmise. *Canadian Journal of Psychology*, 1989, 43:369–396.

Harris, L. J. and Carlson, D. F. Hand preference for visually guided reaching in human infants and adults. *Behavioral and Brain Sciences*, 1988, 11:726–727.

Harris, L. J. and Carlson, D. F. Hand preference for visually guided reaching in human

410

infants and adults. In J. P. Ward (ed.), Current Behavioral Evidence of Primate Asymmetries. New York: Springer, 1992, pp. 285–306.

Harrison, G. A.; Tanner, J. M.; Pilbeam, D. R.; and Baker, P. T. Human Biology: An Introduction to Human Evolution, Variation, Growth and Adaptability. 3rd edition. Oxford: Oxford University Press, 1988.

Hartley, R. J. and Suthers, R. A. Lateralization of syringeal function during song production in the canary. *Journal of Neurobiology*, 1990, 21:1236–1248.

Hartwig-Scherer, S. and Martin, R. D. Was "Lucy" more human than her "child"? Observations on early hominid postcranial skeletons. *Journal of Human Evolution*, 1991, 21:439–449.

Harvey, P. H. Allometric analysis and brain size. In H. J. Jerison and I. Jerison (eds.), Intelligence and Evolutionary Biology. New York: Springer, 1986, pp. 199–210.

Harvey, P. H. and Nee, S. How to live like a mammal. *Nature*, 1991, 350:23–24.

Hauser, M. D. Invention and social transmission: New data from wild vervet monkeys. In R. Byrne and A. Whiten (eds.), Machiavellian Intelligence: Social Expertise and the Evolution of Intellect in Monkeys, Apes and Humans. Oxford: Oxford University Press, 1988, pp. 327–343.

Hauser, M.; Perry, S.; Manson, J. H.; Ball, H.; Williams, M.; Pearson, E.; and Berard, J. It's all in the hands of the beholder: New data on free ranging rhesus monkeys. *Behavioral and Brain Sciences*, 1991, 14:342–347.

Hayhow, W. R.; Sefton, A.; and Webb, C. Primary optic centers of the rat in relation to the terminal distribution of the crossed and uncrossed optic nerve fibers. *Journal of Comparative Neurology*, 1962, 118:295–322.

Heath, C. J. and Jones, E. G. Interhemispheric pathways in the absence of a corpus callosum. *Journal of Anatomy*, 1971, 109:253–270.

Hedges, S. B.; Kumar, S.; Tamura, K.; and Stoneking, M. Human origins and analysis of mitochondrial DNA sequences. *Science*, 1992, 255:737–739.

Heestand, J. Asymmetric hand preference in four species of captive apes. *Society for Neuroscience Abstracts*, 1987, 13 (1):44.

Heffner, H. E. and Heffner, R. S. Temporal lobe lesions and perception of species-specific vocalizations by macaques. *Science*, 1984, 226:75–76.

Heffner, H. E. and Heffner, R. S. Effect of unilateral and bilateral auditory cortex lesions on the discrimination of vocalizations by Japanese macaques. *Journal of Neurophysiology*, 1986, 56:683–701.

Hegstrom, R. A. and Kondepudi, D. K. The handedness of the universe. *Scientific American*, 1990, January, 98–105.

Heilbroner, P. L. and Holloway, R. L. Anatomical brain asymmetries in New World and Old World monkeys: Stages of temporal lobe development in primate evolution. *American Journal of Physical Anthropology*, 1988, 76:39–48.

Heilbroner, P. L. and Holloway, R. L. Anatomical brain asymmetry in monkeys: Frontal, temporoparietal and limbic cortex in *Macaca*. *American Journal of Physical Anthropology*, 1989, 80:203–211.

Heilman, K. M.; Watson, R. T.; and Valenstein, E. Neglect and related disorders. In K. M. Heilman and E. Valenstein (eds.), Clinical Neuropsychology. Oxford: Oxford University Press, 1985, pp. 243–293.

Heine, O. and Galaburda, A. M. Olfactory asymmetry in the rat brain. *Experimental Neurology*, 1986, 91:392–398.

Helmkamp, R. C. and Falk, D. Age and sex-associated variations in the directional asymmetry of rhesus macaque forelimb bones. *American Journal of Physical Anthropology*, 1990, 83:211–218.

Henson, R. A. Amusia. In J. A. M. Frederiks (ed.), Handbook of Clinical Neurology Vol. 1 (45): Clinical Neuropsychology. New York: Elsevier, 1985, pp. 483–490.

Hernandez-Mesa, N. and Bureš, J. Lateralized rewarding brain stimulation affects forepaw preference in rats. *Physiology and Behavior*, 1985, 34:495–499.

Hewes, G. W. The current status of the gestural theory of language origin. *Annals of the New York Academy of Sciences*, 1976, 280:482–504.

Hewes, G. W. Language origin theories. In D. M. Rumbaugh (ed.), Language Learning by a Chimpanzee: The Lana Project. New York: Academic, 1977, pp. 3–53.

Hinde, R. A. and Fisher, J. Further observations on the opening of milk bottles by birds. British Birds, 1951, 44:393–396.

Hochberg, F. H. and LeMay, M. Arteriographic correlates of handedness. *Neurology*, 1975, 25:218–222.

Hodoba, D. Paradoxic sleep facilitaion by interictal epileptic activity of right temporal origin. *Biological Psychiatry*, 1986, 21:1267–1278.

Hodos, W. Comparative neuroanatomy and the evolution of intelligence. In H. J. Jerison and I. Jerison (eds.), Intelligence and Evolutionary Biology. New York: Springer, 1986, pp. 93–107.

Holden, C. Briefings: Brains: Is bigger better? *Science*, 1991, 254:1584.

Holliday, R. Ambidextrous RNA. *Nature*, 1990, 348:491–492.

Holloway, R. L. Volumetric and asymmetry determinants on recent hominid endocasts: Spy I and II, Djebel Ihroud I and the Sale *Homo erectus* specimens, with some notes on Neanderthal brain size. *American Journal of Physical Anthropology*, 1981, 55:385–393 (a).

Holloway, R. L. The Indonesian *Homo erectus* brain endocasts revisited. *American Journal of Physical Anthropology*, 1981, 55:503–521 (b).

Holloway, R. L. Human paleoneurological evidence relevant to language behavior. *Human Neurobiology*, 1983, 2:105–114.

Holloway, R. L. and de LaCoste-Lareymondie, M. C. Some preliminary findings on the palaeontology of cerebral dominance. *American Journal of Physical Anthropology*, 1982, 58:101–110.

Holman, S. D. and Hutchison, J. B. Lateralized action of androgen and development of behavior and brain sex differences. *Brain Research Bulletin*, 1991, 27:261–265.

Hoogmartens, M. Left–right ratio of the stretch reflex amplitude in the human masseter muscles: A methodological approach. Thesis submitted in partial fulfilment of the requirements for the degree of "Doctor in de medische wetenschappen," at the University of Leuven, 1986.

Hopkins, W. D. and Morris, R. D. Laterality for visuospatial processing in two language-trained chimpanzees (*Pan Troglodytes*). *Behavioral Neuroscience*, 1989, 103:227–239.

Hopkins, W. D.; Morris, R. D.; and Savage-Rumbaugh, E. S. Evidence for asymmetrical hemispheric priming using known and unknown warning stimuli in two language trained chimpanzees (*Pan troglodytes*). *Journal of Experimental Psychology: General*, 1991, 120:46–56.

Hopkins, W. D.; Washburn, D.; and Rumbaugh, D. Note on hand use in the manipulation of joysticks by rhesus monkeys (*Macaca mulatta*) and chimpanzees (*Pan troglodytes*). *Journal of Comparative Psychology*, 1989, 103:91–94.

Hopkins, W. D.; Washburn, D. A.; and Rumbaugh, D. M. Processing of form stimuli presented unilaterally in humans, chimpanzees (*Pan troglodytes*), and monkeys (*Macaca mulatta*). *Behavioral Neuroscience*, 1990, 104:577–582.

Horn, G. Memory Imprinting and the Brain. Oxford: Clarendon Press, 1985.

Horn, G. Neural basis of recognition memory investigated through an analysis of imprinting. *Philosophical Transactions of the Royal Society Lond. B.*, 1990, 329:133–142.

412

References

Horn, G. and Johnson, M. H. Memory systems in the chick: Dissociations and neuronal analysis. *Neuropsychologia*, 1989, 27:1–22.

Horn, G. and McCabe, B. J. The time course of N-methyl aspartate (NMDA) receptor binding in the chick brain after imprinting. *Journal of Physiology*, 1990, 423:92P.

Horn, G.; Rose, S. P. R.; and Bateson, P. P. G. Monocular imprinting and regional incorporation of tritiated uracil into the brains of intact and "split-brain" chicks. *Brain Research*, 1973, 56:227–237.

Hörster, W. and Ettlinger, G. An association between hand preference and tactile discrimination performance in the rhesus monkey. *Neuropsychologia*, 1985, 21:411–413.

Howard, K. J.; Rogers, L. J.; and Boura, A. L. A. Functional lateralization of the chicken forebrain revealed by use of intracranial glutamate. *Brain Research*, 1980, 188:369–382.

Howe, M. A. J. The strange achievements of idiots savants. In A. M. Colman and J. G. Beaumont (eds.), Psychological Survey Seven. London: Routledge, 1989, pp. 194–210.

Humphrey, D. E. and Humphrey, G. K. Sex differences in infant reaching. *Neuropsychologia*, 1987, 25:971–975.

Humphrey, N. K. The social function of intellect. In P. P. G. Bateson and R. A. Hinde, Growing Points in Ethology. Cambridge: Cambridge University Press, 1976, pp. 307–317.

Humphreys, G. W. and Riddoch, M. J. Visual Object Processing: A Cognitive Neuropsychological Approach. Hillsdale, N. J.: Erlbaum, 1987.

Hupfer, K.; Jürgens, U.; and Ploog, D. The effect of superior temporal lesions on the recognition of species-specific calls in the squirrel monkey. *Experimental Brain Research*, 1977, 30:75–87.

Hutchison, J. B.; Steimer, T. J.; and Hutchison, R. E. Formation of behaviorally active oestrogen in the dove brain: Induction of preoptic aromatase by intracranial testosterone. *Neuroendocrinology*, 1986, 43:416–427.

Ifune, C. K., Vermeire, B. A. and Hamilton, C. R. Hemispheric differences in split–brain monkeys viewing and responding to videotape recordings. *Behavioral and Neural Biology*, 1984, 41:231–235.

Innocenti, G. M. and Clarke, S. Multiple sets of visual cortical neurons projecting transitorily through the corpus callosum. *Neuroscience Letters*, 1983, 41:27–32.

Innocenti, G. M. and Frost, D. O. The postnatal development of visual callosal connections in the absence of visual experience or of the eyes. *Experimental Brain Research*, 1980, 39:365–375.

Jahnke, H. J. Binocular visual field differences among various breeds of chickens (Columbia livia). *Bird Behaviour*, 1984, 5:86–102.

James, S. R. Hominid use of fire in the Lower and Middle Pleistocene. *Current Anthropology*, 1989, 30:1–26.

Jarman, P. J. The development of the dermal shield in impala. *Journal of Zoology* (London), 1972, 166:349–356.

Jason, G. W.; Cowey, A.; and Weiskrantz, L. Hemispheric superiority for a visuospatial task in monkeys. *Neuropsychologia*, 1984, 22:777–784.

Jaynes, J. The Origin of Consciousness in the Breakdown of the Bicameral Mind. Boston: Houghton Mifflin, 1977.

Jeeves, M. A. J. Agenesis of the corpus callosum. In F. Boller and J. Grafman (eds.), Handbook of Neuropsychology, Vol. 4. New York: Elsevier, 1990, pp. 99–114.

Jefferies, R. P. S. Calcichordates. In R. W. Fairbridge and D. Jablonski (eds.), Encyclopedia of Earth Sciences, Vol. VII: The Encyclopedia of Palaeontology. Strondsburg, Pennsylvania: Dowden, Hutchinson & Ross, Inc., 1979, pp. 161–167.

Jefferies, R. P. S. The Ancestry of the Vertebrates. London: British Museum (Natural History), 1986.

Jefferies, R. P. S. and Lewis, D. N. The English Silurian fossil *Placocystites forbesianus* and the ancestors of the vertebrates. Philosophical *Transactions of the Royal Society of London B*, 1978, 282:204–323.

Jerison, H. J. Evolution of the Brain and Intelligence. New York: Academic Press, 1973.

Jerison, H. J. The evolution of biological intelligence. In R. Sternberg (ed.), Handbook of Intelligence. Cambridge: Cambridge University Press, 1982, pp. 723–791.

Jerison, H. J. Evolutionary biology of intelligence: The nature of the problem. In H. J. Jerison and I. Jerison (eds.), Intelligence and Evolutionary Biology. New York, Springer, 1986, pp. 1–12.

Jerussi, T. P. and Glick, S. D. Drug-induced rotation in rats without lesions: Behavioral and neurochemical indices of a normal asymmetry in nigrostriatal function. *Psychopharmacology*, 1976, 47:249–260.

Jerussi, T. P.; Glick, S. D.; and Johnson, C. L. Reciprocity of pre- and postsynaptic mechanisms involved in rotation as revealed by dopamine metabolism and adenylate cyclase stimulation. *Brain Research*, 1977, 129:385–388.

Johnson, J. S. and Newport, E. L. Critical period effects in second language learning: The influence of maturational state on the acquisition of English as a second language. *Cognitive Psychology*, 1989, 21:60–99.

Johnson, M. S. Polymorphism for direction of coil in *Portula suturalis:* Behavioral isolation and positive frequency dependent selection. *Heredity*, 1982, 49:145–151.

Johnston, A. N.; Rogers, L. J.; and Johnston, G. A. R. Glutamate and imprinting memory: The role of glutamate receptors in the encoding of imprinting memory. *Behavioural Brain Research*, In press.

Jolly, A. Conscious chimpanzees? A review of recent literature. In C. A. Ristau (ed.), Cognitive Ethology: The Minds of Other Animals. Hillsdale, N. J.: Erlbaum, 1991, pp. 231–254.

Jolly, C. J. and Plog, F. Physical Anthropology and Archeology. New York: Knopf, 1987.

Jones, R. East of Wallace's line: Issues and problems in the colonization of the Australian continent. In P. Mellars and C. Stringer (eds.), The Human Revolution: Behavioral and Biological Perspectives on the Origins of Modern Humans. Princeton: Princeton University Press, 1989, pp. 743–782.

Joseph, R. Neuropsychology, Neuropsychiatry and Behavioral Neurology. New York: Plenum, 1990.

Joubert, D. Eyewitness to an elephant wake. *National Geographic*, 1991, 179 (5):39–41.

Jungers, W. L. Relative joint size and hominoid locomotor adaptations with implications for the evolution of hominid bipedalism. *Journal of Human Evolution*, 1988, 17:247–265.

Kaas, J. H. The organization and evolution of the neocortex. In S. P. Wise (ed.), Higher Brain Functions: Recent Explorations of the Brain's Emergent Properties. New York: Wiley, 1987, pp. 347–378.

Karten, H. J. and Hodos, W. Telencephalic projections of the nucleus rotundus in the pigeon (*Columbia livia*). *Journal of Comparative Neurology*, 1970, 140:35–52.

Karten, H. J.; Hodos, W.; Nauta, W. J. H.; and Revzin, A. M. Neural connections of the "visual Wulst" of the avian telencephalon. Experimental studies in the pigeon (*Columbia livia*) and owl (*Speotyto cunicularia*). *Journal of Comparative Neurology*, 1973, 150:253–278.

Kawai, M. Newly acquired pre-cultural behavior of the natural troop of Japanese monkeys on Koshima islet. *Primates*, 1965, 6:1–30.

Kawai, M. Catching behavior in the Koshima troop: A case of newly acquired behavior. *Primates*, 1967, 8:181–186.

414

Kean, W. F.; Lock, C. J. L.; and Howard-Lock, H. E. Chirality in antirheumatic drugs. *The Lancet*, 1991, 338, Dec. 21/28, 1565–1568.

Keeley, L. H. The functions of Paleolithic flint tools. *Scientific American*, 1977, November, 237:108–126.

Kellog, R. J. Consciousness . . . again? *Contemporary Psychology*, 1990, 35:1152–1153.

Kelly, D. B. and Nottebohm, F. Projections of a telencephalic auditory nucleus–field L—in the canary. *Journal of Comparative Neurology*, 1979, 183:455–470.

Kemali, M. Morphological relationship established through the habenulo-interpeduncular system between the right and left portions of the frog brain. In I. Steele Russell, M. W. van Hof and G. Berlucchi (eds.), Structure and Function of Cerebral Commissures. Baltimore: University Park Press, 1977, pp. 15–33.

Kemali, M. and Agrelli, I. The habenulo-interpeduncular nuclear system of a reptilian representative *Lacerta sicula*. *Zeitscrift Microkeskopische—Anatomie Forschung*, 1972, 82:325–333.

Kemali, M.; Guglielmotti, V.; and Fiorino, L. The asymmetry of the habenular nuclei of female and male frogs in spring and winter. *Brain Research*, 1990, 517:251–255.

Kendon, A. Some considerations for a theory of language origins. *Man*, 1991, 26:199–221.

Kendrick, K. M. and Baldwin, B. A. Cells in temporal cortex of conscious sheep can respond preferentially to the sight of faces. *Science*, 1987, 236:448–450.

Kerr, R. A. Dinosaurs and friends snuffed out? *Science*, 1991, 251:160–162.

Kertesz, A. Aphasia. In J. A. M. Frederiks (ed.), Handbook of Clinical Neurology, Vol. 1. (45): Clinical Neuropsychology (gen. eds. P. Vinken, G. Bruyn and G. Klawans). Amsterdam: Elsevier, 1985, pp. 287–331.

Kertesz, A. and Geschwind, N. Patterns of pyramidal decussation and their relationship to handedness. *Archives of Neurology*, 1971, 24:326–332.

Kertesz, A.; Polk, M.; Black, S. E.; and Howell, J. A. Sex, handedness and the morphometry of cerebral asymmetries on magnetic resonance imaging. *Brain Research*, 1990, 530: 40–48.

Kien, J. The need for data reduction may have paved the way for evolution of language ability in hominids. *Journal of Human Evolution*, 1991, 29:157–165.

Kimura, D. Functional asymmetry of the brain in dichotic listening. *Cortex*, 1967, 3:163–178.

Kimura, D. The neural basis of language *qua* gesture. In H. Whitaker and H. A. Whitaker (eds.), Studies in Neurolinguistics, Vol. 1. New York: Academic, 1976, pp. 145–156.

Kimura, D. Left hemisphere control of oral and brachial movements and their relation to communication. *Philosophical Transactions of the Royal Society of London B*, 1982, 298:135–149.

Kimura, D. Sex differences in cerebral organization for speech and praxic functions. *Canadian Journal of Psychology*, 1983, 37:19–35.

Kimura, D. and Harshman, R. A. Sex differences in brain organization for verbal and nonverbal functions. *Progress in Brain Research*, 1984, 61:423–444.

King, J. E. and Landau, V. I. Manual preference in varieties of reaching in squirrel monkeys. In J. Ward (ed.), Current Behavioral Evidence of Primate Asymmetries. New York: Springer, 1992, pp. 107–124.

King, J. E.; Landau, V. I.; Scott, A. G; and Berning, A. L. Hand preference during capture of live fish by squirrel monkeys. *International Journal of Primatology*, 1987, 8:540.

Kinsbourne, M. Evolution of language in relation to lateral action. In M. Kinsbourne (ed.), Asymmetrical Function of the Brain. Cambridge: Cambridge University Press, 1978, pp. 553–565.

Kirn, J. R.; Alvarez-Buylla, A.; and Nottebohm, F. Production and survival of projection neurones in a forebrain vocal center of adult male canaries. *Journal of Neuroscience*, 1991, 11:1756–1762.

Kirzinger, A. and Jürgens, U. Cortical lesion effects and vocalization in the squirrel monkey. *Brain Research*, 1982, 233:299–315.

Klein, R. G. The Human Career: Human Biological and Cultural Origins. Chicago: University of Chicago Press, 1989.

Knusden, E. I. The hearing of the barn owl. *Scientific American*, 1981, 245:83–91.

Knusden, E. I. and Konishi, M. Sound localization by the barn owl measured with the search coil technique. *Journal of Comparative Physiology*, 1979, 133:1–11.

Koff, E.; Naeser, M. A.; Pieniadz, J. M.; Foundas, A. C.; and Levine, H. L. Computed tomographic scan hemispheric asymmetries in right- and left-handed male and female subjects. *Archives of Neurology*, 1986, 43:487–491.

Kolb, B.; Milner, B.; and Taylor, L. Perception of faces by patients with localized cortical lesions. *Canadian Journal of Psychology*, 1983, 37:8–18.

Kolb, B.; Sutherland, R. J.; Nonneman, A. J.; and Whishaw, I. Q. Asymmetry in the cerebral hemispheres of the rat, mouse, rabbit and cat: The right hemisphere is larger. *Experimental Neurology*, 1982, 78:348–359.

Kolb, B. and Whishaw, I. Q. Fundamentals of Human Neuropsychology, 3rd edition. New York: Freeman, 1990.

Kondepudi D. K.; Kaufman, R. J.; and Singh, N. Chiral symmetry breaking in sodium chlorate crystallization. *Science*, 1990, 250:975–976.

Kooistra, C. A. and Heilman, K. M. Motor dominance and lateral asymmetry of the globus pallidus. *Neurology*, 1988, 38:388–390.

Kortlandt, A. The use of stone tools by wild-living chimpanzees, In P. G. Heltne and L. A. Marquardt (eds.), Understanding Chimpanzees. Cambridge, Mass.: Harvard University Press, 1989, pp. 146–147.

Kosslyn, S. M. Seeing and imagining in the cerebral hemispheres. *Psychological Review*, 1987, 94:148–175.

Kosslyn, S. M. Mental imagery. In D. N. Osherson, S. M. Kosslyn and J. M. Hollerbach (eds.), Visual Cognition and Action, Vol. 2: An Invitation to Cognitive Science. Cambridge, MA: MIT Press, 1990, pp. 73–97.

Kozlowski, J. K. A multiaspectual approach to the origins of the Upper Palaeolithic in Europe. In P. Mellars (ed.), The Emergence of Modern Humans: An Archaeological Perspective. Edinburgh: Edinburgh University Press, 1990, pp. 419–439.

Kramer, A. Modern human origins in Australasia: Replacement or evolution? *American Journal of Physical Anthropology*, 1991, 86:455–473.

Kruper, D. C.; Boyle, B. E.; and Patton, R. A. Eye and hand preferences in rhesus monkeys. *Psychonomic Science*, 1966, 5:277–278.

Kubos, K. L. and Robinson, R. G. Asymmetrical effects of cortical island lesions in the rat. *Behavioural Brain Research*, 1984, 11:89–93.

Kuenzel, W. J. and Masson, M. A Sterotaxic Atlas of the Brain of the Chick (*Gallus domesticus*). Baltimore and London: The Johns Hopkins University Press, 1988.

Kuhl, P. K. On handedness in primates and human infants. *Behavioral and Brain Sciences*, 1988, 11:727–729 (a).

Kuhl, P. K. Auditory perception and the evolution of speech. *Human Evolution*, 1988, 3:19–43 (b).

LaBerge, D. and Samuels, S. J. Toward a theory of automatic information processing in reading. *Cognitive Psychology*, 1974, 6:293–323.

LaHoste, G. J.; Neveu, P. J.; Mormède, P.; and Le Moal, M. Hemispheric asymmetry in the effects of cerebral cortical ablations on mitogen-induced lymphoproliferation and plasma prolactin levels in female rats. *Brain Research*, 1989, 483:123–129.

Lambert. A. J. Right hemisphere language ability: 1. Clinical evidence. *Current Psychological Reviews*, 1982, 2:77–94 (a).

416

Lambert, A. J. Right hemisphere language ability: 2. Evidence from normal subjects. *Current Psychological Reviews*, 1982, 2:139–152 (b).

Langlois, J. H.; Ritter, J. M.; Roggman, L. A.; and Vaughn, L. S. Facial diversity and infant preferences for attractive faces. *Developmental Psychology*, 1991, 27:79–84.

Langlois, J. H.; Roggman, L. A.; Casey, R. J.; Ritter, J. M.; Rieser-Danner, L. A.; and Jenkins, V. Y. Infant preferences for attractive faces: Rudiments of a stereotype? *Developmental Psychology*, 1987, 23:363–369.

Laplane, D.; Talairach, J.; Meninger, V.; Bancaud, J.; and Orgogozo, J. M. Clinical consequences of corticectomies involving the supplementary motor cortex in man. *Journal of Neurological Science*, 1977, 34:301–314.

Larson, C. F.; Dodson, D. L.; and Ward, J. P. Hand preferences and whole body turning biases of lesser bush babies (*Galago senegalensis*). *Brain, Behaviour and Evolution*, 1989, 33:261–267.

Larson, R. Why jellyfish stick together. *Natural History*, 1991, 10:66–71.

Lassonde, M.; Bryden, M. P.; and Demers, P. The corpus callosum and cerebral speech lateralization. *Brain and Language*, 1990, 38:195–206.

Latimer, B. and Lovejoy, C. O. The calcaneus of *Australopithecus afarensis* and its implications for the evolution of bipedality. *American Journal of Physical Anthropology*, 1989, 78:369–386.

Latimer, B. and Lovejoy, C. O. Hallucal tarsometatarsal joint in *Australopithecus afarensis*. *American Journal of Physical Anthropology*, 1990, 82:125–133 (a).

Latimer, B. and Lovejoy, C. O. Metatarso-phalangeal joints of *Australopithecus afarensis*. *American Journal of Physical Anthropology*, 1990, 83:13–23 (b).

Leakey, M. D. and Hay, R. L. Pliocene footprints in the Laetolil beds at Laetoli, northern Tanzania. *Nature*, 1979, 278:317–323.

Lehman, R. A. W. Distribution and changes in strength of hand preference of cynomolgus monkeys. *Brain, Behaviour and Evolution*, 1980, 17:209–217.

Lehman, R. A. W. Hand preferences of rhesus monkeys on differing tasks. *Neuropsychologia*, 1989, 27:1193–1196.

Lehman, R. A. W. Manual preference in prosimians, monkeys and apes. In J. P. Ward (ed.), Current Behavioral Evidence of Primate Asymmetries. New York: Springer, 1992, pp. 149–182.

LeMay, M. Morphological cerebral asymmetries of modern man, fossil man and nonhuman primate. *Annals of the New York Academy of Sciences*, 1976, 280:349–366.

LeMay, M. Asymmetries of the skull and handedness. *Journal of Neurological Science*, 1977, 32:243–253.

LeMay, M. Radiological, developmental and fossil asymmetries. In N. Geschwind and A. M. Galaburda (eds.), Cerebral Dominance: The Biological Foundations. Cambridge, Mass.: Harvard University Press, 1984, pp. 26–42.

LeMay, M. Asymmetries of the brains and skulls of nonhuman primates. In S. D. Glick (ed.), Cerebral lateralization in Nonhuman Species. New York: Academic Press, 1985, pp. 233–245.

LeMay, M.; Billig, S.; and Geschwind, N. Asymmetries of the brains and skulls of nonhuman primates. In E. Armstrong and D. Falk (eds.), Primate Brain Evolution: Methods and Concepts. New York: Plenum Press, 1982, pp. 263–277.

LeMay, M. and Culebras, A. Human brain: Morphological differences in the hemispheres demonstrable by carotid arteriography. *New England Journal of Medicine*, 1972, 287:168–170.

LeMay, M. and Geschwind, N. Hemispheric difference in the brains of great apes. *Brain, Behaviour and Evolution*, 1975, 11:48–52.

Lemon, R. E. Nervous control of the syrinx of white-throated sparrows (*Zonotrichia albicollis*), *Journal of Zoology*, 1973, 171:131–140.

Leonard, G. and Milner, B. Contribution of the right frontal lobe to the encoding and recall of kinesthetic distance information. *Neuropsychologia*, 1991, 29:47–58.

Leslie, A. M. Pretence, autism and the basis of "theory of mind." *The Psychologist*, 1990, 3:120–123.

Leslie, A. M. The "theory of mind" impairment in autism: Evidence for a modular mechanism of development? In A. Whiten (ed.), Neural Theories of Mind: Evolution, Development and Simulation of Everyday Mindreading. Oxford: Blackwell, 1991, pp. 63–78.

Lesser, F. and Bown, W. Asthmatics at risk from "right-handed" drug. *New Scientist*, 1991, November 9, 8.

Leutenegger, W. Origin of hominid bipedalism. *Nature*, 1987, 325:305.

Lewin, R. Family relationships are a biological conundrum. *Science*, 1988, 242:671 (a).

Lewin, R. Linguistics in search of the mother tongue. *Science*, 1988, 242:1128–1129 (b).

Lewin, R. Human Evolution: An Illustrated Introduction. Oxford: Blackwell Scientific Publications, 1989 (a).

Lewin, R. Species questions in modern human origins. *Science*, 1989, 243:1666–1667 (b).

Lewin, R. Neanderthals puzzle the anthropologists. *New Scientist*, 1991, April 20, 17 (a).

Lewin, R. Look who's talking now. *New Scientist*, 1991, April 27, 39–42 (b).

Lewin, R. Stone age psychedelia. *New Scientist*, 1991, 1772, June 8, 24–28 (c).

Lewis-Williams, J. D. and Dowson, T. A. The signs of all times: Entoptic phenomena in Upper Palaeolithic art. *Current Anthropology*, 1988, 29:201–245.

Liberman, A. M.; Cooper, F. S.; Shankweiler, D. P.; and Studdert-Kennedy, M. Perception of the speech code. *Psychological Review*, 1967, 74:431–461.

Lieberman, P. On the evolution of human syntactic ability: Its pre-adaptive bases motor control and speech. *Journal of Human Evolution*, 1985, 14:657–668.

Lieberman, P. Language, intelligence, and rule-governed behavior. In H, J. Jerison and I. Jerison (eds.), Intelligence and Evolutionary Biology, New York: Springer, 1986, pp. 143–156.

Lieberman, P. Voice in the wilderness. *The Sciences*, 1988, July/August, 23–29.

Lieberman, P. The origins of some aspects of human language and cognition. In P. Mellars and C. Stringer (eds.), The Human Revolution: Behavioral and Biological Perspectives on the Origins of Modern Humans. Princeton: Princeton University Press, 1989, pp. 391–414.

Lieberman, P. Uniquely Human: The Evolution of Speech, Thought and Selfless Behavior. Cambridge, Mass.: Harvard University Press, 1991.

Lieberman, P.; Laitman, J. T.; Reidenberg, J. S.; Landahl, K.; and Gannon, P. J. Folk physiology and talking hyoids. *Nature*, 1989, 342:486.

Liederman, J. Is there a stage of left sided precocity during early manual specialization? In G. Young, S. J. Segalowitz, C. M. Corter and S. E. Trehub (eds.), Manual Specialization and the Developing Brain. New York: Academic Press, 1983, pp. 321–330.

Lindly, J. M. and Clark, C. A. Symbolism and modern human origins. *Current Anthropology*, 1990, 31:233–261.

Lipp, H-P.; Collins, R. L.; and Nauta, W. J. H. Structural asymmetries in brains of mice selected for strong lateralization. *Brain Research*, 1984, 310:393–396.

Lipsey, J. R. and Robinson, R. G. Sex dependent behavioral response to frontal cortical suction lesions in the rat. *Life Sciences*, 1986, 38:2185–2192.

Lloyd, G. E. R. Polarity and Analogy: Two Types of Argument in Early Greek Thought. Cambridge: Cambridge University Press, 1966.

Lockard, J. S. Handedness in a captive group of lowland gorillas. *International Journal of Primatology*, 1984, 5:356.

Lovejoy, C. O. Biomechanical perspectives on the lower limb of early hominids. In R. H. Tuttle (ed.), Primate Functional Morphology and Evolution. The Hague: Mouton, 1975, pp. 291–326.

Lovejoy, C. O. The origin of Man. *Science*, 1981, 211:341–350.

Lovejoy, C. O. Evolution of human walking. *Scientific American*, 1988, 11:82–89.

Loy, T. H. When is a stone a tool? *Australian Natural History*, 1990–1991, 23:584.

Luine, V.; Nottebohm, F.; Harding, C.; and McEwan, B. S. Androgen affects cholinergic enzymes in syringeal motor neurones and muscle. *Brain Research*, 1980, 192:89–107.

Lund, R. D. Uncrossed visual pathways of hooded and albino rats. *Science*, 1965, 149:1506–1507.

Lund, R. D.; Chang, F.-L. F.; and Land, P. W. The development of callosal projections in normal and one-eyed rats. *Development Brain Research*, 1984, 14:139–142.

McCabe, B. J. Hemispheric asymmetry of learning-induced changes. In R. J. Andrew (ed.), Neural and Behavioural Plasticity: The Use of the Domestic Chick as a Model. Oxford: Oxford University Press, 1991, pp. 262–276.

McCabe, B. J. and Horn, G. Learning and memory: Regional changes in N-methyl-D-aspartate receptors in the chick brain after imprinting. *Proceedings of the National Academy of Science*, USA, 1988, 85:2849–2853.

McCabe, B. J. and Horn, G. Synaptic transmission and recognition memory: Time course of changes in N-methyl-D-aspartate receptors after imprinting. *Behavioral Neuroscience*, 1991, 105:289–294.

McCasland, J. S. Neuronal control of bird song production. *The Journal of Neuroscience*, 1987, 7:23–39.

McCasland, J. S. and Konishi, M. Interaction between auditory and motor activities in an avian song control nucleus. *Proceedings of the National Academy of Science, U.S.A.*, 1981, 78:7815–7819.

McClelland, J.; Rumelhart, D. E.; and the PDP Research Group (eds.), Parallel Distributed Processing: Explorations in the Microstructure of Cognition, Vol. 2. Cambridge, Mass: MIT Press, 1986.

McDonnell, P. Patterns of eye–hand coordination in the first year of life. *Canadian Journal of Psychology*, 1978, 33:253–267.

McGlone, J. The neuropsychology of sex differences in human brain organization. In G. Goldstein and R. E. Tarter (eds.), Advances in Clinical Neuropsychology, Vol. 3. New York: Plenum Press, 1986, pp. 1–30.

MacFadden, S. A. and Salmelainen, P. The effect of experimentally induced myopia on the normal ground–horizon refractive difference in the pigeon. *Proceedings of the Australian Neuroscience Society*, 1990, 1:108.

McHenry, H. M. Sexual dimorphism in *Australopithecus afarensis*. *Journal of Human Evolution*, 1991, 20:21–32 (a).

McHenry, H. M. Femoral lengths and stature in Plio-Pleistocene hominids. *American Journal of Physical Anthropology*, 1991, 85:149–158 (b).

McManus, I. C. Scrotal asymmetry in man and ancient sculpture. *Nature*, 1976, 259:426.

McMullen, P. A. and Bryden, M. P. The effect of word imageability and frequency on hemsipheric asymmetry in lexical decisions. *Brain and Language*, 1987, 31:11–25.

McNaughton, B. L.; Barnes, C. A.; Rao, G.; Baldwin, J.; and Rasmussen, M. Long-term enhancement of hippocampal synaptic transmission and the acquisition of spatial information. *Journal of Neuroscience*, 1986, 6:563–571.

MacNeilage, P. F. The evolution of hemispheric specialization for manual function and language. In S. P. Wise (ed.), Higher Brain Functions: Recent Explorations of the Brain's Emergent Properties. New York: Wiley, 1987, pp. 285–309.

MacNeilage, P. F. Grasping in modern primates: The evolutionary context. In M. A. Goodale (ed.), Vision and Action: The Control of Grasping. Norwood, New Jersey: Ablex Publishing Corporation, 1990, pp. 1–13.

MacNeilage, P. F.; Studdert-Kennedy, M. G.; and Lindblom, B. Functional precursors of language and its lateralization. *American Journal of Physiology: Regulatory, Integrative and Comparative Physiology, Vol.15,* 1984, 246:R 912–914.

MacNeilage, P. F.; Studdert-Kennedy, M. G.; and Lindblom, B. Primate handedness reconsidered. *Behavioral and Brain Sciences,* 1987, 10:247–303.

MacNeilage, P. F.; Studdert-Kennedy, M. G.; and Lindblom, B. Primate handedness: A foot in the door. *Behavioral and Brain Sciences,* 1988, 11:737–746.

MacNeilage, P. F.; Studdert-Kennedy, M. G.; and Lindblom, B. Primate handedness: The other theory, the other hand, and the other attitude. *Behavioral and Brain Sciences,* 1991, 14:344–349.

Mann, A.; Lampl, M.; and Monge, J. Patterns of ontogeny in human evolution: Evidence from dental development. *Yearbook of Physical Anthropology,* 1990, 33:111–150.

Manning, J. T. and Chamberlain, A. T. The left side cradling preference in great apes. *Animal Behaviour,* 1990, 39:1224–1227.

Marchant, L. F. and McGrew, W. C. Laterality of function in apes. A meta-analysis of methods. *Journal of Human Evolution,* 1991, 21:425–438.

Marchant, L. F. and Steklis, H. D. Hand preference in a captive island group of chimpanzees (*Pan troglodytes*). *American Journal of Primatology,* 1986, 10:301–313.

Margoliash, D. and Konishi, M. Auditory representation of autogenous song in the song system of white-crowned sparrows. *Proceedings of the National Academy of Science, U.S.A.,* 1985, 82:5997–6000.

Mark, R. F. and Marotte, L.R. Australian marsupials as models for the developing mammalian visual system. *Trends in Neurosciences,* 1991, 15:51–57.

Marshack, A. Hierarchical evolution of the human capacity: The Paleolithic evidence Fifty–fourth James Arthur lecture on the evolution of the human brain. New York: American Museum of Natural History, 1985, pp. 1–52.

Marshack, A. The Neanderthals and the human capacity for symbolic thought: Cognitive and problem-solving aspects of Mousterian symbol. Paper presented at the international colloquium L'Homme de Néandertal, Liège, 1988, pp. 57–91 (a).

Marshack, A. Paleolithic calendar . . . Paleolithic images. In I. Tattersall, E. Delson, J. van Couvering (eds.), Encyclopedia of Human Evolution and Prehistory. New York: Garland Publishing, 1988, pp. 419–429 (b).

Marshack, A. Evolution of the human capacity: The symbolic evidence. *Yearbook of Physical Anthropology,* 1989, 32:1–34 (a).

Marshack, A. Methodology in the analysis and interpretation of Upper Palaeolithic image: Theory versus contextual analysis. *Rock Art Research,* 1989, 6:17–38 (b).

Marshack, A. Deliberate engravings on bone artifacts of *Homo erectus*. Further comment: A reply to Davidson on Mania and Mania. *Rock Art Research,* 1991, 8:47–58.

Marshack, A. The origin of language: An anthropological approach. In B. Chiarelli (ed.), The Origins of Language. The Hague: Mouton, in press.

Marshall, E. Paleoanthropology gets physical. *Science,* 1990, 247:798–801.

Marshall, J. C. The descent of the larynx. *Nature,* 1989, 338:702–703.

Martin, G. R. Total panoramic vision in the mallard duck, *Anas platyrhynchus*. *Vision Research,* 1986, 26:1303–1305.

Martin, R. D. Allometric approaches to the evolution of the primate nervous system. In E.

Armstrong and D. Falk (eds.), Primate Brain Evolution: Methods and Concepts. New York: Plenum, 1982, pp. 39–56.

Marzke, M. W.; Longhill, J. M.; and Rasmussen, S. A. Gluteus maximus muscle function and the origin of hominid bipedality. *American Journal of Physical Anthropology*, 1988, 77:519–528.

Masataka, N. Population level asymmetry of hand preference in lemurs. *Behaviour*, 1989, 110:1–4.

Masataka, N. Handedness of capuchin monkeys. *Folia Primatologica*, 1990, 55:189–192.

Mason, S. The left hand of nature. *New Scientist*, 1984, 101, January 19, 10–14.

Mathews, L. J. The Natural History of the Whale. London: Weidenfeld and Nicholson, 1978.

Matoba, M.; Masataka, N.; and Tanioka, Y. Cross generational continuity of hand use preference in marmosets. *Behaviour*, 1991, 117:281–286.

Maurus, J. A cometary source for cock-eyed molecules? *Science*, 1991, 254:1110–1111.

Maurus, M.; Barclay, D.; and Streit, K.-M. Acoustic patterns common to human communication between monkeys. *Language and Communication*, 1988, 8:87–94.

Mebert, C. J. and Michel, G. F. Handedness in artists. In J. Herron (ed.), Neuropsychology of Left Handedness. New York: Academic Press, 1980, pp. 273–280.

Megirian, D.; Weller, L.; Martin, G. F.; and Watson, C. R. R. Aspects of laterality in the marsupial *Trichosurus vulpecula* (brush-tailed possum). *Annals of the New York Academy of Sciences*, 1977, 299:197–212.

Mehta, Z. and Newcombe, F. A role for the left hemisphere in spatial processing. *Cortex*, 1991, 27:153–167.

Meier, R. P. Language acquisition by deaf children. *American Scientist*, 1991, 79:60–70.

Melekian, B. Lateralization in the newborn at birth: Asymmetry of the stepping reflex. *Neuropsychologia*, 1981, 19:707–711.

Mellars, P. Major issues in the emergence of modern humans. *Current Anthropology*, 1989, 30:349–385 (a).

Mellars, P. Technological changes at the Middle-Upper Palaeolithic transition: Economic, social and cognitive perspectives. In P. Mellars and C. Stringer (eds.), The Human Revolution: Behavioral and Biological Perspectives on the Origins of Modern Humans. Princeton: Princeton University Press, 1989, pp. 338–365 (b).

Melone, J. H.; Teitelbaum, S. A.; Johnson, R. E.; and Diamond, M. C. The rat amygdaloid nucleus: A morphometric right–left study. *Experimental Neurology*, 1984, 86:293–302.

Melosik, M. and Szuba, Z. Morphometric studies of the bovine eyeball during early fetal period. *Folia Morphologica (Warsz.)*, 1986, 45:19–28.

Melsbach, G.; Hahmann, U.; Waldmann, C.; Wörtwein, G.; and Güntürkün, O. Morphological asymmetries of the optic tectum of the pigeon. In N. Elsner and H. Penzlin (eds.), Synapse-Transmission-Modulation. Stuttgart: Thiem, 1991, p. 553.

Meltzoff, A. N. and Moore, M. K. Newborn infants imitate adult facial gestures. *Child Development*, 1983, 54:702–709.

Mench, J. A. and Andrew, R. J. Lateralization of a food search task in the domestic chick. *Behavioral and Neural Biology*, 1986, 46:107–114.

Menzel, E. W. A group of young chimpanzees in a 1-acre field: Leadership and communication. In R. Byrne and A. Whiten (eds.), Machiavellian Intelligence: Social Expertise and the Evolution of Intellect in Monkeys, Apes and Humans. Oxford: Oxford University Press, 1988, pp. 155–159.

Mercier, N.; Valladas, H.; Joron, J.-L.; Reyss, J.-L.; Lévêque, F.; and Vandermeersch, B. Thermoluminescence dating of the late Neanderthal remains from Saint-Césaire. *Nature*, 1991, 351:737–739.

Merzenich, M. M. Dynamic neocortical processes and the origins of higher brain functions. In J.-P. Changeux and M. Konishi (eds.), The Neural and Molecular Bases of Learning. Chichester: Wiley, 1987, pp. 337–358.

Mesulam, M.-M. Large-scale neurocognitive networks and distributed processing for attention, language, and memory. Annals of Neurology, 1990, 28:597–613.

Meulenbrock, R. G. J. and Van Galen, G. P. The acquisition of skilled handwriting: Discontinuous trends in kinematic variables. In A. M. Colley and J. R. Beech (eds.), Cognition and Action in Skilled Behaviour. Amsterdam: North Holland, 1988, pp. 273–281.

Miceli, D.; Peyrichoux, J.; Repérant, J.; and Weidner, C. Étude anatomique des afférences du teléncéphale rostral chez le poussin de Gallus domesticus L. Journal Hirnforschung, 1980, 21:627–646.

Miceli, D. and Repérant, J. Thalamo-hyperstriatal projections in the pigeon Columa livia as demonstrated by retrograde double-labeling with fluorscent tracers. Brain Research, 1982, 245:365–371.

Michel, G. F. and Harkins, D. A. Concordance of handedness between teacher and student facilitates learning manual skills. Journal of Human Evolution, 1985, 4:597–601.

Michel, G. and Harkins, D. A. Postural and lateral asymmetries in the ontogeny of handedness during infancy. Developmental Psychobiology, 1986, 19:247–258.

Miller, G. Language and Speech. San Francisco: Freeman, 1981.

Miller, J. A. Does brain size variability provide evidence of multiple species in Homo habilis? American Journal of Physical Anthropology, 1991, 84:385–398.

Milliken, G. W.; Forsythe, C.; and Ward, J. P. Multiple measures of hand-use lateralization in the ring-tailed lemur (Lemur catta). Journal of Comparative Psychology, 1989, 103:262–268.

Milliken, G. W.; Stafford, D. K.; Dodson, D. L.; Pinger, C. D.; and Ward, J. P. Analyses of feeding lateralization in the small-eared bush baby (Otolemur gurnettii): A comparison with the ring-tailed lemur (Lemur catta). Journal of Comparative Psychology, 1991, 105:274–285.

Milton, K. Foraging behaviour and the evolution of primate intelligence. In R. Byrne and A. Whiten (eds.), Machiavellian Intelligence: Social Expertise and the Evolution of Intellect in Monkeys, Apes and Humans. Oxford: Oxford University Press, 1988, pp. 284–305.

Mitchell, I. J.; Brotchie, J. M.; Brown, G. D. A.; and Crossman, A. R. Modeling the functional organization of the basal ganglia. A parallel distributed processing approach. Movement Disorders, 1991, 6:189–204.

Mittleman, G.; Whishaw, I. Q.; and Robbins, T. W. Cortical lateralization of function in rats in a visual reaction time task. Behavioural Brain Research, 1988, 31:29–36.

Mittwoch, U. To be right is to be born male. New Scientist, 1977, 73:74–77.

Mittwoch, U. Erroneous theories of sex determination. Journal of Medical Genetics, 1985, 22:164–170.

Mittwoch, U. One gene may not make a man. New Scientist, 1990, November 10, 32–34.

Miyamoto, M.; Slightom, J. L.; and Goodman, M. Phylogenetic relations of humans and African apes from DNA sequences in the psi-eta-globin region. Science, 1987, 238:369–373.

Molfese, D. L. and Molfese, V. J. Hemispheric specialization in infancy. In G. Young, S. J. Segalowitz, C. M. Corter, and S. Trehub (eds.), Manual Specialization and the Developing Brain. New York: Academic Press, 1983, pp. 375–394.

Moody, D. B.; Stebbins, W. C.; and May, B. J. Auditory perception of communicatory signals by Japanese monkeys. In W. C. Stebbins and M. A. Berkley (eds.), Comparative Perception, Vol. II: Complex Signals. New York: Wiley, 1990, pp. 311–344.

Moore, C. L. Maternal behavior of rats is affected by hormonal condition of pups. *Journal of Comparative and Physiological Psychology*, 1982, 96:123–129.

Moore, C. L. Maternal contributions to the development of masculine sexual behavior in laboratory rats. *Developmental Psychobiology*, 1984, 17:347–356.

Morgan, E. The aquatic hypothesis. *New Scientist*, 1984, April 12, 11–13.

Morgan, M. J. and Corballis, M. C. On the biological basis of human laterality: II. The mechanisms of inheritance. *Behavioral and Brain Sciences*, 1978, 2:270–277.

Morgan, M. J. and McManus, I. C. The relationship between braindedness and handedness. In F. C. Rose, R. Whurr and M. A. Wyke (eds.), Aphasia. London: Whurr Publishers, 1988, pp. 85–130.

Morgan, M. J.; O'Donnell, J. M.; and Oliver, R. F. Development of left–right asymmetry in the habenular nuclei of *Rana temporaria*. *Journal of Comparative Neurology*, 1973, 149:203–214.

Morris, R. D. and Hopkins, W. D. Perception of human chimeric faces by chimpanzees: Evidence for a right-hemisphere asymmetry. Submitted.

Morris, R. D.; Hopkins, W. D.; Gilmore, L. B.; and Washburn, D. A. Behavioral lateralization in language-trained chimpanzees. In J. Ward (ed.), Current Behavioral Evidence of Primate Asymmetries. New York: Springer, 1992, pp. 207–234.

Morris, R. D.; Hopkins, W. D.; Washburn, D. A.; and Bolser, L. Hemispheric activation in the chimpanzee: Evidence for a left hemispheric asymmetry for known symbols. In preparation.

Morse, P. A.; Molfese, D.; Laughlin, N. K.; Linnville, S.; and Wetzel, F. Categorical perception for voicing contrasts in normal and lead treated rhesus monkeys: Electrophysiological indices. *Brain and Language*, 1987, 30:63–80.

Moscovitch, M. Information processing and the cerebral hemispheres. In M. S. Gazzaniga (ed.), Handbook of Behavioral Neurobiology, Vol. 2.: Neuropsychology. New York: Plenum Press, 1979, pp. 379–446.

Moscovitch, M.; Scullion, D.; and Christie, D. Early versus late stages of processing and their relation to functional hemispheric asymmetries in face recognition. *Journal of Experimental Psychology: Human Perception and Performance*, 1976, 2:401–416.

Mukhametov, L. M.; Lyamin, O. I.; and Polyyakova, I. G. Interhemispheric asynchrony of the sleep EEG in northern fur seals. *Experientia*, 1985, 41:1034–1035.

Mukhametov, L. M.; Supin, A. Y.; and Polyakova, I. G. Interhemispheric asymmetry of the electroencephalographic sleep patterns in dolphins. *Brain Research*, 1977, 134:581.

Myhrer, T. and Iversen, E. G. Changes in retention of a visual discrimination task following unilateral and bilateral transections of temporo-entorhinal connections in rats. *Brain Research Bulletin*, 1990, 25:293–298.

Myslobodsky, M. (ed.), Hemisyndromes: Psychobiology, Neurology, Psychiatry. New York: Academic Press, 1983.

Nance, W. E. Anencephaly and spina bifida: A possible example of cytoplasmic inheritance in man. *Nature*, 1969, 224:373–375.

Narang, H. K. Right-left asymmetry of myelin development in epiretinal portion of rabbit optic nerve. *Nature*, 1977, 266:855–856.

Narang, H. K. and Wisniewski, H. M. The sequence of myelination in the epiretinal portion of the optic nerve in the rabbit. *Neuropathology and Applied Neurobiology*, 1977, 3:15–27.

Nathan, P. W.; Smith, M. C.; and Deacon, P. The corticospinal tracts in man: Course and location of fibers at different segmental levels. *Brain*, 1990, 113:303–324.

Nelson, J. M.; Phillips, R.; and Goldstein, L. Interhemispheric EEG laterality relationships following psychoactive agents and during operant performance in rabbits. In S. Harnard, R. W. Doty, L. Goldstein, J. Jaynes and G. Krauthamer, (eds.), Lateralization in the Nervous System. New York: Academic Press, 1977, pp. 451–470.

Neveu, P. J. Cerebral neocortex modulation of immune functions. *Life Sciences*, 1988, 42:1917–1923.

Neveu, P. J. Asymmetrical brain modulation of immune reactivity in mice: A model for studying interindividual differences and physiological population heterogeneity? *Life Sciences*, 1992, 50:1–6.

Neveu, P. J.; Barnéoud, P.; Vitiello, S.; Betancur, C.; and LeMoal, M. Brain modulation of the immune system: Association between lymphocyte responsiveness and paw preference in mice. *Brain Research*, 1988, 157:392–394.

Neveu, P. J.; Betancur, C.; Barnéoud, P.; Preud'homme, J. L.; Aucouturier, P.; LeMoal, M.; and Vitiello, S. Functional brain asymmetry and murine systemic lupus erythematosus. *Brain Research*, 1989, 498:159–162.

Neveu, P. J.; Betancur, C.; Barnéoud S.; Vitiello S.; and LeMoal, M. Functional brain asymmetry and lymphocyte proliferation in female mice: Effects of right- and left-cortical ablation. *Brain Research*, 1991, 550.125–128.

Neville, A. C. Animal asymmetry; The Institute of Biology's Studies on Biology No. 67. London: Edward Arnold, 1976.

Newton, P. N. and Nishida, T. Possible buccal administration of herbal drugs by wild chimpanzees, *Pan troglodytes*. *Animal Behaviour*, 1990, 39:798–801.

Noble, S. B. The emergence of human intelligence and of civilization. *Language Origins Society*, Volendam, Holland, 1990.

Noble, W. and Davidson, I. The evolutionary emergence of modern human behavior: Language and its archaeology. *Man*, 1991, 26:223–253.

Noller, N. J. Eye enlargement in the chick following visual deprivation. M.Sc. dissertation (unpublished). University of Melbourne, Australia, 1984.

Noonan, M. and Axelrod, S. Improved acquisition of left-right response differentiation in the rat following section of the corpus callosum. *Behavioural Brain Research*, 1991, 46:135–142.

Norberg, R. Å. Occurrence and independent evolution of bilateral ear asymmetry in owls and implications on owl taxonomy. *Philosophical Transactions of the Royal Society of London*, B., 1977, 280(973):375–408.

Norberg, R. Å. Skull asymmetry, ear structure and function, and auditory localization in Tengmalm's owl, *Aegolius funereus* (Linné). *Philosophical Transactions of the Royal Society of London*, B., 1978, 282(991):325–410.

Nordeen, E. J. and Nordeen, K. W. Neurogenesis and sensitive periods in avian song learning. *Trends in Neuroscience*, 1990, 13:31–36.

Nordeen, E. J. and Yahr, P. Hemispheric asymmetries in the behavioral and hormonal effects of sexually differentiating mammalian brain. *Science*, 1982, 218:391–394.

Nosten, M.; Roubertoux, P.; Degrelle, H.; and Leoyer, M. Effect of the TFM mutation on handedness in mice. *Journal of Endocrinology*, 1989, 121:R5–R7.

Nottebohm, F. Ontogeny of bird song. *Science*, 1970, 167:950–956.

Nottebohm, F. Neural lateralization of vocal control in a Passerine bird. I. Song. *The Journal of Experimental Zoology*, 1971, 177:229–261.

Nottebohm, F. Neural lateralization of vocal control in a passerine bird. II. Sub-song, calls, and a theory of vocal learning. *The Journal of Experimental Zoology*, 1972, 179:35–50.

Nottebohm, F. Vocal behavior in birds. In D.S. Farner and J.R. King (eds), Avian Biology, Vol. 5. New York: Academic Press, 1975, pp. 287–221.

Nottebohm, F. Phonation in the orange-winged amazon parrot, *Amazona amazonica*. *Journal of Comparative Physiology*, 1976, 108:157–170.

Nottebohm, F. Asymmetries in neural control of vocalization in the canary. In S. Harnard, R. W. Doty, L. Goldstein, J. Jaynes and G. Krauthamer (eds), Lateralization in the Nervous System. New York: Academic Press, 1977, pp. 23–44.

Nottebohm, F. Origins and mechanisms in the establishment of cerebral dominance. In M. S. Gazzaniga (ed.), Handbook of Behavioral Neurobiology, Vol. 2, Neuropsychology. New York: Plenum, 1979, pp. 295–344.

Nottebohm, F. Plasticity in adult avian central nervous system: Possible relation between hormones, learning, and brain repair. In V.B. Mountcastle, F. Plum and S. R. Geiger (eds.), Handbook of Physiology V; Section 1. The Nervous System. Maryland: American Physiological Society, 1987, pp. 85–108.

Nottebohm, F. From bird song to neurogenesis. Scientific American, 1989, February, 55–61.

Nottebohm, F.; Alvarez-Buylla, A.; Cynx, J.; Kirn, J.; Ling, C. Y.; and Nottebohm, M. Song learning in birds: The relation between perceptions and production. Philosophical Transactions of the Royal Society of London B, 1990, 329:115–124.

Nottebohm, F. and Arnold, A. P. Sexual dimorphism in vocal control areas of the songbird brain. Science, 1976, 194:211–213.

Nottebohm, F.; Manning, E.; and Nottebohm, M. E. Reversal of hypoglossal dominance in canaries following unilateral syringeal denervation. Comparative Physiology, 1979, 134:227–240.

Nottebohm, F. and Nottebohm, M. E. Left hypoglossal dominance in the control of canary and white-crowned sparrow song. Journal of Comparative Physiology, 1976, 108:171–192.

Nottebohm, F. and Nottebohm, M. E. Relationship between song repertive and age in the canary, Serinus canarius. Zeitschrift für Tierpsychologie, 1978, 46:298–305.

Nottebohm, F.; Stokes, T. M.; and Leonard, C. M. Central control of song in the canary, Serinus canarius. Journal of Comparative Neurology, 1976, 165:457–486.

Nowicki, S. and Capranica, R. R. Bilateral syringeal interaction in vocal production of an oscine bird sound. Science, 1986, 231:1297–1299.

Oakley, K. P. Skill as a human possession. In S. L. Washburn and P. Dolhinow (eds.), Perspectives on Human Evolution, Vol. 2. New York: Holt, Rinehart & Winston, Inc., 1972, pp. 14–50.

Oakley, K. P. Emergence of higher thought, 3.0—0.2 Ma B.P. Philosophical Transactions of the Royal Society of London B, 1981, 292:205–211.

Ojemann, G. A. Brain organization for language from the perspective of electrical stimulation mapping. Behavioral and Brain Sciences, 1983, 6:189–230.

Olavarria, J. and van Sluyters, C. Organization and postnatal development of callosal connections in the visual cortex of the rat. Journal of Comparative Neurology, 1985, 239:1–26.

Oldfield, R. C. Handedness in musicians. British Journal of Psychology, 1969, 60:91–99.

O'Leary, D. D. M.; Stanfield, B. B.; and Cowan, W. M. Evidence that the early postnatal restriction of the cells of origin of the callosal projection is due to the elimination of axonal collaterals rather than to the death of neurons. Developmental Brain Research, 1981, 1:607–617.

Olson, D. A.; Ellis, J. E.; and Nadler, R. D. Hand preferences in captive gorillas, orangutans and gibbons. American Journal of Primatology, 1990, 20:83–94.

Overman, W. H. and Doty, R. W. Hemispheric specialization displayed by man but not macaques for analysis of faces. Neuropsychologia, 1982, 20:113–128.

Pandya, D. N. and Yeterian, E. H. Architecture and connections of cerebral cortex: Implications for brain evolution and function. In A. B. Scheibel and A. F. Wechsler (eds.), Neurobiology of Higher Cognitive Function. Hillsdale, N.J.: Erlbaum, 1990, pp. 53–84.

Parker, S. T. and Gibson, K. R. Object manipulation, tool use and sensorimotor intelligence as feeding adaptations in cebus monkeys and great apes. Journal of Human Evolution, 1977, 6:623–641.

Parker, S. T. and Gibson, K. R. A developmental model for the evolution of language and intelligence in early hominids. Behavioral and Brain Sciences, 1979, 2:367–408.

Parsons, C. and Rogers, J. L. Role of the tectal and posterior commissures in lateralization in the avian brain, *Behavioural Brain Research* submitted.

Patel, S. N. and Stewart, M. G. Changes in the number and structure of dendritic spines, 25h after passive avoidance training in the domestic chick, *Gallus domesticus. Brain Research*, 1988, 449:34–46.

Patterson, K. and Besner, D. Is the right hemisphere literate? *Cognitive Neuropsychology*, 1984, 1:315–341 (a).

Patterson, K. and Besner, D. Reading from the left: A reply to Rabinowicz and Moscovitch and to Zaidel and Schweiger. *Cognitive Neuropsychology*, 1984, 1:365–380 (b).

Patterson, K. and Bradshaw, J. L. Differential hemispheric mediation of nonverbal visual stimuli. *Journal of Experimental Psychology: Human Perception and Performance*, 1975, 1:246–252.

Patterson, T. A.; Alvarado, M. C.; Warner, I. T.; Bennett, E. L.; and Rosenzweig, M. R. Memory stages and brain asymmetry in chick learning. *Behavioral Neuroscience*, 1986, 100:856–865.

Patterson, T. A.; Gilbert, D. B.; and Rose, S. P. R. Pre- and post-training lesions of the intermediate medial hyperstriatum ventrale and passive avoidance learning in the chick. *Experimental Brain Research*, 1990, 80:189–195.

Pearlson, G. D. and Robinson, R. G. Suction lesions of the frontal cortex in the rat induce asymmetrical behavioral and catecholaminergic responses. *Brain Research*, 1981, 213:233–242.

Pepperberg, I. M. Tool use in birds: An avian monkey wrench? *Behavioral and Brain Sciences*, 1989, 12:604–605.

Pepperberg, I. M. Some cognitive capacities of an African Grey parrot (*Psittacus erithacus*). *Advances in the Study of Behavior*, 1990, 19:357–409.

Pepperberg, I. M. A communicative approach to animal cognition: A study of conceptual abilities of an African grey parrot. In C. A. Ristau (ed.), Cognitive Ethology: The Minds of Other Animals. Hillsdale, N. J.: Erlbaum, 1991, pp. 153–178.

Pepperberg, I. M. and Kozak, F. A. Object permanence in the African Grey parrot (*Psittacus erithacus*). *Animal Learning and Behavior*, 1986, 14:322–330.

Perello, J. Digressions on the biological foundations of language. *Journal of Communication Disorders*, 1970, 3:140–149.

Peretz, I. and Morais, J. Determinants of laterality for music: Towards an information processing account. In K. Hugdhal (ed.), Handbook of Dichotic Listening: Theory, Methods and Research. New York: Wiley, 1988, pp. 323–358.

Pérez, H.; Ruiz, S.; Hernández, A.; and Soto-Moyano, R. Asymmetry of interhemispheric responses evoked in the prefrontal cortex of the rat. *Journal of Neuroscience Research*, 1990, 25:139–142.

Perrett, D. I.; Mistlin, A. J.; Chitty, A. J.; Smith, P. A. J.; Potter, D. D.; Broennimann, R.; and Harries, M. Specialized face processing and hemispheric asymmetry in man and monkeys: Evidence from single unit and reaction time studies. *Behavioural Brain Research*, 1988, 29:245–258.

Peters, M. The primate mouth as an agent of manipulation and its relation to human handedness. *Behavioral and Brain Sciences*, 1988, 11:729.

Peters, M. and Petrie, B. F. Functional asymmetry in the stepping reflex of human neonates. *Canadian Journal of Psychology*, 1979, 33:198–200.

Petersen, M. R.; Beecher, M. D.; Zoloth, S. R.; Green, S.; Marler, P. R.; Moody, D. B.; and Stebbins, W. C. Neural lateralization of vocalizations by Japanese macaques: Communicative significance is more important than acoustic structure. *Behavioral Neuroscience*, 1984, 98:779–790.

Petersen, M.; Beecher, M.; Zoloth, S.; Moody, D.; and Stebbins, W. Neural lateralization of species-specific vocalizations by Japanese macaques (*Macaca fuscata*). *Science*, 1978, 202:324–327.

Petersen, S.; Fox, P.; Snyder, A.; and Raichle, M. Activation of extra-striate and frontal cortical areas by visual words and wordlike stimuli. *Science*, 1990, 249:1041–1043.

Peterson, G. M. Mechanisms of handedness in the rat. *Comparative Psychology Monographs*, 1934, 9(6):1–67.

Petito, L. A. and Marentette, P. F. Babbling in the manual mode: Evidence for the ontogeny of language. *Science*, 1991, 251:1493–1496.

Phillips, R. E. and Youngren, O. M. Unilateral kainic acid lesions reveal dominance of right archistriatum in avian fear behavior. *Brain Research*, 1986, 377:216–220.

Pickford, M. The evolution of intelligence: A palaeontological perspective. In H. J. Jerison and I. Jerison (eds.), Intelligence and Evolutionary Biology. New York: Springer, 1986, pp. 175–198.

Pinker, S. Rules of language. *Science*, 1991, 253:530–535.

Pinker, S. and Bloom, P. Natural language and natural selection. *Behavioral and Brain Sciences*, 1990, 13:707–784.

Plante, E. MRI findings in the parents and siblings of language impaired boys. *Brain and Language*, 1991, 41:67–80.

Plato, C. C.; Wood, J. L.; and Norris, A. H. Bilateral asymmetry in bone measurements of the hand and lateral hand dominance. *American Journal of Physical Anthropology*, 1980, 52:27–31.

Pohl, P. Central auditory processing V: Ear advantages for acoustic stimuli in baboons. *Brain and Language*, 1983, 20:44–53.

Pohl, P. Ear advantages for temporal resolution in baboons. *Brain and Cognition*, 1984, 3:438–441.

Poizner, H.; Klima, E. S.; and Bellugi, U. What the Hands Reveal About the Brain. Cambridge, MA: MIT/Bradford, 1987.

Policansky, D. Flatfishes and the inheritance of asymmetry. *Behavioral and Brain Sciences*, 1982, 5:262–266 (a).

Policansky, D. The asymmetry of flounders. *Scientific American*, 1982, 246, May, 96–102 (b).

Pons, T. P.; Garraghty, P. E.; Ommaya, A. K.; Kaas, J. H.; Taub, E.; and Mishkin, M. Massive cortical reorganization after sensory deafferentation in adult macaques. *Science*, 1991, 252:1857–1860.

Pope, G. G. Bamboo and human evolution. *Natural History*, 1989, October, 48–59.

Pope, G. G. Evolution of the zygomatico–maxillary region in the genus *Homo* and its relevance to the origin of modern humans. *Journal of Human Evolution*, 1991, 21:189–213.

Post, D. G.; Hausfater, G.; and McCuskey, S. A. Feeding behavior of yellow baboons (*Papio cynocephalus*): Relationship to age, gender and dominance rank. *Folia Primatologica*, 1980, 34:170–195.

Potts, R. Early Hominid Activities at Olduvai. New York: A. de Gruyter, 1988.

Povinelli, D. J. Failure to find self-recognition in Asian elephants (*Elephas maximus*) in contrast to their use of mirror cues to discover hidden food. *Journal of Comparative Psychology*, 1989, 103:122–131.

Preilowski, B.; Reger, M.; and Engele, H. C. Handedness and cerebral asymmetry in nonhuman primates. In D. M. Taub and F. A. King (eds.), Current Perspectives in Primate Biology. New York: Van Nostrand & Reinhold, 1986, pp. 270–282.

Premack, D. Gavagai or the Future History of the Animal Language Controversy. Cambridge, Mass.: M.I.T. Press, 1986.

Premack, D. "Does the chimpanzee have a theory of mind?" revisited. In R. Byrne and A.

Whiten (eds.), Machiavellian Intelligence: Social Expertise and the Evolution of Intellect in Monkeys, Apes and Humans. Oxford: Oxford University Press, 1988, pp. 160–179.

Proudfoot, R. E. Hemiretinal differences in face recognition: Accuracy versus reaction time. *Brain and Cognition*, 1983, 2:25–31.

Provins, K. A. Motor skills, handedness and behavior. *Australian Journal of Psychology*, 1967, 19:137–150.

Puech, P.-F. and Cianfarani, F. Interproximal grooving of teeth: Additional evidence and interpretation. *Current Anthropology*, 1988, 29:665–668.

Radner, D. and Radner, M. Animal Consciousness. Buffalo, N. Y.: Prometheus Books, 1989.

Raijmakers, G. M. J. Het onstaan van spraak en taal bij de primaten. Unpublished Ph.D. thesis, Free University of Amsterdam, 1990.

Rajendra, S. and Rogers, L. J. Asymmetry is present in the thalamofugal visual projections of female chicks. *Experimental Brain Research*, submitted.

Rak, Y. On the differences between two pelvises of Mousterian context from the Qafzeh and Kebara caves, Israel. *American Journal of Physical Anthropology*, 1990, 81:323–332.

Rak, Y. Lucy's pelvic anatomy: Its role in bipedal gait. *Journal of Human Evolution*, 1991, 20:283–290.

Rapacz, J.; Chen, L.; Butler-Brunner, E.; Wu, M.-J.; Hasler-Rapacz, J. U.; Butler, R.; and Schumaker, V. N. Identification of the ancestral haplotype for apolipoprotein B suggests an African origin of *Homo sapiens sapiens* and traces their subsequent migration to Europe and the Pacific. *Proceedings of the National Academy of Science*, 1991, 88:1403–1406.

Rashid, N. and Andrew, R. J. Right hemisphere advantage for topographical orientation in the domestic chick. *Neuropsychologia*, 1989, 27:937–948.

Rauschecker, J. P. and Hahn, S. Ketamine oxylazine anaesthesia blocks consolidation of ocular dominance changes in the kitten visual cortex. *Nature*, London, 1987, 326:138–185.

Reist, J. D.; Bodaly, R. A.; Fudge, R. J. P.; Cash, K. J.; and Stevens, T. V. External scarring of whitefish, *Coregonus nasus* and *C. clupeaformis* complex, from the western Northwest Territories, Canada. *Canadian Journal of Zoology*, 1987, 65:1230–1239.

Renfrew, C. Archaeology and Language: The Puzzle of Indo-European Origins. London: J. Cape, 1987.

Renoux, G.; Biziere, K.; Renoux, M.; Guillaumin, S. M.; and Degenne, D. A. A balanced brain asymmetry modulates T cell mediated events. *Journal of Neuroimmunology*, 1983, 5:227–238.

Reynolds, G. P.; Czudek, C.; Browej, N.; and Seeman, P. Dopamine receptor asymmetry in schizophrenia. *The Lancet*, 1987, April 25, 978–979.

Ridgway, S. H. Physiological observation on dolphins' brains. In R. J. Schusterman, J. A. Thomas and F. G. Wood (eds.), Dolphin Cognition and Behavior: A Comparative Approach. Hillsdale, N. J. Erlbaum, 1986, pp. 31–60.

Rightmire, G. P. *Homo erectus* and later Middle Pleistocene humans. *Annual Review of Anthropology*, 1988, 17:239–259.

Rightmire, G. P. The Evolution of *Homo erectus*: Comparative Anatomical Studies of an Extinct Human Species. Cambridge: Cambridge University Press, 1990.

Rightmire, G. P. and Deacon, H. J. Comparative studies of Late Pleistocene human remains from Klasies River Mouth, South Africa. *Journal of Human Evolution*, 1991, 20:131–156.

Ristau, C. A. Before mindreading: Attention, purposes and deception in birds. In A. Whiten (ed.), Natural Theories of Mind: Evolution, Development and Simulation of Everyday Mindreading. Oxford: Blackwell, 1991, pp. 195–207.

Roberts, R. G.; Jones, R.; and Smith, M. A. Thermoluminescence dating of a 50,000-year-old human occupation site in northern Australia. *Nature*, 1990, 345:153–156.

Robinson, R. G. Differential behavioral effects of right versus left hemispheric cerebral infarction: Evidence for cerebral lateralization in the rat. *Science*, 1979, 205:707–710.

Robinson, T. E. Variation in the pattern of behavioral and brain asymmetries due to sex differences. In S. D. Glick (ed.), Cerebral Lateralization in Nonhuman Species. New York: Academic Press, 1985, pp. 185–231.

Robinson, T. E. and Becker, J. B. Variation in lateralization: Selected examples do not a population make. *Brain Science*, 1981, 4:34–35.

Robinson, T. E.; Becker, J. B.; and Camp, D. M. Sex differences in behavioral and brain asymmetries. In M. Myslobodsky (ed.), Hemisyndromes: Psychobiology, Neurology, Psychiatry. New York: Academic Press, 1983, pp. 91–128.

Robinson, T. E.; Becker, J. B.; Camp, D. M.; and Mansour, A. Variation in the pattern of behavioral and brain asymmetries due to sex differences. In S. D. Glick (ed.), Cerebral Lateralization in Nonhuman Species. New York: Academic Press, 1985, pp.185–231.

Rodman, P. S. and McHenry, H. M. Bioenergetics and the origin of hominid bipedalism. *American Journal of Physical Anthropology*, 1980, 52:103–106.

Rogers, L. J. Functional lateralization in the chicken forebrain revealed by cycloheximide treatment. In R. Nohring (ed.), Acta XVI Congressus Ornithologici, Berlin, 1978, Vol. 1. Berlin: Deutsche Ornithologen-Gesellschaft, 1980, pp. 653–659 (a).

Rogers, L. J. Lateralization in the avian brain. *Bird Behaviour*, 1980, 2:1–12 (b).

Rogers, L. J. Environmental influences on brain lateralization. *Behavioral and Brain Sciences*, 1981, 4:35–36.

Rogers, L. J. Light experience and asymmetry of brain function in chickens. *Nature* (London), 1982, 297:223–225.

Rogers, L. J. Lateralization of learning in chicks. *Advances in the Study of Behavior*, 1986, 16:147–189.

Rogers, L. J. Laterality in animals. *The International Journal of Comparative Psychology*, 1989, 3:5–25.

Rogers, L. J. Light input and the reversal of functional lateralization in the chicken brain. *Behavioural Brain Research*, 1990, 38:211–221.

Rogers, L. J. Development of lateralization. In R.J. Andrew (ed.), Neural and Behavioural Plasticity: The Use of the Domestic Chick as a Model. Oxford: Oxford University Press, 1991, pp. 507–535.

Rogers, L. J. and Anson, J. M. Lateralization of function in the chicken forebrain. *Pharmacology, Biochemistry and Behavior*, 1979, 10:679–686.

Rogers, L. J. and Bolden, S. W. Light-dependent development and asymmetry of visual projections. *Neuroscience Letters*, 1991, 121:63–67.

Rogers, L. J.; Drennen, H. D.; and Mark, R. F. Inhibition of memory formation in the imprinting of memory formation in the imprinting period: Irreversible action of cycloheximide in young chickens. *Brain Research*, 1974, 79:213–233.

Rogers, L. J. and Ehrlich, D. Asymmetry in the chicken forebrain during development and the possible role of the supraoptic decussation. *Neuroscience Letters*, 1983, 37:123–127.

Rogers, L. J. and Hambley, J. W. Specific and nonspecific effects of neuro-excitatory amino acids on learning and other behaviours in the chicken. *Behavioural Brain Research*, 1982, 4:1–18.

Rogers, L. J. and Kaplan, G. Observations of the orangutan at Sepilok rehabilitation center, June/July 1991. Paper presented at the 22nd International Ethological Conference, Kyoto, Japan, August 1991.

Rogers, L. J. and Rajendra, S. Modulation of the development of light-initiated asymmetry in chick thalamofugal visual projections by estradiol. *Experimental Brain Research*, submitted.

Rogers, L. J. and Sink, H. S. Transient asymmetry in the projections of the rostral thalamus to the visual hyperstriatum of the chicken, and reversal of its direction by light exposure. *Experimental Brain Research*, 1988, 70:378–384.

Rogers, L. J. and Workman, L. Light exposure during incubation affects competitive behavior in domestic chicks. *Applied Animal Behaviour Science*, 1989, 23:187–198.

Rogers, L. J. and Workman, L. Footedness in birds. *Animal Behaviour*, In press.

Rogers, L. J.; Zappia, J. V.; and Bullock, S. P. Testosterone and eye–brain asymmetry for copulation in chickens. *Experientia*, 1985, 1:1447–1449.

Rose, S. P. R. Can memory be the brain's rosetta stone? In R. M. J. Cotterill, (ed.), Models of Brain Function. Cambridge: Cambridge University Press, 1989, pp.1–13.

Rose, S. P. R. How chicks make memories, the cellular cascade from c-fos to dendritic modelling. *Trends in Neurosciences*, 1991, 14:390–397.

Rose, S. P. R. and Csillag, A. Passive avoidance training results in lasting changes in deoxyglucose metabolism in left hemisphere regions of chick brain. *Behavioral and Neural Biology*, 1985, 44:315–324.

Rose, S. P. R. and Jork, R. Long-term memory formation in chicks is blocked by 2-deoxygalactose, a fucose analog. *Behavioral and Neural Biology*, 1987, 48:246–258.

Rosen, G. D.; Berrebi, A. S.; Yutzey, D. A.; and Denenberg, V. H. Prenatal testosterone causes shift of asymmetry in neonatal tail posture of the rat. *Developmental Brain Research*, 1983, 9:99–101.

Rosen, G. D.; Galaburda, A. M.; and Sherman, G. F. The ontogeny of anatomic asymmetry: Constraints derived from basic mechanisms. In A. B. Scheibel and A. F. Wechsler (eds.), Neurobiology of Higher Cognitive Function. New York: Guilford Press, 1990, pp. 215–238.

Rosen, G. D.; Sherman, G. F.; and Galaburda A. M. Interhemispheric connections differ between symmetrical and asymmetrical brain regions. *Neuroscience*, 1989, 3:525–533.

Rosen, G. D.; Sherman, G. F.; and Galaburda, A. M. Ontogenesis of neocortical asymmetry: a [3H]thymidine study. *Neuroscience*, 1991, 41:779–790.

Rosen, G. D.; Sherman, G. F.; Mehler, C.; Emsbo, K.; and Galaburda, A. M. The effect of developmental neuropathology on neocortical asymmetry in New Zealand black mice. *International Journal of Neuroscience*, 1989, 45:247–254.

Ross, D. A. and Glick, S. D. Lateralized effects of bilateral frontal cortex lesions in rats. *Brain Research*, 1981, 210:379–382.

Ross, D. A.; Glick, S. D.; and Melbach, R. C. Sexually dimorphic brain and behavioral asymmetries in the neonatal rat. *Proceedings of the National Academy of Science, U.S.A.*, 1981, 78:1958–1961.

Ross, E. D. Modulation of affect and nonverbal communication by the right hemisphere. In M. M. Mesulam (ed.), Principles of Behavioral Neurology. Philadelphia: F. A. Davis, 1985, pp. 239–258.

Rubens, A. B. Aphasia with infarction in the territory of the anterior cerebral artery. *Cortex*, 1975, 11:239–250.

Rubens, A. B.; Mahowold, M. W.; and Hutton, J. T. Asymmetry of the lateral sylvian fissures in man. *Neurology*, 1976, 26:620–624.

Ruff, C. B. Climate and body shape in hominid evolution. *Journal of Human Evolution*, 1991, 21:81–105.

Ruhlen, M. Voices from the past. *Natural History*, 1987, 96:6–10.

Rumbaugh, D. M. and Savage-Rumbaugh, E. S. Chimpanzees, competence for language, and numbers. In W. C. Stebbins and M. A. Berkley (eds.), Comparative Perception, Vol. II: Complex Signals. New York: Wiley, 1990, pp. 409–439.

Ruvolo, M.; Disotell, T. R.; Allard, M. W.; Brown, W. M.; and Honeycutt, R. L. Resolution of the African hominoid trichotomy by use of a mitochondrial gene sequence. *Proceedings of the National Academy of Science*, 1991, 88:1570–1574.

Saitou, N. Reconstruction of molecular phylogeny of extant hominoids from DNA sequence data. *American Journal of Physical Anthropology*, 1991, 84:75–85.

Saling, M. M. and Kaplan-Solms, K. L. On lateral asymmetry in cradling behavior and breast sensitivity. *Current Anthropology*, 1989, 30:210–211.

Salzen, E. A.; Parker, D. M.; and Williamson, A. J. A forebrain lesion preventing imprinting in domestic chicks. *Experimental Brain Research*, 1975, 24:145–157.

Sandhu, S.; Cook, P.; and Diamond, M. C. Rat cerebral cortical estrogen receptors: Male–female, right–left. *Experimental Neurology*, 1986, 92:186–196.

Sandi, C.; Patterson, T. A.; and Rose, S. P. R. Visual input and lateralization of brain function in learning in the chick, *Neuroscience*, 1992, In press.

Sanford, C.; Guin, K.; and Ward, J. P. Posture and laterality in the bush baby (*Galago senegalensis*). *Brain, Behaviour and Evolution*, 1984, 25:217–224.

Savage-Rumbaugh, E. S. Ape Language: From Conditioned Response to Symbol. New York: Columbia University Press, 1986.

Savage-Rumbaugh, E. S. and McDonald, K. Deception and social manipulation in symbol-using apes. In R. Byrne and A. Whiten (eds.), Machiavellian Intelligence: Social Expertise and the Evolution of Intellect in Monkeys, Apes and Humans. Oxford: Oxford University Press, 1988, pp. 224–237.

Savage-Rumbaugh, E. S.; Romski, M. A.; Hopkins, W. D.; and Sevcik, R. A. Symbol acquisition and use by *Pan troglodytes, Pan paniscus, Homo sapiens*. In P. G. Heltne and L. A. Marquardt (eds.), Understanding Chimpanzees. Cambridge, Mass.: Harvard University Press, 1989, pp. 266–295.

Savage-Rumbaugh, E. S.; Sevcik, R. A.; Brakke, K. E.; Rumbaugh, D. M.; and Greenfield, P. M. Symbols: Their communicative use, comprehension, and combination by bonobos (*Pan paniscus*). In C. Rovee-Collier and L. P. Lipsitt (eds.), Advances in Infancy Research, Vol. 6. Norwood, N.J.: Ablex, 1990, pp. 221–278.

Schaeffel, F. and Howland, H. C. Plasticity of refractive development in chick eyes. In J. Erber, R. Menzel, H-J. Pflüger and D. Todt (eds.), Neural Mechanisms of Behavior: Proceedings of the 2nd International Congress of Neuroethology, Stuttgart/New York: Georg Thieme Verlag, 1989, p. 193.

Schaeffel, F.; Howland, H. C.; and Farkas, L. Natural accommodation in the growing chicken. *Vision Research*, 1986, 26:1977–1993.

Schaeffer, A. A. Spiral movement in man. *Journal of Morphology*, 1928, 45:293–398.

Schaller, G. B. The Mountain Gorilla: Ecology and Behavior. Chicago: Chicago University Press, 1963.

Schell, L. M.; Johnston, F. E.; Smith, D. R.; and Paolone, A. M. Direction asymmetry of body dimensions among white adolescents. *American Journal of Physical Anthropology*, 1985, 67:317–322.

Schiff, H. B.; Sabin, T. D.; Geller, A.; Alexander, L.; and Mark, V. Lithium in aggressive behavior. *Americam Journal of Psychiatry*, 1982, 139:1346–1348.

Schmidt, R. A. Motor Control and Learning. Champaign, Illinois: Human Kinetics Publishers, 2nd edition, 1988.

Schmidt, S. L. and Caparelli-Dáquer, E.M. The effects of total and partial callosal agenesis on the development of morphological brain asymmetries in the BALB/cCF mouse. *Experimental Neurology*, 1989, 104:172–180.

Schmidt, S. L.; Manhães, A.; and de Moraes, V. Z. The effects of total and partial callosal agenesis on the development paw preference performance in the BALB/cCF mouse. *Brain Research*, 1991, 545:123–130.

Schneider, L. H.; Murphy, R. B.; and Coons, E. E. Lateralization of striatal dopamine (D2) receptors in normal rats. *Neuroscience Letters*, 1982, 33:281–284.

Schulte, von E-H. Unterschiede im Lern-und Abstraktionsvermögen von binokular und monokular sehenden Hühnern. *Zeitschrift für Tierpsychologie*, 1970, 27:946–970.

Schwarz, I. M. and Rogers, L. J. Testosterone: A role in the development of brain asymmetry in the chick. *Neuroscience Letters*, 1992, In press.

Scott, J. P.; Bradt, D.; and Collins, R. L. Fighting in female mice in lines selected for laterality. *Aggressive Behavior*, 1985, 12:41–44.

Searleman, A. Language capabilities and the right hemisphere. In A. W. Young (ed.), Functions of the Right Cerebral Hemisphere. New York: Academic Press, 1983, pp. 87–111.

Seccombe, A. and Rogers, L. J. Handedness in the Australian sugar glider, *Petaurus breviceps*, In preparation.

Sedláček, J. Development of the optic afferent system in chick embryos. In G. Newton and A. Reisen (eds), Advances in Psychobiology, Vol. 1. New York: Wiley, 1972, 129–170.

Sefton, A. J.; Dreher, B.; and Lim, W.-L. Interactions between callosal, thalamic and associational projections to the visual cortex of the developing rat. *Experimental Brain Research*, 1991, 84:142–158.

Selfe, L. Nadia: A Case of Extraordinary Drawing Ability in an Autistic Child. New York: Academic, 1978.

Seltzer, C.; Forsythe, C.; and Ward, J. P. Multiple measures of motor lateralization in human primates (*Homo sapiens*). *Journal of Comparative Psychology*, 1990, 104:159–166.

Seth, G. Eye–hand coordination and handedness: A developmental study of visuomotor behavior in infancy. *British Journal of Educational Psychology*, 1973, 43:35–49.

Sherman, G. F. and Galaburda, A. M. Neocortical asymmetry and open-field behavior in rat. *Experimental Neurology*, 1984, 86:473–487.

Sherman, G. F. and Galaburda, A. M. Asymmetries in anatomy and pathology in the rodent brain. In S. D. Glick (ed.), Cerebral Lateralization in Nonhuman Species. New York: Academic Press, 1985, pp. 185–231.

Sherman, G. F.; Garbanati, J. A.; Rosen, G. D.; Hofmann, M.; Yutzey, D. A.; and Denenberg, V. H. Lateralization of spatial preference in the female rat. *Life Sciences*, 1983, 33:189–193.

Sherman, G. F.; Garbanati, J. A.; Rosen, G. D.; Yutzey, D. A.; and Denenberg, V. H. Brain and behavioral asymmetries for spatial preference in rats. *Brain Research*, 1980, 192:61–67.

Sherman, G. F.; Morrison, L.; Rosen, G. D.; Behan, P. O.; and Galaburda, A. M. Brain abnormalities in immune defective mice. *Brain Research*, 1990, 532:25–33.

Shevoroshkin, V. The mother tongue. *The Sciences*, 1990, May/June, 20–27.

Shubin, N. H. and Alberch, P. A morphogenetic approach to the origin and basic organization of the tetrapod limb. In M. K. Hecht, B. Wallace and G. T. Prance (eds.), Evolutionary Biology, Vol. 20. New York: Plenum, 1986, pp. 319–387.

Sidtis, J. Music, pitch perception and the mechanisms of cortical hearing. In M. S. Gazzaniga (ed.), Handbook of Cognitive Neuroscience. New York: Plenum, 1984, pp. 91–114.

Signore, P.; Nosten-Bertrand, M.; Chaoui, M.; Roubertoux, P. L.; Marchland, C.; and Perez-Diaz, F. An assessment of handedness in mice. *Physiology and Behavior*, 1991, 49:701–704.

Sillen, A. and Brain, C. K. Old flame. *Natural History*, 1990, April, 6–10.

Simons, E. L. African origins, characteristics and context of earliest higher primates. In P. V. Tobias (ed.), Hominid Evolution: Past Present and Future. New York: Alan R. Liss Inc., 1985, pp. 101–106.

Simons, E. L. Human origins. *Science*, 1989, 245:1343–1350.

Sinclair, A. R. E.; Leakey, M. D.; and Norton-Griffiths, M. Migration and hominid bipedalism. *Nature*, 1986, 324:307–308.

Singhaniyom, W.; Wreford, N. G. M.; and Güldner, F.-H. Asymmetric distribution of catecholamine-containing neural perikarya in the upper cervical spinal cord of the rat. *Neuroscience Letters*, 1983, 41:91–97.

Skinner, B. F. About Behaviorism. New York: Alfred A. Knopf, 1974.

Smart, J. L.; Tonkiss, J.; and Massey, R. F. A phenomenon: Left-biased asymmetrical eye-opening in artificially reared rat pups. *Developmental Brain Research*, 1986, 28:134–136.

Smith, B. H. The cost of a large brain. *Behavioral and Brain Sciences*, 1990, 13:265–266.

Smith, F. H.; Falsetti, A. B.; and Donnelly, S. M. Modern human origins. *Yearbook of Physical Anthropology*, 1989, 32:35–68.

Smith, M. Behaviour of the koala, *Phascolarctos cinereus* Goldfuss, in captivity, 1. Non-social behaviour. *Australian Wildlife Research*, 1979, 6:117–129.

Smith, M. L. and Milner, B. Differential effects of frontal lobe lesions on cognitive estimation and spatial memory. *Neuropsychologia*, 1984, 22:697–705.

Smotherman, W. P.; Brown, C. P.; and Levine, S. Maternal responsiveness following differential pup treatment and mother–pup interactions. *Hormones and Behavior*, 1977, 8:242–253.

Snowdon, C. T. Language capacities of nonhuman animals. *Yearbook of Physical Anthropology*, 1990, 33:215–243.

Spennemann, D. Right- and left-handedness in early South-East Asia: The graphic evidence of the Borobudur. *Bijdragen: Tot de Taal-, Land- en Volkenkunde* (Foris Publications), 1984, 140:163–166 (a).

Spennemann, D. R. Handedness data on the European Neolithic. *Neuropsychologia*, 1984, 22:613–615 (b).

Spuhler, J. N. Evolution of mitochondrial DNA in monkeys, apes and humans. *Yearbook of Physical Anthropology*, 1988, 31:15–48.

Stafford, D. K.; Milliken, G. W.; and Ward, J. P. Lateral bias in feeding and brachiation in *Hylobates*. *Primates*, 1990, 31:407–414.

Starkstein, S. E.; Ginsberg, S.; Shnayder, L.; Bowersox, J.; Mersey, J. H.; Robinson, R. G.; and Moran, T. H. Developmental and hormonal factors in the sexually dimorphic, asymmetrical response to focal cortical lesions. *Brain Research*, 1989, 478:16–23.

Starr, M. S. and Kilpatrick, I. C. Bilateral asymmetry in brain GABA function? *Neuroscience Letters*, 1981, 25:167–172.

Steele, J. Hominid evolution and primate social cognition. *Journal of Human Evolution*, 1989, 18:421–432.

Sternberg, R. J. Beyond IQ: A Triarchic Theory of Human Intelligence. Cambridge: Cambridge University Press, 1985.

Stewart, M. G. Changes in dendritic and synaptic structure in chick forebrain consequent on passive avoidance learning. In R. J. Andrew (ed.), Neural and Behavioural Plasticity: The Use of the Domestic Chick as a Model. Oxford: Oxford University Press, 1991, pp. 305–328.

Stewart, M. G.; Csillag, A.; and Rose, P. R. Alterations in synaptic structure in the paleostriatal complex of the domestic chick, *Gallus domesticus*, following passive avoidance training. *Brain Research*, 1987, 426:69–81.

Stewart, M. G.; Rose, S. P. R.; King, T. S.; Gabbot, P. L. A.; and Bourne, R. Hemispheric asymmetry of synapses in the chick medial hyperstriatum ventrale following passive avoidance training: A stereological investigation. *Developmental Brain Research*, 1984, 12:261–269.

Stiles-Davis, J.; Kritchevsky, M.; and Bellugi, U. Spatial Cognition: Brain Bases and Development. Hillsdale, N. J.: Erlbaum, 1988.

Stokes, K. A. and McIntyre, D. C. Lateralized asymmetrical state-dependent learning produced by kindled convulsions from the rat hippocampus. *Physiology and Behavior*, 1981, 26:163–169.

Stoneking, M. and Cann, R. L. African origin of human mitochondrial DNA. In P. Mellars and C. Stringer (eds.), The Human Revolution: Behavioral and Biological Perspectives on the Origins of Modern Humans. Princeton: Princeton University Press, 1989, pp. 17–30.

Straus, L. G. Age of the modern Europeans. *Nature*, 1989, 342:476–477.

Stringer, C. B. The origin of early modern humans: A comparison of the European and non-European evidence. In P. Mellars and C. Stringer (eds.), The Human Revolution: Behavioral and Biological Perspectives on the Origins of Modern Humans. Princeton: Princeton University Press, 1989, pp. 232–244.

Stringer, C. B. The emergence of modern humans. *Scientific American*, 1990, December, 68–74 (a).

Stringer, C. B. The Asian connection. *New Scientist*, 1990, November 17, 23–27 (b).

Stringer, C. B. and Andrews, P. Genetic and fossil evidence for the origin of modern humans. *Science*, 1988, 239:1263–1268.

Stringer, C. B. and Grün, R. Time for the last Neanderthals. *Nature*, 1991, 351:701–702.

Stringer, C. B.; Grün, R.; Schwarcz, H. P.; and Goldberg, P. ESR dates for the hominid burial site of Es Skhul in Israel. *Nature*, 1989, 338:756–758.

Studdert-Kennedy, M. This view of language. *Behavioral and Brain Sciences*, 1990, 13:758–759.

Suarez, S. D. and Gallup, G. G. Face touching in primates: A closer look. *Neuropsychologia*, 1986, 24:597–600.

Susman, R. L. Evidence for tool behavior in *Paranthropus robustus* from Member 1, Swartkrans: Fossil evidence for tool behavior. *Science*, 1988, 240:781–784.

Sutton, D.; Trachy, R. E.; and Lindeman, R. C. Primate phonation: Unilateral and bilateral cingulate lesion effects. *Behavioural Brain Research*, 1981, 3:99–114.

Suzuki, Y. and Arai, Y. Laterality associated with sexual dimorphism in the volume of the mouse hypogastric ganglion. *Experimental Neurology*, 1986, 94.241–245.

Szuba, Z. Asymetria Budowy Komor Bocznych Mózgu owcy. (The asymmetry of the lateral ventricles of the sheep's brain.) *Zeszyty Naukowe Wyż szej Szkoły Rolniczej w Szczecinie*, 1965, 20:3–12.

Tan, Ü. Paw preferences in dogs. *International Journal of Neuroscience*, 1987, 32:825–329.

Tan, Ü. and Çalişkan , S. Allometry and asymmetry in the dog brain: The right hemisphere is heavier regardless of paw preference. *International Journal of Neuroscience*, 1987, 35:189–194 (a).

Tan, Ü. and Çalişkan , S. Asymmetries in the cerebral dimensions and fissures of the dog. *International Journal of Neuroscience*, 1987, 32:943–952 (b).

Tan, Ü. and Kutlu, N. The distribution of paw preference in right-, left-, and mixed pawed male and female cats: The role of a female right-shift factor in handedness. *International Journal of Neuroscience*, 1991, 59:219–229.

Tan, Ü.; Yaprak, M.; and Kutlu, N. Paw preference in cats: Distribution and sex differences. *International Journal of Neuroscience*, 1990, 50:195–208.

Tanabe, Y.; Nakamura, T.; Fujioka, K.; and Doi, O. Production and secretion of sex steroid hormones by the testes, the ovary and adrenal glands of embryonic and young chickens. *General and Comparative Ednocrinology*, 1979, 39:26–33.

Tattersall, I and Eldredge, N. Fact, theory and fantasy in human paleontology. *American Scientist*, 1977, 65:204–211.

Tauber, H.; Waehneldt, T. V.; and Neuhoff, V. Myelination in rabbit optic nerves is accelerated by artificial eye opening. *Neuroscience Letters*, 1980, 16:239–244.

Taylor, A. M. and Warrington, E. K. Visual discrimination in patients with localized cerebral lesions. *Cortex*, 1973, 9:82–93.

Tees, R. C. Visual experience, unilateral cortical lesions, and lateralization of function in rats. *Behavioral Neuroscience*, 1984, 98:969–978

Templeton, A. R. Human origins and analysis of mitochondrial DNA sequences. *Science*, 1992, 255:737.

ten Cate, C. Population lateralization in zebra finch courtship: A reassessment. *Animal Behaviour*, 1991, 41:900–901.

ten Cate, C.; Baauw, A.; Ballintijn, M.; Majoor, M.; and van der Horst, I. Lateralization of orientation in sexually active zebra finches: Eye use asymmetry or locomotor bias? *Animal Behaviour*, 1990, 39:992–994.

Terrace, H. S. "Language" in apes. In R. Harré and V. Reynolds (eds.), The Meaning of Primate Signals. Cambridge: Cambridge University Press, 1984, pp. 179–207.

Thanos, S. and Bonhoeffer, F. Axonal arborization in the developing chick retinotectal system. *Journal of Comparative Neurology*, 1987, 261:155–164.

Thomson, K. S. Spiral discord. *Nature*, 1987, 327:196–197.

Tillier, A-M. The evolution of modern humans: Evidence from young Mousterian individuals. In P. Mellars and C. Stringer (eds.), The Human Revolution: Behavioral and Biological Perspectives on the Origins of Modern Humans. Princeton: Princeton University Press, 1989, pp. 286–297.

Tobias, P. V. The brain of *Homo habilis:* A new level of organization in cerebral evolution. *Journal of Human Evolution*, 1987, 16:741–761.

Todor, J. I. and Cisneros, J. Accommodation to increased accuracy demands by the right and left hands. *Journal of Motor Behavior*, 1985, 17:355–372.

Tomasello, M. Cultural transmission in the tool use and communicatory signaling of chimpanzees? In S. T. Parker and K. R. Gibson (eds.), Language and Intelligence in Monkeys and Apes: Comparative and Developmental Perspectives. Cambridge: Cambridge University Press, 1990, pp. 274–311.

Toth, N. Archaeological evidence for preferential right-handedness in the Lower and Middle Pleistocene, and its possible implications. *Journal of Human Evolution*, 1985, 14:607–614.

Toth, N. Behavioral inferences from Early Stone Age artifact assemblages: An experimental model. *Journal of Human Evolution*, 1987, 16:763–787.

Treffert, D. A. Extraordinary People. London: Bantam Press, 1989.

Trinkaus, E. Cladistics and the hominid fossil record. *American Journal of Physical Anthropology*, 1990, 83:1–11.

Tulving, E. and Thomson, D. M. Encoding specificity and retrieval processes in episodic memory. *Psychological Review*, 1973, 80:352–373.

Turner II, C. G. Teeth and prehistory in Asia. *Scientific American*, 1989, February, 70–76.

Turner II, C. G. Major functions of Sundadonty and Sinodonty, including suggestions about east Asia microevolution, population history and Late Pleistocene relations with Australian aborigines. *American Journal of Physical Anthropology*, 1990, 82:295–317.

Tuttle, R. H. The pitted pattern of Laetoli feet. *Natural History*, 1990, March, 61–64.

Ubaghs, G. Early Paleozoic echinoderms. *Annual Review of Earth and Planetary Science*, 1975, 3:79–98.

Uhrbrock, R.S. Bovine laterality. *Journal of Genetic Psychology*, 1969, 115:77.

Uhrbrock, R. S. Laterality in art. *Journal of Aesthetics and Art Criticism,* 1973, 32:27–35.

Valdes, J. J.; Mactutus, C. F.; and Cory, R. N. Lateralization of norepinephrine, serotonin and choline uptake into hippocampal synaptosomes of sinistral rats. *Brain and Behavior,* 1981, 27:381–383.

Vallortigara, G. Behavioral asymmetries in visual learning of young chickens. *Physiology and Behavior,* 1989, 44:797–800.

Vallortigara, G. and Andrew, R. J. Lateralization of response by chicks to change in a model partner. *Animal Behaviour,* 1991, 41:187–194.

Vallortigara, G.; Zandforlin, M.; and Cailotto, M. Right–left asymmetry in position learning of male chicks. *Behavioural Brain Research,* 1988, 27:189–191.

Vandermeersch, B. The evolution of modern humans: Recent evidence from southwest Asia. In P. Mellars and C. Stringer (eds.), The Human Revolution: Behavioral and Biological Perspectives on the Origins of Modern Humans. Princeton: Princeton University Press, 1989, pp. 155–164.

Vandiver, P. B.; Soffer, O.; Klima, B.; and Svoboda, J. The origins of ceramic technology at Dolni Věstonice, Czechoslovakia. *Science,* 1989, 246:1002–1008.

Van Hof, M. W. Interocular transfer and interhemispheric communication. In I. S. Russell, M. W. Van Hof and G. Berlucci (eds.), Structure and Function of the Cerebral Commissures. New York: Macmillan, 1979, pp. 224–235.

Vargo, J. M., Richard-Smith, M. and Corwin, J. V. Spiroperidol reinstates asymmetries in rats recovered from left or right oorsomedial prefrontal cortex lesions. *Behavioural Neuroscience,* 1989, 103:1017–1027.

Vauclair, J. Phylogenetic approach to object manipulation in human and ape infants. *Human Development,* 1984, 27:321–328.

Vauclair, J. Tool use, hand cooperation and the development of object manipulation in human and nonhuman primates. In A. F. Kalverboer, B. Hopkins and R. H. Geuze (eds.), Motor Development in Early and Later Childhood: Longitudinal Approaches. Cambridge: Cambridge University Press, in press.

Vauclair, J. and Fagot, J. Spontaneous hand usage and handedness in a troop of baboons. *Cortex,* 1987, 23:265–274.

Vauclair, J. and Fagot, J. Manual specialization in gorillas and baboons. In J. Ward (ed.), Current Behavioral Evidence of Primate Asymmetries, New York: Springer, 1992, pp 193–207.

Verhaegen, M. Origin of bipedalism. *Nature,* 1987, 325:305–306.

Vermeij, G. J. Left-asymmetry in the animal kingdom. *Behavioral and Brain Sciences,* 1978, 2:320–321.

Vernon, P. A. Speed of Information Processing and Intelligence. Norwood, N. J.: Ablex, 1987.

Vigilant, L.; Stoneking, M.; Harpending, H.; Hawkes, K.; and Wilson, A. C. African populations and the evolution of human mitochondrial DNA. *Science,* 1991, 253:1503–1507.

Vince, M. A. Use of the feet in feeding by the Great Tit, *Parus major. Ibis,* 1964, 106:500–529.

Vince, M. A. and Toosey, F. M. Posthatching effects of repeated prehatching stimulation with an alien sound. *Behaviour,* 1980, 72: 65–76.

Vockel, A.; Pröve, E.; and Balthazart, J. Sex- and age-related differences in the activity of testosterone-metabolizing enzymes in microdissected nuclei of the zebra finch brain. *Brain Research,* 1990, 511:291–302.

Von Frisch, K. The Dance Language and Orientation of Bees. Cambridge, MA: Harvard University Press, 1967.

Vrensen, G., and De Groot, D. The effect of dark rearing and its recovery on synaptic terminals in the visual cortex of rabbits: A quantitative electron microscopic study. *Brain Research*, 1974, 78:263–278.

Wada, J. A.; Clarke, R.; and Hamm, A. Cerebral hemispheric asymmetry in humans. *Archives of Neurology*, 1975, 32:239–246.

Wagnon, K. A. and Rollin, W. C. Bovine laterality. *Journal of Animal Science*, 1972, 35:486.

Wakshlak, A. and Weinstock, M. Neonatal handling reverses behavioral abnormalities induced in rats by prenatal stress. *Physiology and Behavior*, 1990, 48:289–292.

Walcott, C. Show me the way you go home. *Natural History*, 1989, 11:40–46.

Walker, A.; Leakey, R. E.; Harris, J. M.; and Brown, F. H. 2.5-Myr *Australopithecus boisei* from west of Lake Turkana, Kenya. *Nature*, 1986, 322:517–522.

Walker, A. and Teaford, M. The hunt for *Proconsul*. *Scientific American*, 1989, January, 58–64.

Walker, M. and Bitterman, M. E. Attached magnets impair magnetic field discrimination by honeybees. *Journal of Experimental Biology*, 1989, 141:447–451.

Walker, S. F. Lateralization of function in the vertebrate brain: A review. *British Journal of Psychology*, 1980, 71:329–367.

Walker, S. Animal Thought. London: Routledge, 1985.

Wallman, J. and Pettigrew, J. D. Conjugate and disjunctive saccades in two avian species with contrasting oculomotor strategies. *Journal of Neuroscience*, 1985, 5:1418–1428.

Walls, G. L. The Vertebrate Eye and its Adaptive Radiation. New York: Hafner, 1942.

Walls, G. L. The lateral geniculate nucleus and visual histophysiology. *University of California Publications in Physiology*, 1953, 9:1–100.

Ward, I. L. and Weisz, J. Maternal stress alters plasma testosterone in fetal males. *Science*, 1980, 207:328–329

Ward, J. P.; Milliken, G. W.; Dodson, D. L.; Stafford, D. K.; and Wallace, M. Handedness as a function of sex and age in a large population of lemur. *Journal of Comparative Psychology*, 1990, 104:167–173.

Ward, J. P.; Milliken, G. W.; and Stafford, D. K. Patterns of lateralized behavior in prosimians. In J. P. Ward (ed.), Current Behavioral Evidence of Primate Asymmetry. New York: Springer, 1992, pp. 43–74.

Ward, R. and Collins, R. L. Brain size and shape in strongly and weakly lateralized mice. *Brain Research*, 1985, 328:243–249.

Ward, R.; Tremblay, L.; and Lassonde, M. The relationship between callosal variation and lateralization in mice is genotype-dependent. *Brain Research*, 1987, 424:84–88.

Warren, J. M. Handedness and laterality in humans and other animals. *Physiological Psychology*, 1980, 8:351–359.

Warren, J. M.; Abplanalp, J. M.; and Warren, H. B. The development of handedness in cats and rhesus monkeys. In A. W. Stevensen (ed.), Early Behavior Comparative Developmental Approaches. New York: Rheingold Wiley, 1967.

Warren, J. M.; Cornwell, P. R.; Webster, W. G.; and Pubols, B. H. Unilateral cortical lesions and paw preferences in cats. *Journal of Comparative and Physiological Psychology*, 1972, 81:410–422.

Watanabe, S. Interocular transfer of learning in the pigeon: Visuo-motor integration and separation of discriminanda and manipulanda. *Behavioural Brain Research*, 1986, 19:227–232.

Watanabe, S. Effects of ectostriatal lesions on natural concept, psuedoconcept, and artificial pattern discrimination in pigeons. *Visual Neuroscience*, 1991, 6:497–506.

Watanabe, S.; Hodos, W.; and Bessette, B. B. Two eyes are better than one: Superior binocular discrimination learning in pigeons. *Physiology and Behavior*, 1984, 32:847–850.

Webster, K. E. Changing concepts of the organization of the central visual pathways in birds. In R. Bellairs and E. G. Gray (ed.), Essays on the Nervous System. Oxford: Clarendon Press, 1974, pp. 258–298.

Webster, W. G. Functional asymmetry between the cerebral hemispheres of the cat. *Neuropsychologia*, 1972, 10:75–87.

Webster, W. Hemispheric asymmetry in cats. In S. Harnard, R. W. Doty, L. Goldstein, J. Jaynes and G. Krauthamer (eds.), Lateralization in the Nervous System. New York: Academic Press, 1977, pp. 471–480.

Webster, W. G. Morphological asymmetries of the cat brain. *Brain Behaviour and Evolution*, 1981, 18:72–79.

Webster, W. G. and Webster, I. H. Anatomical asymmetry of the cerebral hemispheres of the cat brain. *Physiology and Behavior*, 1975, 14:867–869.

Weidner, C.; Repérant, J.; Miceli, D.; Haby, M.; and Rio, J. P. An anatomical study of ipsilateral retinal projections in the quail using radioautographic, horseradish peroxidase, fluorescence and degenerative techniques. *Brain Research*, 1985, 340:99–108.

Weinberger, D. R.; Luchins, D. J.; Morihisa, J.; and Wyatt, R. J. Asymmetrical volumes of the right and left frontal and occipital regions of the human brain. *Annals of Neurology*, 1982, 11:97–100.

Westergard, G. C. Hand preference in the use and manufacture of tools by tufted capuchin (*Cebus apella*) and lion-tailed macaques (*Macaca silenus*) monkeys. *Journal of Comparative Psychology*, 1991, 105:172–176.

Wheeler, P. E. The evolution of bipedality and loss of functional body hair in hominids. *Journal of Human Evolution*, 1984, 13:91–98.

White R. Visual thinking in the Ice Age. *Scientific American*, 1989, July, 74–81.

Whiten, A. Causes and consequences in the evolution of hominid brain size. *Behavioral and Brain Sciences*, 1990, 13:367.

Whiten, A. The emergence of mindreading. Steps towards an interdisciplinary enterprise. In A. Whiten (ed.), Natural Theories of Mind: Evolution, Development and Simulation of Everyday Mindreading. Oxford: Blackwell, 1991, pp. 195–207.

Whiten, A. and Byrne, R. W. The Machiavellian intelligence hypothesis. In R. W. Byrne and A. Whiten (eds.), Machiavellian Intelligence. Oxford, Clarendon Press, 1988, pp. 1–9.

Wilson, D. Paleolithic dextrality. *Royal Society of Canada, Proceedings and Transactions III*, 1885, 2:119–133.

Wind, J. The evolutionary history of the human speech organs. In J. Wind and E. Pulleyblank (eds.), Studies in Language Origins. Amsterdam: Benjamins, 1989, pp. 173–197.

Witelson, S. F. Hand preference: Basis or reflection of hemispheric specialization? *Behavioral and Brain Sciences*, 1988, 11:35–36.

Witelson, S. F. Hand and sex differences in the isthmus and genu of the human corpus callosum. *Brain*, 1989, 112:799–832.

Witelson, S. F. and Kigar, D. L. Asymmetry in brain function follows asymmetry in anatomical form: Gross, microscopic, postmorten and imaging studies. In F. Boller and J. Grafman (eds.), Handbook of Neuropsychology, Vol.1. New York: Elsevier, 1988, pp. 111–142.

Witelson, S. F. and Pallie, W. Left hemisphere specialization for language in the newborn: Neuroanatomical evidence of asymmetry. *Brain*, 1973, 96:641–646.

Wolpoff, M. H. Multiregional evolution: The fossil alternative to Eden. In P. Mellars and C. Stringer (eds.), The Human Revolution: Behavioral and Biological Perspectives on the Origins of Modern Humans. Princeton: Princeton University Press, 1989, pp. 62–108.

Wood, W. B. Evidence from reverse handedness in *Caenorhabditis elegans* embryos for early cell interactions determining cell fates. *Nature*, 1991, 349:536–538.

Woods, J. E.; Simpson, R. M.; and Moore, P. L. Plasma testosterone levels in the chick embryo. *General and Comparative Endocrinology*, 1975, 27:543–547.

Workman, L. and Andrew, R. J. Asymmetries of eye use in birds. *Animal Behaviour*, 1986, 34:1582–1584.

Workman, L. and Andrew, R. J. Simultaneous changes in behaviour and in lateralization during the development of male and female domestic chicks. *Animal Behaviour*, 1989, 38:596–605.

Workman, L. and Andrew, R. J. Population lateralization in zebra finch courtship: An unresolved issue. *Animal Behaviour*, 1991, 41:545–546.

Workman, L.; Kent, J. P.; and Andrew, R. J. Development of behaviour in the chick. In R. J. Andrew (ed.), Neural and Behavioural Plasticity: The Use of the Domestic Chick as a Model. Oxford: Oxford University Press, 1991, pp. 157–176.

Wrangham, R. W. Bipedal locomotion as a feeding adaptation in gelada baboons, and its implications for hominid evolution. *Journal of Human Evolution*, 1980, 9:329–331.

Wundram, I. J. Cortical motor asymmetry and hominid feeding strategies. *Human Evolution*, 1986, 1:183–188.

Wynn, T. Tools and the evolution of human intelligence. In R. Byrne and A. Whiten (eds.), Machiavellian Intelligence: Social Expertise and the Evolution of Intellect in Monkeys, Apes and Humans. Oxford: Oxford University Press, 1988, pp. 270–284.

Wynn, T. The Evolution of Spatial Competence: Illinois Studies in Anthropology Number 17. Urbana & Chicago: University of Illinois Press, 1989.

Wynn, T. and McGrew, W. C. An ape's view of the Oldowan. *Man*, 1989, 24:383–398.

Yakovlev, P. I. A proposed definition of the limbic system. In C. H. Hockman (ed.), Limbic System Mechanisms and Autonomic Function. Springfield, Ill.: Charles C. Thomas, 1972.

Yakovlev, P. I. and Rakic, P. Patterns of decussation of bulbar pyramids and distribution of pyramidal tracts on two sides of the spinal cord. *Transactions of the American Neurological Association*, 1966, 91:366–367.

Yamada, J. E. Laura: A Case for the Modularity of Language. Cambridge, MA: Bradford/ MIT, 1990.

Yamane, S.; Kaji, S.; and Kawano, K. What facial features activate face neurons in the inferotemporal cortex of the monkey? *Experimental Brain Research*, 1988, 73:209–214.

Yeni-Komshian, G. H. and Benson, D. A. Anatomical study of cerebral asymmetry in the temporal lobe of humans, chimpanzees and rhesus monkeys. *Science*, 1976, 192:387–389.

Young, G. Manual specialization in infancy: Implications for lateralization of brain function. In S. J. Segalowitz and F. A. Gruber (eds.), Language, Development and Neurological Theory. New York: Academic Press, 1977, pp. 289–311.

Young, J. Z. A Model of the Brain. Oxford: The Clarendon Press, 1964.

Youngren, O. M.; Peek, F. W.; and Phillips, R. E. Repetitive vocalizations evoked by local electrical stimulation of avian brains. III. Evoked activity in the tracheal muscles of the chicken (*Gallus gallus*). *Brain, Behaviour and Evolution*, 1974, 9:393–421.

Zaidel, E. and Schweiger, A. On wrong hypotheses about right hemisphere: Commentary on K. Patterson and D. Besner, "Is the right hemisphere literate?" *Cognitive Neuropsychology*, 1984, 1:351–364.

Zappia, J. V. and Rogers, L. J. Light experience during development affects asymmetry of forebrain function in chickens. *Developmental Brain Research*, 1983, 11:93–106.

Zappia, J. V. and Rogers, L. J. Sex differences and reversal of brain asymmetry by testosterone in chickens. *Behavioural Brain Research*, 1987, 23:261–267.

Zatorre, R. J. Musical perception and cerebral function: A critical review. *Music Perception*, 1984, 2:196–221.

Zihlman, A. Women as shapers of the human adaptation. In F. Dahlberg (ed.), Woman the Gatherer. New Haven: Yale University Press, 1981, pp. 75–119.

Zihlman, A. and Tanner, N. M. Gathering and the hominid adaptation. In L. Tiger and H. Fowler (eds.), Female Hierarchies. Chicago: Beresford Books, 1978, pp. 163–194.

Zimmerberg, B.; Glick, S. D.; and Jerussi, T. P. Neurochemical correlate of a spatial preference in rats. *Science*, 1974, 185:623–625.

Zimmerberg, B. and Reuter J. M. Sexually dimorphic behavioral and brain asymmetries in neonatal rats: Effects of prenatal alcohol exposure. *Developmental Brain Research*, 1989, 46:281–290.

Zimmerberg, B.; Stumpf, A. J.; and Glick, S. D. Cerebral asymmetry and left–right discrimination. *Brain Research*, 1978, 140:194–196.

Author Index

Wheeler, P. E., 241
Whishaw, I. Q., 21, 123, 131, 156, 170
Whitaker, H. A., 368
White, A. C., 193
White, P. J., 181, 190
White, R., 302, 308
Whiten, A., 242, 373, 374, 376, 388
Williams, A. C., 193
Williams, M., 187
Williamson, A. J., 62
Wilson, A. C., 267, 271
Wilson, D., 291
Wind, J., 318, 333, 336, 345
Wisniewski, H. M., 145, 160
Witelson, S. F., 139, 197, 214, 216, 219, 224
Wolpoff, M. H., 251, 258, 261, 264, 269, 271, 272
Wood, J. L., 173
Wood, W. B., 26, 27
Woods, J. E., 83
Woodward, D. J., 217
Workman, L., 50, 51, 53, 55, 91–93, 95
Wörtwein, G., 80
Wrangham, R. W., 240
Wreford, N. G. M., 148
Wu, M.-J., 259
Wundram, I. J., 372
Wyatt, R. J., 213
Wynn, T., 285, 287, 313, 372

Y

Yahr, P., 127, 128
Yakovlev, P. I., 171, 172
Yamada, J. E., 322
Yamane, S., 206, 207
Yamanouchi, K., 128
Yaprak, M., 162
Yeni-Komshian, G. H., 214, 221
Yeterian, E. H., 356
Young, D., 129
Young, G., 190
Young, J. Z., 364
Youngren, O. M., 49, 75
Yutzey, D. A., 113, 114, 116, 119, 120, 136, 141, 169

Z

Zaidel, E., 201
Zandforlin, M., 47
Zanelli, G., 193
Zappia, J. V., 39, 42, 43, 49–52, 54, 55, 59, 97
Zatorre, R. J., 359
Zeidner, L., 141, 169
Zihlman, A., 240
Zimmerberg, B., 108–111, 113, 129, 143, 144
Zoccolotti, P., 244
Zoloth, S. R., 210, 211, 224

Subject Index

A

Acheulian, 235, 253, 286–288, 292, 293, 304
  see also Homo erectus
Aegyptopithecus, 229, 233
Aesthetics, 357
Age difference
  ape, 181
  bird, 44, 51–59, 82–86, 89
  human, 191
  monkey, 174, 180, 189
  prosimian, 186, 187
  rat, 115, 132
Agnosia, 350–352
Albinism, 101, 117
Allometry, 234, 367
Altriciality, 371, 373
Alzheimer's disease, 370
America, 266, 302, 362
American Sign Language, 323, 324, 346
Amino acid asymmetry, 12–15
Amphetamine, 108–110, 150, 192
Amphioxus, 27, 226
Amusia, see Music
Amygdaloid nucleus, 132, 134
Analytic/holistic processing dichotomy 41,
  73, 122–127, 281, 357, 359
Anatomically modern humans, see Homo
  sapiens sapiens
Anaxagoras, 8
Animal carving, 302
Aphasia, 338, 384
Apraxia, see Praxis
Aquatic hypothesis, 241
Arago cave, 252
Arboreal brachiation, see Brachiation
Aromatase, 51,128
Art, 257, 260, 264, 286, 290–313, 349–352,
  358, 378, 381–385
Asymmetric tonic neck reflex, 190

Attack, see Emotional behavior
Auditory processing
  bird, 66–75, 87, 88
  monkey, 209–212
  nonhuman primates, 209, 210
  rat, 122
Australia, 259, 264–267, 271, 308, 311
Australian brush turkey, 59
Australopithecus aethiopicus, 235
Australopithecus afarensis, 233–245
Australopithecus africanus, 219, 233, 234,
  236
Australopithecus boisei, 233–235, 241, 242
Australopithecus robustus, 219, 233–235, 241,
  242
Autism, 316, 366, 376, 384, 385

B

Baboon
  auditory discrimination, 210, 212
  brain morphology, 218, 221
  hand preference, 180, 190, 195
Bacillus subtilis, 17
Bamboo, 251, 253, 265
Barbiturate, 166
Basal ganglia, 108–112, 133, 143, 146, 172,
  192, 193, 217, 342
Bible, see Myth
Bipedalism, 233, 237–245
Bird, 37–97, 282, 283, 332, 358, 364, 375, 382
Birdsong, 66–75, 332, 382
Black-capped chickadee, 67, 68
Blue Jay, 282
Bonobo, 323, 324, 385
Border cave, 258
Borobudur, 291
Bower bird, 358,
Brachiation, 239, 243
Brain size, 233, 242, 247, 251, 366